英汉地学专业术语集注系列
丛书主编　董元兴　张红燕

英汉矿物学专业术语集注

江敏　杨红燕　丘晓娟　编著

WUHAN UNIVERSITY PRESS
武汉大学出版社

图书在版编目(CIP)数据

英汉矿物学专业术语集注/江敏,杨红燕,丘晓娟编著.—武汉:武汉大学出版社,2017.1
英汉地学专业术语集注系列/董元兴 张红燕主编
ISBN 978-7-307-12119-5

Ⅰ.英… Ⅱ.①江… ②杨… ③丘… Ⅲ.矿物学—术语—英、汉 Ⅳ.P57-61

中国版本图书馆CIP数据核字(2017)第013035号

责任编辑:李 琼　　责任校对:李孟潇　　版式设计:马 佳

出版发行:**武汉大学出版社** (430072 武昌 珞珈山)
(电子邮件:cbs22@whu.edu.cn 网址:www.wdp.com.cn)
印刷:湖北金海印务有限公司
开本:787×1092 1/16 印张:38.5 字数:959千字 插页:1
版次:2017年1月第1版 2017年1月第1次印刷
ISBN 978-7-307-12119-5 定价:70.00元

序

矿物学是地质学的重要分支学科，是研究矿物的化学成分、晶体结构、形态、性质、成因、产状、共生组合、变化条件、时间与空间上的分布规律、形成与演化的历史和用途以及它们之间关系的一门学科。许多生产部门，如采矿、选冶化工等都离不开矿物原料。因此，矿物学研究不仅有理论意义，而且对矿物资源的开发和应用有重要的实际意义。近些年来，中外矿产企业交流合作日益深入，矿物学研究发展迅猛并日益趋向国际化，地质工作者和学生迫切需要有一本矿物学方面的专业术语英汉集注，为此我们编写了这本集注。

本集注根据矿物学的学科特点，共分为三个部分，分别是结晶学词汇、矿物名称词汇、矿物学词汇。其中结晶学词汇 527 条，矿物名称词汇 524 条，矿物学词汇 401 条。词汇选择严格参照矿物学专业术语库。在体例上，以英语音标、英语释义、对应汉语名、汉语拼音注音、词源、例句（含汉语译文）和拓展词组为主要组成部分。在词汇释义和例句选择上紧扣专业主题，部分较为复杂的复合词术语中的单词也进行了单独释义。词源部分除了提供部分术语在该专业领域的缘起，还包含许多普通词汇的由来和演变，有助于了解该词汇的来龙去脉。该集注兼顾专业和语言，融合了英汉科技专业词典和英语语言词典的双重特色。

本集注的例句不少来自国内外有关著作和文献，限于体例，未予注明出处；本集注的编写得到了中国地质大学（武汉）材化学院相关专家的支持和帮助；外语学院 2012 届研究生何颖、周佳、徐倩、李寐竹等做了大量资料收集整理工作，在此表示诚挚的谢意。

由于时间和经验所限，本书在选词和例句选用等方面难免有不足或错误之处，我们诚恳地希望广大读者提出批评和建议，以利于我们进一步改正和修订。

目　录

第一部分　结晶学词汇

第二部分　矿物名称词汇

第三部分　矿物学词汇

第一部分
结晶学词汇

Abbe Refractometer [ˈæbi riˌfrækˈtɔmitə]

Definition: An Abbe Refractometer is a bench-top device for the high-precision measurement of an index of refraction. 阿贝折射仪 ā bèi zhé shè yí

abbe：阿贝 ā bèi

refractometer：A refractometer is a laboratory or field device for the measurement of an index of refraction. The index of refraction is calculated from Snell's law and can be calculated from the composition of the material using the Gladstone-Dale relation. 折射计 zhé shè jì

Example:

A method of verifying the formula of index of refraction of a solution is given by measuring its index of refraction and percentage concentration with Abbe Refractometer.

提供了一种通过阿贝折射仪测量溶液折射率和百分比浓度来验证其折射率公式的方法。

Extended Terms:

high-precision abbe refractometer 高精阿贝折射仪

abbe refractometer with light source 带光源阿贝折射计

---

abelian group [əˈbiːljən gruːp]

Definition: An abelian group, also called a commutative group, is a group in which the result of applying the group operation to two group elements does not depend on the order in which they are written. This means that the order in which the binary operation is performed does not matter, and any two elements of the group commute. Groups that are not commutative are called non-abelian (rather than non-commutative). 阿贝尔群 ā bèi ěr qún

Origin: named after Niels Henrik Abel

Example:

In some finiteness conditions, we prove that there exists a natural abelian group homomorphism from the Grothendieck group of R to the Grothendieck group of A.

在适当的有限性条件下,我们证明了一个从 R 的格罗腾迪克群到 A 的格罗腾迪克群的自然阿贝尔群同态。

Extended Terms:

discrete abelian group 离散阿贝尔群

A

compact abelian group 紧阿贝尔群
free abelian group 自由交换群;自由阿贝尔群
topological abelian group 拓扑阿贝尔群
primary abelian group 准素阿贝尔群

absolute structure [ˈæbsəluːt ˈstrʌktʃə]

Definition: The spatial arrangement of the atoms of a physically identified non-centrosymmetric crystal and its description by way of unit-cell dimensions, space group, and representative coordinates of all atoms. 绝对构造 jué duì gòu zào

absorption edge [əbˈsɔːpʃən edʒ]

Definition: An absorption edge is a sharp discontinuity in the absorption spectrum of X-rays by an element that occurs when the energy of the photon corresponds to the energy of a shell of the atom. 吸收边 xī shōu biān

Example:

It also showed that the optical absorption edge of the annealed film appeared shifted towards the longer wavelength side and the band gap decreased by 0.
光学性质显示退火处理的薄膜吸收边缘明显向长波的方向移动,发生红移现象,而且禁带宽度减少了 0。

Extended Terms:

X-ray absorption near edge structure (XANES) X 射线吸收近边结构
optical absorption edge 光学吸收限;光吸收限
fundamental absorption edge 基本吸收边缘
X-ray absorption edge X 射线吸收带边缘
main absorption edge 主吸收边限

absorption index [əbˈsɔːpʃən ˈindeks]

Definition: A measure of the attenuation caused by absorption of energy per unit of distance that occurs in an electromagnetic wave of given wavelength propagating in a material medium of given refractive index. 吸收指数 xī shōu zhǐ shù

Example:

When salinity of formation water is rather high, the volumetric photoelectric absorption index of water and that of oil (gas) are quite different.
当地层水含盐量较高时,水的体积光电吸收指数与油(气)的体积光电吸收指数相差较大。

Extended Terms:

pseudo-absorption index 伪吸收系数
tensile energy absorption index 抗张能量吸收指数
index of absorption 吸收指数;吸收率

absorption [əbˈsɔːpʃən] *n.*

Definition: (*Chemistry*) Absorption is a physical or chemical phenomenon or a process in which atoms, molecules, or ions enter some bulk phase-gas, liquid or solid material. This is a different process from adsorption, since molecules undergoing absorption are taken up by the volume, not by the surface (as in the case for adsorption). A more general term is "sorption", which covers absorption, adsorption and ion exchange. Absorption is basically where something takes in another substance.吸收性 xī shōu xìng

Origin: from absorb

Example:

This article makes on classified explanation to absorption spectrum on analyzing and researching lignin structure, application of lignin.
本文对吸收光谱在分析和研究木素结构及应用等方面进行了分类综述。

Extended Terms:

absorption rate 吸收率;分摊率;摊配率;吸收速率
absorption ratio 吸收系数
absorption band 吸收带;吸收谱带;吸收频带;吸收光带
absorption capacity 吸收能力;吸收容量;吸收本领;吸收力

acceptance domain [əkˈseptəns dəuˈmein]

Definition: When an aperiodic crystalline point set is obtained by the intersection method, as the intersection of a periodic array of finite, disjoint components in superspace and the physical space, then there is, for each point, a component in the higher-dimensional unit cell. These components are called *acceptance domains*, *atomic surfaces*, *atomic domains*, or *windows* in the literature. The positions of atoms in aperiodic crystals (or vertices in the case of a tiling) are the intersection of the atomic surfaces with the physical space. This construction of the points is called the intersection method. 接受域 jiē shòu yù

Example:

This general procedure leads to acceptance domain or motif identical to those discussed in literature for primitive orthogonal hyperlattices.

这种普通的程序会形成与相关文献中原始正交点阵相同的接受域或结构基元。

A

affine isomorphism [əˈfain ˌaisəuˈmɔːfizəm]

Definition: Each symmetry operation of crystallographic group in E^3 may be represented by a 3×3 matrix W (*the linear part*) and a vector w. Two crystallographic groups $G1 = \{(W_{1i}, w_{1i})\}$ and $G2 = \{(W_{2i}, w_{2i})\}$ are called affine isomorphic if there exists a non-singular 3×3 matrix A and a vector a such that: $G2 = \{(A,a)(W_{1i}, w_{1i})(A,a)^{-1}\}$. Two crystallographic groups are affine isomorphic if and only if their arrangement of symmetry elements may be mapped onto each other by an affine mapping of E^3. Two affine isomorphic groups are always isomorphic. 类质同晶 lèi zhì tóng jīng

affine: related by marriage 姻亲的 yīn qīn de

isomorphism: similarity or identity of form or shape or structure 类质同晶 lèi zhì tóng jīng

affine mapping [əˈfain ˈmæpiŋ]

Definition: An affine mapping is any mapping that preserves collinearity and ratios of distances: if three points belong to the same straight line, their images under an affine transformation also belong to the same line. 仿射映射 fǎng shè yìng shè

Example:

Its characteristic is also in the capability of describing complex surface texture of a model. It uses affine mapping from a rectangular texture image to each facet to describe simple texture mapping on a surface, and it describes complex texture mapping by simply storing the result image of the complex texture mapping onto each facet.

该格式的另一特点在于它对模型表面复杂纹理的描述能力:对于表面上的简单纹理映射,用从矩形纹理图案到三角片的仿射映射方法进行描述,而对于复杂的纹理映射,则采用了直接保存每个三角片上的复杂纹理映射结果图案的简单方法。

Extended Term:

quasi affine mapping 拟仿射映射

albite twin law [ˈælbait twin lɔː]

Definition: It is a rule specifying the orientation of alternating lamellae in multiple twin feldspar crystals; the twinning plane is brachypinacoid and is common in albite. 钠长石双晶律 nà cháng shí shuāng jīng lǜ

albite: a widely distributed rock-forming feldspar 钠长石 nà cháng shí

twin: either of two offspring born at the same time from the same pregnancy 孪生的 luán shēng de

law: a generalization that describes recurring facts or events in nature 定律 dìng lǜ

amorphous substance [əˈmɔːfəs ˈsʌbstəns]

Definition: An amorphous substance is something that has a random molecular formation in its natural form (when solidified). The opposite would be semi-cristaline which has random molecular structure when heated but returns to an organised uniform state when solidified. 非晶质 fēi jīng zhì;无定形物质 wú dìng xíng wù zhì

Example:

Fluid is a continuous, amorphous substance whose molecules move freely past one another and that has the tendency to assume the shape of its container: a liquid or gas.

流体是一种分子能自由移动并随容器形状而变化的连续的无定形物质,如液体或气体。

Extended Term:

amorphous ground substance 无结构基质;无定型基质

angle of polarization [ˈæŋgl əv ˌpəulərai'zeiʃən]

Definition: An angle of polarization (also known as the Brewster's angle) is an angle of incidence at which light with a particular polarization is perfectly transmitted through a surface, with no reflection. 偏振角 piān zhèn jiǎo

Example:

The performance of passive OCT is analyzed when the angle of polarization and phase shift of the incident light and the different initial birefringence are introduced.

(本文)分析了入射光偏转角、入射光相移与初始线性双折射对光学电流互感器输出的共同影响。

Extended Term:

angle of restored polarization 复偏振角

anisotropic body [ænˌaisəu'trɔpik ˈbɔdi]

Definition: objects of which optical properties vary with respect to direction 各向异性体 gè xiàng yì xìng tǐ

anisotropic: not invariant with respect to direction 各向异性的 gè xiàng yì xìng de

body: an individual 3-dimensional object that has mass and that is distinguishable from other objects 物体 wù tǐ

Example:

Using the complex potential method in the plane theory of elasticity of an anisotropic body, the series solution of finite anisotropic thin plate containing an elliptical inclusion is proposed with the help of Faber series.
采用各向异性体平面弹性理论中的复势方法,以 Faber 级数为工具,给出了有限大含椭圆核各向异性板弹性问题的级数解形式。

Extended Terms:

anisotropic elastic body 异方性弹性体
orthogonal anisotropic body 正交各向异性体

---

anisotropy [ˌænaiˈsɔtrəpi]

Definition: (*Physics*) Anisotropy is a physical property (of an object or substance) which has a different value when measured in different directions. An example is wood, which is stronger along the grain than across it. 各向异性(异向性)gè xiàng yì xìng (yì xiàng xìng)

Origin: late 19th century: from Greek anisos "unequal" + tropos "turn" +-ic

Example:

The objective is to explicit the dynamic changes of average diffusion coefficient (DCav), Isotropic image (Iso), fractional anisotropy (FA), relative anisotropy (RA), anisotropy index (AI) value in acute cerebral ischemia.
目的在于明确急性脑缺血平均扩散系数(DCav)、各向同性图像(Iso)、部分各向异性(FA)、相对各向异性(RA)、各向异性指数 (AI) 值的动态变化。

Extended Terms:

growth anisotropy 生长各向异性
macroscopic anisotropy 宏观各向异性;巨视异向性

---

anomalous absorption [əˈnɔmələs əbˈsɔːpʃən]

Definition: Anomalous absorption is absorption that takes place when radiation is dynamically diffracted by a perfect or nearly perfect crystal. 反常吸收 fǎn cháng xī shōu

anomalous: inconsistent with or deviating from what is usual, normal, or expected 反常 fǎn cháng

Example:

It is recognized that anomalous absorption is the major energy absorption mechanism for an intense relativistic electron in the target matter. With the two-stream instability and magnetic enhancement effect taken into account, experimental results can be interpreted rather well.

(本文)认为反常吸收是靶对强流相对论电子束的主要能量吸收机制,考虑双流不稳定性效应及磁增强效应就可以比较好地解释实验结果。

anomalous dispersion [əˈnɔmələs disˈpəːʃən]

Definition: extraordinary behavior in the curve of refractive index versus wavelength which occurs in the vicinity of absorption lines or bands in the absorption spectrum of a medium 反常色散 fǎn cháng sè sàn

dispersion: In optics, dispersion is the phenomenon in which the phase velocity of a wave depends on its frequency, or alternatively when the group velocity depends on the frequency. Media having such a property are termed dispersive media. Dispersion is sometimes called chromatic dispersion to emphasize its wavelength-dependent nature, or group-velocity dispersion (GVD) to emphasize the role of the group velocity. 色散 sè sàn

Example:

Bright soliton pairs propagating in the birefringent fiber with the anomalous dispersion exchange the energy periodically by the couple between them.

在反常色散的双折射光纤中传输的亮孤子对之间由于耦合周期性地交换能量,一个亮孤子可把能量完全传给另一个亮孤子。

Extended Terms:

anomalous rotatory dispersion 异常旋光色散

anomalous dispersion of X-rays X射线的异常折散

anomalous scattering [əˈnɔmələs ˈskætəriŋ]

Definition: a change in a diffracting X-ray's phase that is unique from the rest of the atoms in a crystal due to strong X-ray absorbance 反常散射 fǎn cháng sǎn shè

scattering: A general physical process where some forms of radiation, such as light, sound, or moving particles, are forced to deviate from a straight trajectory by one or more localized non-uniformities in the medium through which they pass. 散射 sǎn shè

Example:

The influence of anomalous scattering on retained austenite determination in steel by X-ray diffraction method was investigated.

采用 X 射线校对法研究了反常散射对钢中残余奥氏体含量的影响。

Extended Terms:

anomalous fluorescence scattering 反常荧光散射

anomalous atomic scattering method 反常原子散射法;反常原子反射法

A

anomalous transmission [əˈnɔmələs trænzˈmiʃən]

Definition: transmission deviating from the normal rule 反常透射 fǎn cháng tòu shè

transmission: the fraction of radiant energy that passes through a substance 透射 tòu shè

Example:

It places the stress on analyzing the anomalous transmission valley which occurs in the passbands, and a simple inhibiting procedure is described.
着重分析了在通带区域内出现反常透射深谷的原因及其抑制方法。

Extended Terms:

Borrmann anomalous-transmission technique 博曼反常透射技术
anomalous transmission method 异常透射法

---

antiphase boundary [ˈæntiˌfeis ˈbaundəri]

Definition: Each side of the boundary has an opposite phase.反相畴界 fǎn xiāng chóu jiè

antiphase: (*Sciences*) describing a boundary between an ordered phase and a disordered or random phase 反相 fǎn xiāng

boundary: the dividing line or location between two areas 界限 jiè xiàn

Example:

It is found by transmission electron microscope that the form of antiphase domain boundary has a strong effect on the ductility of ordered alloys.
利用透射电镜观察发现,反相畴界形态严重影响合金的塑性。

Extended Terms:

antiphase domain boundary 反相畴界
antiphase boundary energy 反相边界能

---

antiphase domain (APD) [ˈæntiˌfeis dəuˈmein]

Definition: Antiphase domain is the region where the atomic arrangements are of the opposite to that of perfect lattice system or simply said, APD is the region of anti-sites in a parent lattice. Generally APD forms anti-phase boundary (APB) with the parent lattice. 反相畴 fǎn xiāng chóu

Example:

The model not only can describe the evolution of the antiphase domains, morphology and microstructure of coherent precipitate successfully, but also take all stages into one physical model, including nucleation, growth and coarsening.

该模型可成功地描述共格沉淀过程中反相畴、沉淀相形貌和结构的演化，并将形核、长大和粗化在同一物理模型内加以考虑。

aperiodic crystal [ˌeipiəriˈɔdik ˈkristəl]

Definition: Aperiodic crystal is a structure with, ideally, sharp diffraction peaks on the positions of a reciprocal lattice. The structure then is invariant under the translations of the direct lattice. Periodicity here means lattice periodicity. Any structure without this property is aperiodic. 非周期性晶体 fēi zhōu qī xìng jīng tǐ

aperiodic: occurring without periodicity; irregular 非周期性的 fēi zhōu qī xìng de

crystal: A crystal or crystalline solid is a solid material, whose constituent atoms, molecules, or ions are arranged in an orderly repeating pattern extending in all three spatial dimensions. 晶体 jīng tǐ

aphanite [æˈfenɑit] *n.*

Definition: Aphanite, or aphanitic as an adjective, is a name given to certain igneous rocks which are so fine grained that their component mineral crystals are not detected by the unaided eye (as opposed to phaneritic igneous rocks, where the minerals are visible to the unaided eye). This texture results from rapid cooling in volcanic or hypabyssal (shallow subsurface) environments. As a rule, the texture of these rocks are not quite the same volcanic glass (e.g. obsidian), with volcanic glass being even finer grained (or more accurately, non-crystalline) than aphanitic rocks, and having a glass-like appearance. Aphanites are commonly porphyritic, having large crystals embedded in the fine groundmass or matrix. The large inclusions are called phenocrysts. They consist essentially of very fine grained minerals, such as plagioclase feldspar, with hornblende or augite, and may contain also biotite, quartz, and orthoclase. 隐晶质 yǐn jīng zhì

Origin: from the Greek αφανης, "invisible"

Example:

The result shows that the content of C in the aphanitic graphite can reach 96% by the method.
结果表明:用该方法提纯隐晶质石墨,可使石墨的含碳量达到96%以上。

Extended Terms:

aphanitic texture 隐晶结构
aphanitic basalt 非显晶玄武岩;隐晶玄武岩
aphanitic variolitic texture 隐晶球颗结构
dense (aphanitic) dolomite 致密白云岩

A

aristotype [ə'ristəˌtaip] n.

Definition: An aristotype is a high-symmetry structure type that can be viewed as an idealized version of a lower-symmetry structure. 珂罗酊氯化银印相法(一种旧式的印相术) kē luó dǐng lǜ huà yín yìn xiāng fǎ

---

arithmetic crystal class [ə'riθmətik 'kristəl klaːs]

Definition: The arithmetic crystal classes are obtained in an elementary fashion by combining the geometric crystal classes and the corresponding types of Bravais lattices. For instance, in the monoclinic system, there are three geometric crystal classes, 2, *m* and 2/*m*, and two types of Bravais lattices, *P* and *C*. There are therefore six monoclinic arithmetic crystal classes. Their symbols are obtained by juxtaposing the symbol of the geometric class and that of the Bravais lattice, in that order: 2*P*, 2*C*, *mP*, *mC*, 2/*mP*, 2/*mC* (note that in the space group symbol the order is inversed: *P*2, *C*2, etc.). In some cases, the centring vectors of the Bravais lattice and some symmetry elements of the crystal class may or may not be parallel; for instance, in the geometric crystal class *mm* with the Bravais lattice *C*, the centring vector and the two-fold axis may be perpendicular or coplanar, giving rise to two different arithmetic crystal classes, *mm*2*C* and 2*mmC* (or *mm*2*A*, since it is usual to orient the two-fold axis parallel to *c*), respectively. There are 13 two-dimensional arithmetic crystal classes and 73 three-dimensional arithmetic crystal classes. Space groups belonging to the same geometric crystal class and with the same type of Bravais lattice belong to the same arithmetic crystal class; these are therefore in one to one correspondence with the symmorphic space groups. 算术晶组 suàn shù jīng zǔ

arithmetic: the branch of pure mathematics dealing with the theory of numerical 算术的 suàn shù de

---

asymmetric unit [ˌæsi'metrik 'juːnit]

Definition: An asymmetric unit is the smallest portion of a crystal structure to which crystallographic symmetry can be applied to generate one unit cell. The symmetry operations most commonly found in biological macromolecular structures are rotations, translations, and screws (combined rotation and translation). The unit cell is the smallest unit in a crystal that when translated in three dimensions makes up the entire crystal. 不对称晶胞 bú duì chèn jīng bāo

asymmetric: something which is asymmetric displays asymmetry 不对称的 bú duì chèn de

Example:

The molecular packing was analyzed with 10 different crystal forms of T4 lysozyme, where

there is only one protein molecule in an asymmetric unit.
以不对称单位中只有一个分子的 10 种不同晶型的 T4 溶菌酶晶体为材料，对晶体中的分子堆积进行研究。

Extended Term:

asymmetric non-linear unit 不对称非线性元件

asymmetry [æ'simətri] *n.*

Definition: lack of equality or equivalence between parts or aspects of something; lack of symmetry 非(不)对称 fēi (bú) duì chèn

Origin: mid-17th century: from Greek asummetria, from a-"without" + summetria (see symmetry)

Example:

The second part involves voltage controlled slow light effects in asymmetry double quantum dots (QDs).
第二部分是在非对称的双量子点中电压调控的慢光效应。

Extended Terms:

asymmetry coefficient 不对称性系数；偏度系数；偏态系数
membrane asymmetry 膜不对称性；膜的不对称性
functional asymmetry 功能性偏利
asymmetry cut 不对称修剪
baryon asymmetry 重子不对称

atomic scattering factor [ə'tɔmik 'skætəriŋ 'fæktə]

Definition: In physics, the atomic form factor, or atomic scattering factor, is a measure of the scattering amplitude of a wave by an isolated atom. The atomic form factor depends on the type of scattering, which in turn depends on the nature of the incident radiation, typically X-ray, electron or neutron. The common feature of all form factors is that they involve a Fourier transform of a spatial density distribution of the scattering object from real space to k-space (also known as reciprocal space). For an object that is spherically symmetric, the spatial density distribution can be expressed as, $\rho(r)$, a function of radius, so that the form factor $f(Q)$ is defined as: $f(Q) = \int \rho(r) e^{iq-r} d^3 r$, where $\rho(r)$ is the spatial density of the scatterer about the center of mass of the scatterer ($r=0$), and Q is the momentum transfer. As a result of the nature of the Fourier transform, the broader the distribution of the scatterer ρ in real space r, the narrower the distribution of f in Q; i.e., the faster the decay of the form factor. For crystals, atomic form factors are used to calculate the structure factor for a given Bragg peak of a crystal. 原子散射因子 yuán zǐ sǎn shè yīn zǐ

atomic surface [ə'tɔmik 'səːfis]

Definition: The atomic surface is the (n-m)-dimensional object in superspace corresponding to a point atom in physical space. For incommensurately modulated crystals with a smooth modulation function, these atomic surfaces are unbounded. They have the same dimension as the internal space and they have lattice periodicity. For other aperiodic crystals, the atomic surfaces consists of disjoint components, the atomic domains (also called windows, atomic surfaces, or acceptance domains). 原子面 yuán zǐ miàn

automorphism [ˌɔːtəu'mɔːfizəm] *n.*

Definition: In mathematics, an automorphism is an isomorphism from a mathematical object to itself. It is, in some sense, symmetry of the object, and a way of mapping the object to itself while preserving all of its structure. The set of all automorphisms of an object forms a group, called the automorphism group. It is, loosely speaking, the symmetry group of the object. 自同构 zì tóng gòu

Origin: 1870-1875:auto+morph+ism.

Example:

An automorphism of a map is an isomorphism from the map to itself.
一个地图的自同构就是到它本身的一个同构。

Extended Terms:

inner automorphism 内自同构
central automorphism 中心自同构
generating automorphism 生成自同构
reciprocal automorphism 反向自同构
singular automorphism 奇异自同构

axial ['æksiəl] *adj.*

Definition: along the same line as an axis of rotation in geometry 轴的 zhóu de

Origin: from axis

Example:

On Mercury, for example, the interplay between axial tilt and orbital path would make an

analemma that's a nearly straight line from east to west.
例如,在水星上,轴向倾斜和轨道路径会形成从东到西近乎直线的"8"字形曲线。

Extended Terms:

axial dihedron 轴双面
axial intercept 轴截矩
axial length 轴长

axial angle [ˈæksiəl ˈæŋgl]

Definition: The angle formed by two surfaces of a structure, as of a tooth, in which the line of union is parallel with its axis. 轴角 zhóu jiǎo

Example:

The Chinese encoding method, the key technology and the characteristics of pseudorandom code absolute photo electric axial angle encoder are detailed.
详细介绍了中国光电轴角编码器制造业中新出现的一种绝对式编码器,伪随机码绝对式光电轴角编码器。

Extended Terms:

apparent optic axial angle 假光轴角;视光轴角
axial angle apparatus (结晶)轴角器
crystallographic axial angle 结晶轴角
crystal axial angle 晶轴角
optic axial angle 光轴角

axial plane [ˈæksiəl plein]

Definition: a plane that includes two of the crystallographic axes 轴面 zhóu miàn 又作 pinacoid

Example:

Please note that this syncline is of inclined axial plane which has the same dip direction as the limb with gentle dip angle.
请注意,这个向斜构造的轴面是倾斜的,轴面的倾向与具有较缓一翼岩层的倾向一致。

Extended Terms:

crystal axial plane 晶轴面
axial plane 轴平面
optical axial plane 光轴平面;光轴面
axial plane cleavage 轴面劈理

B

axial ratios [ˈæksiəl ˈreiʃiəuz]

Definition: Axial ratio, for any structure or shape with two or more axes, is the ratio of the length (or magnitude) of those axes to each other — the longer axis divided by the shorter. 轴率 zhóu lǜ

Example:

The integral hydroforming technology of oblate spheroidal shells is investigated experimentally and theoretically with various axial length ratios and petal numbers.
对不同轴长率和瓣数的不锈钢扁球壳体进行了整体无模胀成形实验研究及理论探讨。

axis of symmetry [ˈæksis əv ˈsimitri]

Definition: The axis of symmetry or line of symmetry of a two-dimensional figure is a line such that, for each perpendicular constructed, if the perpendicular intersects the figure at a distance “d” from the axis along the perpendicular, then there exists another intersection of the figure and the perpendicular, at the same distance “d” from the axis, in the opposite direction along the perpendicular. 对称轴 duì chèn zhóu

Example:

Of a source, distribution of luminous intensity has an axis of symmetry or at least one plane of symmetry.
对光源而言,发光强度有一个轴对称或至少有一面对称。

Extended Terms:

twofold axis of symmetry 二重对称轴
screw axis of symmetry 螺旋对称轴
ordinary axis of symmetry 普通对称轴;常对称轴
common axis of symmetry 常对称轴
alternating axis of symmetry 更迭对称轴

Babinet compensator [ˈbæbinit ˈkɔmpenseitə]

Definition: The Babinet compensator is a continuously variable, zero-order retarder. It consists of a birefringent wedge which is movable and another birefringent wedge which is fixed to a compensator plate. The orientation of the long axis of the wedges is perpendicular to the long axis of the compensator plate. 巴比涅补偿器 bā bǐ niè bǔ cháng qì

Origin: named after Babinet, a French physicist, mathematician, and astronomer

Example:

Compensator have some types. Babinet compensator and Solet compensator are used most extensively in mechanical compensator. But they have some flaws.

补偿器有多种类型,在机械补偿器中应用最广泛的是巴比涅补偿器和索累补偿器,但它们各有其缺点。

Extended Terms:

Babinet-Solet compensator 巴比涅—索累补偿器

Babinet-Jamin Compensator 巴比涅—亚门补偿器

---

band gap [bænd gæp]

Definition: In solid state physics, a band gap, also called an energy gap or bandgap, is an energy range in a solid where no electron states can exist. In a graph of the electronic band structure of a solid, the band gap generally refers to the energy difference (in electron volts) between the top of the valence band and the bottom of the conduction band which is found in insulators and semiconductors. It is the amount of energy required to free an outer shell electron from its orbit about the nucleus to become a mobile charge carrier, able to move freely within the solid material. In conductors, the two bands often overlap, so they may not have a band gap. 带隙 dài xì

Example:

The reduction of Si band gap and the enhancement of light intensity under external tensile strain are observed.

在施加外加的伸展应力之下,我们观察到矽的带隙缩减还有光强度的增加。

Extended Terms:

band-gap crystal 带隙晶体

band-gap energy 频带隙能量;禁带能量

PBG photonic band-gap 光能隙

basic theory and analysis of electromagnetic band-gap structure 电磁晶体结构之基本理论分析

---

band theory [bænd 'θiəri]

Definition: (*Chemistry/Physics*) A theoretical model describing the states of electrons in solid materials, which have energy values only within certain specific ranges, is called bands. The band theory accounts for many of the electrical and thermal properties of solids and forms the basis of the technology of devices such as semiconductors, heating elements, and capacitors (see capacitance). 能带理论 néng dài lǐ lùn

Example:

The article deals with the oil and rock mineral property according to the molecular orbit theory and the energy band theory of the solid material.
该文依据分子轨道理论和固体物质的能带理论,对石油及岩石矿物的性质进行了论述。

Extended Terms:

band theory of solid 固体能带理论;固体的能带理论;固态能带理论
band theory of ferromagnetism 铁磁性能带理论
energy band theory 能带理论

---

basal pinacoid [ˈbeisəl ˈpinəkɔid]

Definition: also called basal plane, is a pinacoid of two parallel faces that intersect only the crystallographic axis 底轴面 dǐ zhóu miàn

basal: serving as or forming a base 在底部的 zài dǐ bù de

pinacoid: a plane that includes two of the crystallographic axes 轴面 zhóu miàn

---

base-centered lattice [beis ˈsentəd ˈlætis]

Definition: also c-lattice, a space lattice in which each unit cell has lattice points at the centers of each of two opposite faces as well as at the vertices; in a monoclinic crystal, they are the faces normal to one of the lattice axes 底心格子(C-格子)dǐ xīn gé zi

Example:

The expression of relation between Miller indices and interplanar spacing in base-centered monoclinic lattice is needed here.
这里需要底心单斜格子中密勒指数与晶面间距的关系式。

Extended Term:

base-centered orthorhombic lattice 底心正交格子;底心正交格子

Baveno twin [bəˈviːnəu ˈtwin]

Definition: Minerals have been collected at Baveno for centuries. The locality, which gave its name to the minerals bavenite and to the Baveno law for orthoclase twins, has produced some of the world's finest orthoclase crystals.巴韦诺双晶 bā wéi nuò shuāng jīng

Baveno twin law [bəˈviːnəu ˈtwin lɔː]

Definition: An uncommon twin law applicable in feldspar, in which the twin plane and composition surface are. A Baveno twin usually consists of two individuals. 巴韦诺双晶定律 bā wéi nuò shuāng jīng dìng lǜ

Becke line [bek lain]

Definition: The Becke Line Test is a technique in optical mineralogy that helps determine the relative refractive index of two materials. It is done by lowering the stage (increasing the focal distance) of the petrographic microscope and observing which direction the light appears to "move" toward. This movement will always go into the material of higher refractive index. Typically, this is done by comparing 1) two (or more) minerals, 2) a mineral verses thin section epoxy (such as Canada Balsam which has a moderate refractive index of 1.54), and/or 3) a mineral verses an oil of known refractive index (as in oil immersion studies).贝克线 bèi kè xiàn

Origin: The method was developed by Friedrich Johann Karl Becke (1855-1931)

Extended Term:

Becke line method 贝克线法:一种利用浸液和显微镜确定宝石大致折射率的方法

Berek compensator [ˈberik ˈkɔmpənseitə]

Definition: The Berek compensator is an optical device that is capable of quantitatively determining the wavelength retardation of a crystal, fiber, mineral, plastic film or other birefringent material. 贝累克补偿器 bèi lèi kè bǔ cháng qì

Bernal chart [ˈkɜːnl tʃɑːt]

Definition: a chart used to determine the coordinates in reciprocal space of X-ray reflections that produce 贝尔纳图 bèi ěr nà tú

Origin: named after John Desmond Bernal, who created the diagram for interpreting X-ray photographs

B

biaxial crystal [baiˈæksiəl ˈkristəl]

Definition: a crystal of low symmetry in which the index ellipsoid has three unequal axes 二轴晶 èr zhóu jīng

Example:

Based on geometric construction and theoretical derivation, the directions of the two kinds of optical axes in the biaxial crystal have been deduced.
用几何作图和理论推导方法,求得在双轴晶体中两类光轴的方向。

Extended Terms:

positive biaxial crystal 正双轴晶体
indicatrix of optic biaxial crystal 二轴晶光率体

Bijvoet pair [bəˈdʒɔit pɛə]

Definition: The couple of reflections h, k, l and $\bar{h}$, $\bar{k}$, $\bar{l}$ is called a Friedel pair, or Bijvoet pair. Their intensities are equal if there is no absorption, but differ otherwise. Friedel's law then does not hold. Generally speaking, when absorption is present, equivalent reflections generated by the symmetry elements in the crystal have intensities different from those of equivalent reflections generated by the introduction of an additional inversion centre in normal scattering. Friedel, or Bijvoet pairs are used in the resolution of the phase problem for the solution of crystal structures. 贝弗特构型(绝对构型) bèi fú tè gòu xíng (jué duì gòu xíng)

Origin: The use of Friedel's pairs for helping in phase determinations was suggested by Bijvoet J. M., Peerdeman A. F. & van Bommel A. J. (1951), *Determination of the absolute configuration of optically active compounds by means of X-rays. Nature* (London), 168, 271-272.

---

binary operation [ˈbainəri ˌɔpəˈreiʃən]

Definition: In mathematics, a binary operation is a calculation involving two operands, in other words, an operation whose arity is two. Examples include the familiar arithmetic operations of addition, subtraction, multiplication and division. More precisely, a binary operation on a set S is a ternary relation that maps elements of the Cartesian product $S \times S$ to S: $f: S \times S \rightarrow S$. If f is not a function, but is instead a partial function, it is called a partial operation. 二元运算 èr yuán yùn suàn

Example:

In current computer application, the 0 binary operation mode with 1 composition is most

frequently used.
在当今的电脑应用中,采用最多的是 0 和 1 组成的二进制运算模式。

Extended Terms:

binary arithmetic operation 二进制算术运算;二进制运算
binary raster operation code 二元光栅操作码
binary operation relations 二元运算关系

---

biprism ['baiprizm] *n.*

Definition: a prism having a highly obtuse angle to facilitate beam splitting 双棱镜 shuāng léng jìng

Example:

The distribution curves of light intensity of orthogonal biprism interference is made from experimental results.
通过实验结果作出正交双棱镜干涉的光强度分布曲线。

Extended Terms:

biprism interference 双棱镜干涉
Fresnel biprism 菲涅耳双棱镜
Fresnel biprism experiment 弗里司内尔双棱镜实验

---

bipyramid(dipyramid) [bai'pirəmid] *n.*

Definition: An n-gonal bipyramid or dipyramid is a polyhedron formed by joining an n-gonal pyramid and its mirror image base-to-base. The referenced n-gon in the name of the bipyramids is not an external face but an internal one, existing on the primary symmetry plane which connects the two pyramid halves.The face-transitive bipyramids are the dual polyhedra of the uniform prisms and will generally have isosceles triangle faces. A bipyramid can be projected on a sphere or globe as n equally spaced lines of longitude going from pole to pole, and bisected by a line around the equator.双锥 shuāng zhuī

Example:

A trigonal bipyramid has three equatorial and two axial positions.
一个三角双锥体有三个赤道位和两个轴向位。

Extended Terms:

tetragonal bipyramid 正方双锥
trigonal bipyramid 三方双锥;三角双锥;三角双锥体
pentagonal bipyramid 五角双锥

dihexagonal bipyramid 复六方双锥
tetragonal bipyramid class 正方双锥体类

birefringence [ˌbairiˈfrindʒəns] *n.*

Definition: Birefringence, or double refraction, is the decomposition of a ray of light (and other electromagnetic radiation) into two rays (the ordinary ray and the extraordinary ray) when it passes through certain types of material, such as calcite crystals or boron nitride, depending on the polarization of the light. This effect can occur only if the structure of the material is anisotropic (directionally dependent). If the material has a single axis of anisotropy or optical axis, (i.e. it is uniaxial) birefringence can be formalized by assigning two different refractive indices to the material for different polarizations. The birefringence magnitude is then defined by $\Delta n = n_e - n_o$ where n_e and n_o are the refractive indices for polarizations parallel (extraordinary) and perpendicular (ordinary) to the axis of anisotropy respectively. The reason for birefringence is the fact that in anisotropic media the electric field vector $\vec{E}$ and the dielectric displacement $\vec{D}$ can be nonparallel (namely for the extraordinary polarisation), although being linearly related. Birefringence can also arise in magnetic, not dielectric, materials, but substantial variations in magnetic permeability of materials are rare at optical frequencies. Liquid crystal materials as used in Liquid Crystal Displays (LCDs) are also birefringent. 重折率(双折射率)chóng zhé lǜ (shuāng zhé shè lǜ)

Example:

The research results of the intrinsic birefringence of the circular-core side hole fiber are reported. 对圆芯型边孔光纤固有双折射的研究结果进行了报道。

Extended Terms:

ionic birefringence 离子双折射;离子[致]双折射
orientation birefringence 顺向双折射率;取向双折射
amorphous birefringence 非晶[形]双折射;非晶双折射
nonlinear birefringence 非线性双折射
strain birefringence 应变双折射;应变双折射

bisectrix [baiˈsektriks] *n.*

Definition: (*Crystallography*) the line which bisects the angle of the optic axes of a crystal 等分线 děng fēn xiàn

Example:

The paper analyzes the algorithm of building linear buffer zone by angular bisectrix method, then puts forward a new method for getting angular bisectrix and buffer point.

本文分析了用对角平分线法建立缓冲区的算法,提出了一种新的求角平分线和缓冲点的方法。

Extended Terms:

acute bisectrix 锐角等分线;敏锐二等分角
obtuse bisectrix 钝角等分线;钝等分线

body-centered lattice [ˈbɔdi-ˈsentəd ˈlætis]

Definition: This unit cell can be thought as made by stuffing another atom into the center of the simple cubic lattice, slightly spreading the corners. Thus, the corner spheres no longer quite touch one another, but do touch the center. The diagonal through the body of the cube is 4x (sphere radius). The packing efficiency of a body-centered lattice is considerably higher than that of a simple cubic: 69.02%. The higher coordination number and packing efficiency mean that this lattice uses space more efficiently than simple cubic. This lattice can also be thought of as made from layers of square-packed spheres. However, alternating layers stack so that the spheres "nestle down" into the spaces between the spheres in the layer below. In a simple cubic lattice the spheres stack directly atop one another. 体心格子 tǐ xīn gé zi 又称 I-lattice

Example:

The crystal structure of Poly(phthalocyaninatostannoxane), $[Sn(P_c)O]_n$, was studied with X-ray and electron diffraction. The experimental diffraction date can be indexed based on a body-centered tetragonal crystal lattice, with a = 17.99, c = 3.91 and z = 2。
本文用X射线衍射和电子衍射研究了聚酞菁锡氧烷$[Sn(P_c)O]_n$的晶体结构,测定出该晶体属体心四方点阵,晶胞参数为a = 17.99, c = 3.91和z = 2。

Extended Terms:

body-centered tetragonal lattice 体心四方格子
body-centered cubic lattice 体心立方格子;晶格
body-centered orthorhombic lattice 体心正交格子
body-centered cubic lattice (B.C.C.) 体心立方晶格

bond energy [bɔnd ˈenədʒi]

Definition: (*Chemistry*) Bond energy (E) is a measure of bond strength in a chemical bond. It is the heat required to break Avogadro's number of molecules into their individual atoms. For example, the carbon-hydrogen bond energy in methane E (C-H) is the enthalpy change involved with breaking up one molecule of methane into a carbon atom and 4 hydrogen radicals divided by 0.5. 键能 jiàn néng

Example:

The composition of butadiene polyperoxide is analyzed. The energy of explosion reactions is

calculated by bond energy method.
分析了丁二烯聚过氧化物的组成,用键能法计算了爆炸反应能量。

Extended Terms:

bond dissociation energy 键的离解能;键裂解能;键离解能
chemical-bond energy 化学键能
nucleus bond energy 核结合能
phosphate bond energy 磷酸键能
average bond energy 平均键能

bond length [bɔnd leŋθ]

Definition: (*Molecular geometry*) Bond length or bond distance is the average distance between nuclei of two bonded atoms in a molecule. Bond length is related to bond order, when more electrons participate in bond formation, the bond will get shorter. Bond length is also inversely related to bond strength and the bond dissociation energy, as a stronger bond is also a shorter bond. In a bond between two identical atoms half the bond distance is equal to the covalent radius. Bond lengths are measured in molecules by means of X-ray diffraction. A set of two atoms sharing a bond is unique going from one molecule to the next. For example, the carbon to hydrogen bond in methane is different from that in methyl chloride. It is, however, possible to make generalizations when the general structure is the same. 键长 jiàn cháng

Example:

The hydrogen bond length between the soap molecules and molecular bulk modulus of the soap in lubricating greases were also calculated.
采用分子模拟技术对润滑脂滴点进行了初步预测,并计算了润滑脂中各种金属皂分子间氢键键长和分子体积模量。

Extended Terms:

effective bond length 有效键长
bond valence-bond length correlation 键价—键长关联

bond strength [bɔnd streŋθ]

Definition: (*Chemistry*) Bond strength is measured between two atoms joined in a chemical bond. It is the degree to which each atom linked to another atom contributes to the valency of this other atom. Bond strength is intimately linked to bond order. Bond strength can be quantified by bond energy and bond dissociation energy. Another criterion of bond strength is the qualitative relation between bond energies and the overlap of atomic orbitals of the bonds (Pauling and Mulliken). The more these overlap, the more the bonding electrons are to be found between the nuclei and hence stronger will be the bond. This overlap can be calculated and is called the overlap integral. 键强 jiàn qiáng

Example:

Bond strength is an important parameter for wafer bond.
键强是关系到键合好坏的一个重要参数。

Extended Terms:

green bond strength 湿砂结合强度
adhesive bond strength tester 黏合强度试验机
dry bond strength 干态强度;干黏结强度
bond strength test 接合强度测定
shear bond strength 剪切强度

Borrmann effect [ˈbɔmən iˈfekt]

Definition: The irregular transmission of X-rays when a single crystal of high perfection is placed in a monochromatic X-ray beam in a reflecting position. 博曼效应 bó màn xiào yìng

Origin: The Borrmann effect was first discovered in quartz (Borrmann G., 1941, Über Extinktionsdiagramme der Röntgenstrahlen von Quarz. Physik Z., 42, 157-162) and then in calcite crystals (Borrmann G., 1950, Die Absorption von Röntgenstrahlen in Fall der Interferenz. Z. Phys., 127, 297-323), and interpreted by Laue (Laue, M. 1949, Die Absorption der Röntgenstrahlen in Kristallen im Interferenzfall. Acta Crystallogr, 2, 106-113).

Bragg angle [bræg ˈæŋgl]

Definition: The angle between an incident X-ray beam and a set of crystal planes for which the secondary radiation displays maximum intensity as a result of constructive interference. 布拉格角 bù lā gé jiǎo

Origin: named after Sir William Henry Bragg and Sir William Lawrence Bragg

Example:

With Mica crystal as dispersive element, Bragg angle is in the range of 30°~60°.
分光晶体使用云母晶体材料,布拉格角范围为30°~60°。

Extended Terms:

Bragg reflection angle 布拉格反射角
Bragg scattering angle 布拉格散射角

B

Bragg equation [bræg iˈkweiʃ(ə)n]

Definition: relates the angles at which X-rays are scattered from a crystal to the spacing between the layers of molecules 布拉格方程 bù lā gé fāng chéng

Origin:

This is the equation derived by William H. Bragg and his son William L. Bragg to explain the scattering (diffraction) of X-rays from crystals, and show how the angle of incidence between the beam and the layer (θ) is related to the wavelength of the X-rays (λ) and the separation of the scattering layers (d). n is a positive integer. The Braggs received the 1915 Nobel Prize for Physics for the work that included this. This is the fundamental theory behind X-ray crystallography.

Example:

According to the ideal approximation of the coupled-mode theory, the Bragg equation and the peak reflectivity expression of the fiber Bragg grating are derived.
根据理想模展开下的耦合模方程,对光纤布拉格光栅的峰值反射率公式进行了数学推导,得到了布拉格光纤光栅的光谱反射率表达式。

Bragg R-factor [bræg aːˈfæktə]

Definition: The R-factor measuring the agreement between the reflection intensities calculated from a crystallographic model and those measured experimentally. $R_B = \frac{\sum |I_{obs} - I_{calc}|}{\sum |I_{obs}|}$.

In the Rietveld method R_B is useful because it depends on the fit of the structural parameters and not on the profile parameters. 布拉格 R 系数 bù lā gé R xì shù

Bragg's law

Definition: Bragg's law provides the condition for a plane wave to be diffracted by a family of lattice planes:

$$2d \sin \theta = n \lambda.$$

where d is the lattice spacing, θ the angle between the wave vector of the incident plane wave, k_o, and the lattice planes, λ its wave length and n is an integer, the order of the reflection. It is equivalent to the diffraction condition in reciprocal space and to the Laue equations. 布拉格定律 bù lā gé dìng lǜ

B

Bravais class [brə'vei klɑːs]

Definition: A Bravais arithmetic crystal class, or Bravais class for short, is an arithmetic crystal class with matrix group of lattices. Each lattice is associated with a Bravais class, and each matrix group of a Bravais class represents the point group of a lattice referred to an appropriate primitive basis. 布拉维晶组 bù lā wéi jīng zǔ

Bravais flock [brə'vei flɔk]

Definition: Each Bravais flock consists of the space groups with the same Bravais lattice type. 布拉维群 bù lā wéi qún

---

Bravais indices [brə'vei 'indisiːz]

Definition: A modification of the Miller indices; frequently used for hexagonal and trigonal crystalline systems; they refer to four axes: the *c* axis and three others at 120° angles in the basal plane.布拉维指数 bù lā wéi zhǐ shù

Extended Terms:

Milier-Bravais indices 米勒-布拉维指数
Bravais-Miller indices 布拉维-米勒指数

Bravais lattice [brə'vei 'lætis]

Definition: (*Crystallography*) one of the 14 possible arrangements of lattice points in space such that the arrangement of points about any chosen point is identical with that about any other point 布拉维晶格 bù lā wéi jīng gé

Origin: In geometry and crystallography, a Bravais lattice was studied by Auguste Bravais (1850).

Example:

Furtherly the author studied half space configurations of four beams (HCFBs), or also so-called pan-umbrella configurations and proved that any Bravais lattice can be formed by HCFBs.
该作者进一步考察了半空间的四光束配置,或者可以称之为泛伞形配置,证明任意一种布拉维晶格都能由半空间光束配置干涉生成。

Extended Terms:

Bravais space lattice 布拉维空间晶格
Miller-Bravais lattice 米勒-布拉维晶格

Bravais' space lattice 布拉维空间点阵

Bravais-Miller indices [brə'vei-milə 'indisiːz]

Definition: (*Crystallography*) A modification of the Miller indices; frequently used for hexagonal and trigonal crystalline systems; they refer to four axes: the c axis and three others at 120° angles in the basal plane. 布拉维-密勒指数 bù lā wéi mì lè zhǐ shù

Origin: The Miller indices were first introduced, among others, by W. Whevell in 1829 and developed by W. H. Miller, his successor at the Chair of Mineralogy at Cambridge University, in his book A Treatise on Crystallogtaphy (1839) — see Historical Atlas of Crystallography (1990), edited by J. Lima de Faria, published for the International Union of Crystallography by Kluwer Academic Publishers, Dordrecht.

Example:

The orientations of all developing facets in the crystals and their Bravais-Miller indices have been determined by means of X-ray and optical method.
用 X 射线和光学方法测定了各个显露面,并确定其布拉维-密勒指数。

Brazil twin law [brə'zil twin lɔː]

Definition: a form of quartz twinning where right and left hand quartz structures are combined in a single crystal 巴西双晶律 bā xī shuāng jīng lǜ,又称为光学双晶(optical twinning), guāng xué shuāng jīng

Example:

No electrical or brazil twin law (or optical twinning) shall be allowed in usable zones.
可用区内不允许存在电学或光学双晶。

Brillouin zone [brilɔi'in zəun]

Definition: (*Mathematics/Physics*) The first Brillouin zone is a uniquely defined primitive cell in reciprocal space. The boundaries of this cell are given by planes related to points on the reciprocal lattice. It is found by the same method as for the Wigner-Seitz cell in the Bravais lattice. The importance of the Brillouin zone stems from the Bloch wave description of waves in a periodic medium, in which it is found that the solutions can be completely characterized by their behavior in a single Brillouin zone. 布里渊区 bù lǐ yuān qū

Origin: The concept of a Brillouin zone was developed by Léon Brillouin (1889-1969), a French physicist.

Example:

The five possible normal polyhedrons are presented. It is impossible for crystals as well as their Brillouin zones to take the shape of normal dodecahedron.
介绍了可能的五种正多面体,指出正十二面体是晶体结构及其布里渊区不可能具有的正多面体形状。

bubble [ˈbʌbl] *n.*

Definition: an air- or gas-filled spherical cavity in a liquid or a solidified liquid such as glass or amber 气泡 qì pào

Origin: Middle English: partly imitative, partly an alteration of burble

Example:

At the tip of the gas bubble, it becomes long and thin for the second penetration.
在气泡末端,由于气体的二次穿透作用而使气泡变得细而长。

Extended Terms:

bubble trap 除泡器;气泡收集器;气泡陷阱;气泡捕捉器
bubble impression 泡痕;气泡痕迹;气泡痕
bubble pressure 气泡压力;泡压
bubble level 气泡水平仪;水准仪;气泡水准仪;水准器
bubble accumulator 气泡储存器

Canada balsam [ˈkænədə ˈbɔːlsəm]

Definition: Canada balsam, also called Canada turpentine or balsam of fir, is a turpentine which is made from the resin of the balsam fir tree (Abies balsamea). The resin, dissolved in essential oils, is a viscous, sticky, colourless (sometimes yellowish) liquid that turns to a transparent yellowish mass when the essential oils have been allowed to evaporate. Canada balsam is amorphous when dried. It does not crystallize with age, so its optical properties do not deteriorate. However, it has poor thermal and solvent resistance. 加拿大树胶 jiā ná dà shù jiāo

Origin: Old English, via Latin from Greek balsamon

Example:

Canada balsam was substituted by water-soluble acacia, which made the making process more

secure.
采用水溶性的阿拉伯树胶取代加拿大树胶，使试验过程更加安全。

Extended Terms:

bottle for Canada balsam 加拿大香脂瓶
abies balsamea (balsam Canada) extract 香脂冷杉提取物

Carlsbad twin [ˈkaːlzbæd twin]

Definition: form of penetration twinning where two Orthoclase crystals form interpenetrating twins as depicted in the figure below 卡尔斯巴双晶(卡式双晶) kǎ ěr sī bā shuāng jīng (kǎ shì shuāng jīng)

Extended Term:

carlsbad-albite compound twin 卡钠复合双晶

---

cartesian product [kaːˈtiziən ˈprɔdʌkt]

Definition: In mathematics, a Cartesian product (or product set) is the direct product of two sets. Specifically, the Cartesian product of two sets X (for example, the points on an X-axis) and Y (for example the points on a Y-axis), denoted $X \times Y$, is the set of all possible ordered pairs whose first component is a member of X and whose second component is a member of Y (e.g., the whole of the X-Y plane): $X \times Y = \{(x, y) \mid x \in X \text{ and } y \in Y\}$. 笛卡儿积 dí kǎ ér jī

Origin: The Cartesian product is named after René Descartes, whose formulation of analytic geometry gave rise to this concept.

Example:

A join is a Cartesian product of two row sets, with the join predicate applied as a filter to the result.
一个连接是两个行集的笛卡儿乘积，谓词连接作为一个过滤结果应用。

Extended Terms:

fuzzy Cartesian product 模糊笛卡儿积；模糊卡的逊素积
Extended Cartesian Product 笛卡儿积；广义笛卡儿积；扩充的笛卡儿积

---

cell parameter [sel pəˈræmitə]

Definition: parameters of units of crystal organisms 晶胞参数 jīng bāo cān shù

cell: the basic structural and functional unit of all organisms 细胞 xì bāo

parameter: a constant in the equation of a curve that can be varied to yield a family of similar curves 参数 cān shù

Example:

The results demonstrated that the as-prepared lead alloy had the same crystalline structure as pure lead, while the cell parameter was decreased by adding Cd, Sb and Ag.
结果表明,自制的铅合金仍具有纯铅的晶体结构,其晶胞参数由于掺入较小原子半径的合金元素而变小。

Extended Terms:

unit-cell parameter 单个电极参数
parameter of cell kinetics 细胞动力学参数

---

center ['sentə] *n.*

Definition: the middle of an object 中心 zhōng xīn

Origin: from the Greek κεντρον /kentron/

Example:

As contraction continues and the density increases, the dust becomes so thick at the center of the cloud that it absorbs the far-infrared radiation.
当收缩过程继续进行下去,密度不断增大时,气尘云中心的尘埃就会变得极其浓密,因而能够吸收远红外辐射。

Extended Terms:

center buckle 表面中部波皱
instantaneous center 瞬心;瞬时中心

---

centralizer ['sentrəlaizə] *n.*

Definition: In group theory, the centralizer and normalizer of a subset *S* of a group *G* are subgroups of *G* which have a restricted action on the elements of *S* and *S* as a whole, respectively. These subgroups can provide insight into the structure of *G*. 扶正器 fú zhèng qì

Example:

The centralizer has the properties of high tensile strength, impact resistance, no deformation with high temperature, high hardness, low wear rate, oil, water, acid, alkali resistance, anti-aging, etc.
该扶正器具有强度高、耐冲击、高温不变形、硬度高、磨损率低、耐油、耐水、耐酸、耐碱、抗老化等特点。

Extended Terms:

centralizer element 中心化子元素

bottom centralizer 底部扶正器
centralizer blade 定中簧片
tubing centralizer 油管扶正器
star centralizer 星形定心器

centre of symmetry(symmetry centre) [ˈsentə əv ˈsimitri]

Definition: (*Mathematics*) A point, within an object or figure, through which any straight line also passes through two points on the edge of the figure at the same distance from the centre but on opposite sides(also inversion centre). 对称中心,又称对称心、反伸中心 duì chèn zhōng xīn(duì chèn xīn, fǎn shēn zhōng xīn)

symmetry: an attribute of a shape or relation; exact reflection of form on opposite sides of a dividing line or plane 对称 duì chèn

centred lattice [ˈsentəd ˈlætis]

Definition: When the unit cell does not reflect the symmetry of the lattice, it is usual in crystallography to refer to a "conventional", non-primitive, crystallographic basis, a_c, b_c, c_c instead of a primitive basis, a, b, c. This is done by adding lattice nodes at the center of the unit cell or at one or three faces. The vectors joining the origin of the unit cell to these additional nodes are called "centring vectors". In such a lattice a_c, b_c and c_c with all their integral linear combinations are lattice vectors again, but there exist other lattice vectors $t \in L$, $t = t_1 a_c + t_2 b_c + t_3 c_c$; with at least two of the coefficients t_1, t_2, t_3 being fractional. The table below gives the various types of centring vectors and the corresponding types of centring. Each one is described by a letter, called the Bravais letter, which is to be found in the Hermann-Mauguin symbol of a space group. The "multiplicity", m, of the centred cell is the number of lattice nodes per unit cell. The volume of the unit cell, $V_c = (a_c, b_c, c_c)$ is given in terms of the volume of the primitive cell, $V = (a, b, c)$, by: $V_c = m\,V$ 定心格子 dìng xīn gé zi

centred: concentrated on or clustered around a central point or purpose 在中心的 zài zhōng xīn de

lattice: an arrangement of points or particles or objects in a regular periodic pattern in 2 or 3 dimensions 晶格 jīng gé

class of symmetry [klaːs əv ˈsimitri]

Definition: a class of crystals determined by a combination of their symmetry elements, all crystals left unchanged by a given set of symmetry elements being placed in the same class. Also called symmetry class. 对称型 duì chèn xíng

C-lattice *n.* C-格子 see base-centered lattice

cleavage [ˈkliːvidʒ] *n.*

Definition: the splitting of rocks or crystals in a preferred plane or direction 解理 jiě lǐ

Example:

The lay of the strata can imply that the rock splits easily along its direction of cleavage.
成层的地层意味着岩石易于沿着其解理方向裂开。

Extended Terms:

cleavage strength 解理强度;劈裂强度;劈理度;劈理强度
cleavage crack 解理裂纹;解理缝;劈理裂隙;劈裂
spiral cleavage 旋裂;螺旋卵裂;旋卵裂
parallel cleavage 平行劈理
slip cleavage 滑劈理;滑动劈理

Clerici's solution [kləˈritʃiːz səˈljuːʃən]

Definition: Clerici solution is a solution of equal parts of thallium formate ($Tl(CHO_2)$) and thallium malonate ($Tl(C_3H_3O_4)$) in water. It is a freely flowing, odorless liquid which changes from yellowish to colorless upon decreasing the concentration of the thallium salts. At a density of 4.25 g/cm^3 at 20℃, the saturated Clerici solution is one of the heaviest aqueous solutions known. 克列里奇液 kè liè lǐ qí yè

Origin: The solution was invented in 1907 by the Italian chemist Enrico Clerici (1862-1938) and introduced to mineralogy and gemology in 1930s as a valuable liquid, which allowed separating minerals by density with a traditional flotation method. Its advantages include transparency and variable and easily controllable density in the range 1-5 g/cm^3.

closed form [kləuzd fɔːm]

Definition: In mathematics, closed form may refer to: closed-form expression, a finitary expression and closed differential form, a differential form α with the property that $d\alpha = 0$.闭形 bì xíng

Example:

This paper presents a closed form analysis method of nonlinear mirror mode locking of a homogeneously broadened Laser.

对于均匀展宽激光器的非线性镜锁模,该文提出了一种闭形分析方法。

Extended Terms:

closed-form solution 闭型解;闭合解;闭合形解
closed differential form 闭微分(形)式
closed-form expression 解析解
closed form measure 闭式度量;闭合形式测度

---

columnar [kə'lʌmnə] *n.*

Definition: an upright pillar, typically cylindrical and made of stone or concrete, supporting an arch, entablature, or other structure or standing alone as a monument. 柱状 zhù zhuàng

Origin: late Middle English: partly from Old French columpne, reinforced by its source, Latin columna "pillar"

Example:

Snow crystals grow in one of two designs — platelike and columnar.
雪晶是按照两种形状中的一种——板状和圆柱形形成的。

Extended Terms:

columnar structure 柱状组织;柱状构造;柱状结构;常呈现柱状结晶
columnar section 地层柱状剖面图;柱状剖面;柱状图;柱状剖面图
columnar grain 柱状晶粒

---

combination form [ˌkɔmbi'neiʃən fɔːm]

Definition: a combination of two or more crystal forms 聚形 jù xíng

combination: a collection of things that have been combined; an assemblage of separate parts or qualities 组合 zǔ hé

form: a category of things distinguished by some common characteristic or quality 形式 xíng shì

Origin: late Middle English: from late Latin combinatio (noun), from the verb combinare "join two by two"
Middle English: from Old French forme (noun), fo(u)rmer (verb, from Latin formare "to form"), both based on Latin forma "a mould or form"

Example:

In the great majority of cases, a crystal shows a combination of forms rather than a single form.
绝大部分情况下,水晶呈现聚形而非单一形态。

C

compensator [ˈkɔmpənseitə] *n.*

Definition: a device that offsets or counterbalances a destabilising factor, which acts so as to neutralize or correct (a deficiency or abnormality in a physical property or effect) 消色器(补色器)xiāo sè qì(bǔ sè qì)

Origin: mid 17th century (in the sense "counterbalance"): from Latin compensat- "weighed against", from the verb compensare, from com- "together" + pensare (frequentative of pendere "weigh")

Example:

The discussion lay a strong emphasis on the application of all-purpose and gemel corrugated-pipe compensator to different piping section.

本文着重论述了万能型和铰链型波纹补偿器在不同管段中的应用。

Extended Terms:

heave compensator 波浪补偿器;升沉补偿器;升沉补偿装置

starting compensator 起动补偿器;自耦变压起动器

synchronous compensator 同步调相机;同步补偿器

dynamic compensator 动态补偿器

sensitivity compensator 灵敏度补偿器;灵敏补偿器

---

complete extinction [kəmˈpliːt ikˈstiŋkʃən]

Definition: the disappearance of crystal rays 全消光 quán xiāo guāng

complete: perfect and complete in every respect 完全的 wán quán de

extinction:no longer in existence 消失 xiāo shī

Example:

The presence of stray light is indicated by the absence of complete extinction when the phase difference is such that this should be achieved.

若位差达到应当消光的程度,却未能完全消光即表示有杂散光存在。

---

complex [ˈkɔmpleks] *n.*

Definition: any loosely bonded species formed by the association of two molecules 络合物 luò hé wù

Origin: mid 17th century (in the sense "group of related elements"): from Latin complexus, past participle (used as a noun) of complectere "embrace, comprise", later associated with complexus "plaited"; the adjective is partly via French complexe

C

Example:

Starch is a complex carbohydrate which combines with water, and causes the material to swell and change character.
淀粉是一种复杂的碳水化合物,淀粉与水结合后会膨胀并引起性能的变化。

Extended Terms:

complex salt 络盐;复盐;络合盐;错化物
complex amplitude 复振幅;复数幅值
complex ore 复合矿石;复杂矿石;复合矿

---

composition plane [ˌkɔmpəˈziʃən plein]

Definition: (*Crystallography*) a planar composition surface in a crystal uniting two individuals of a contact twin 双晶接合面 shuāng jīng jiē hé miàn

composition:the way in which someone or something is composed 组合 zǔ hé

plane:an unbounded two-dimensional shape 平面 píng miàn

Extended Term:

plane of composition 接合面

---

compound twin [ˈkɔmpaund twin]

Definition: a twin crystal or one seeming to be made up of two or more crystals combined according to regular laws of composition 复合双晶 fù hé shuāng jīng

compound:consisting of two or more substances or ingredients or elements or parts 复合的 fù hé de

twin:crystal individuals related by a reflection (either plane or centre of symmetry) or a rotation 双晶 shuāng jīng

Extended Term:

carlsbad-albite compound twin 卡钠复合双晶

---

conduction band [kənˈdʌkʃən bænd]

Definition: In the field of semiconductors and insulators, the conduction band is the range of electron energies, higher than that of the valence band, sufficient to free an electron from binding with its individual atom and allow it to move freely within the atomic lattice of the material. Electrons within the conduction band are mobile charge carriers in solids, responsible for conduction of electric currents in metals and other good electrical conductors. 导带 dǎo dài

Example:

The highest-energy band containing electrons is called the valence band, and the next higher one is the conduction band.
填有电子而能量最高的能带称为价带，相邻的更高能带称为导带。

Extended Terms:

bottom of conduction band 导带底
effective density of state of conduction band 导带有效状态密度
impurity band conduction 杂质带传导；杂质带电导
impurity-band conduction theory 杂质带导电理论

---

conductivity [ˌkɔndʌkˈtiviti] *n.*

Definition: Conductivity may refer to: 1) electrical conductivity, a measure of a material's ability to conduct an electric current; 2) conductivity (electrolytic), a measurement of an electrolytic solution, such as water; 3) hydraulic conductivity, a property of a porous material's ability to transmit water; 4) thermal conductivity, the intensive property of a material that indicates its ability to conduct heat; 5) rayleigh conductivity, describing the behavior of apertures concerning the flow of a liquid or gas 传导性 chuán dǎo xìng

Example:

The result shows that the material has high strength and good conductivity.
结果表明，处理后材料具有较高的强度和良好的导电性能。

Extended Terms:

heat conductivity 导热系数；热导率；导热度；热传导性
acoustic conductivity 声导率；传声性
earth conductivity 大地电导率
anisotropic conductivity 各向异性传导率；非均质电导
extrinsic conductivity 外质电率；外质导电率；外在传导率

---

conjugacy class [ˈkɔndʒugəsi klɑːs]

Definition: (*Mathematics, especially group theory*) The elements of any group may be partitioned into conjugacy classes; members of the same conjugacy class share many properties, and study of conjugacy classes of non-abelian groups reveals many important features of their structure. In all abelian groups every conjugacy class is a set containing one element (singleton set). Functions that are constant for members of the same conjugacy class are called class functions. 共轭类 gòng è lèi

Example:

This paper gives several ways by knowledge of combinatorial mathematics and abstract algebra, in which determines the number of elements in every conjugacy class of permutation group Sn. 文章利用组合数学和抽象代数的知识给出了确定置换群 Sn 中每个共轭类所含元素数目的不同方法。

constant form [ˈkɔnstənt fɔːm]

Definition: one of the crystal forms that is solidly existent and recurringly observed 定形 dìng xíng

constant: persistent in occurrence and unvarying in nature 恒定的 héng dìng de

form: a category of things distinguished by some common characteristic or quality 形式 xíng shì

Example:

There is, however, one additional constant form that you need to know about.
不过,还有一种定形需要了解。

Extended Terms:

constant applicative form 常数应用形式
form-constant 树形常数
form field constant 表格字段常数

constrained refinement [kənˈstreind riˈfainmənt]

Definition: A refinement is said to be constrained if one or more parameters in the refinement are held fixed or are determined by the value of one or more refined parameters. Constraints related to space group symmetry are not usually counted among the constraints applied to a given refinement as they are always present where applicable. 强制性精修 qiáng zhì xìng jīng xiū

constrained: lacking spontaneity; not natural 强制的 qiáng zhì de

refinement: the result of improving something 改进 gǎi jìn

contact twin [ˈkɔntækt twin]

Definition: a twin characterized by a definite composition surface separating the two individuals 接触双晶 jiē chù shuāng jīng

contact: close interaction 接触 jiē chù

Example:

On the crystal surfaces of the beryl crystal at {0001}, {1010}, {1121} occur universally screw dislocation, constriction screw dislocation, stacking fault, and contact twin that serve as the major step like source for the growth of beryl crystal.

在绿柱石晶体的{0001}、{1010}、{1121}面上，普遍发育螺旋位错、束合螺旋位错、层错及接触双晶，它们共同构成了绿柱石晶体生长的主要台阶源。

Extended Term:

twin contact 双触点；双子接点

---

conventional cell [kən'venʃənəl sel]

Definition: For each lattice, the conventional cell is the cell obeying the following conditions: its basis vectors define a right-handed axial setting; its edges are along symmetry directions of the lattice; it is the smallest cell compatible with the above condition. Crystals having the same type of conventional cell belong to the same crystal family. 惯用晶胞 guàn yòng jīng bāo

conventional: conforming with accepted standards 惯例的 guàn lì de

cell: the basic structural and functional unit of all organisms 细胞 xì bāo

Example:

The method was similar to conventional cell culture ELBA in reproducibility and sensitivity ($P>0.05$).

此法的重复性良好，灵敏度与常规组织培养法相似($P>0.05$)。

Extended Term:

conventional unit cell 惯用单胞，简称单胞

---

convergent [kən'vəːdʒənt] *n.*

Definition: Convergent is related to convergence, which is the approach toward a definite value, a definite point, a common view or opinion, or toward a fixed or equilibrium state. 会聚光 huì jù guāng

Origin: early 18th century: from late Latin convergent- "inclining together", from the verb convergere

Example:

The convergent and divergent velocities of active plate boundaries in the north hemisphere are obtained with space geodetic data.

利用现代空间大地测量技术测出北半球活动板块边缘会聚、扩张和滑动速度。

Extended Terms:

convergent oscillation 减幅振荡；减幅摇荡；衰减振荡

convergent synthesis 汇集合成

convergent juncture 会聚接合带；聚合界线；聚合线

C

convergent beam 收敛射束;收敛光束;会聚光束
convergent reaction 收敛反应;收聚反应;会聚反应
convergent streaks 辐合条纹

convergent light figure [kən'vəːdʒənt lait 'figə]

Definition: A conoscopic interference pattern or interference figure is the best way to determine if a mineral is uniaxial or biaxial and also for determining optic sign in optical mineralogy. The observed interference figure essentially shows all possible birefringence colors at once, including the extinctions (in dark bands called isogyres). 干涉图 gān shè tú

Extended Terms:

monoaxial interference figure 一轴干涉像;单轴干涉图
oblique interference figure 斜交干涉图
pseudo interference figure 假干涉图
biaxial interference figure 二轴晶干涉图;二轴干涉像
interference figure complain 干涉像
bull's-eye interference figure 牛眼干涉图

converse piezoelectricity [kən'vɜːs paɪˌiːzəʊɪlek'trɪsɪtɪ]

Definition: The charge that accumulates in certain solid materials in response to applied mechanical stress also works in the opposite way, with the material deforming slightly when a small electric current is applied to it, encouraging their use in instruments for which great degrees of mechanical control are necessary.反压电性 fǎn yā diàn xìng

converse:a logical implication with the propositions reversed; see conversion (logic) 反向的 fǎn xiàng de

piezoelectricity:Piezoelectricity is the ability of certain crystals to produce a voltage when subjected to mechanical stress. The word is derived from the Greek piezein, which means to squeeze or press. The effect is reversible; piezoelectric crystals, subject to an externally applied voltage, can change shape by a small amount. The effect is of the order of nanometres, but nevertheless finds useful applications such as the production and detection of sound, generation of high voltages, electronic frequency generation, and ultrafine focusing of optical assemblies. 压电性 yā diàn xìng

coordination number [kəuˌɔːdi'neiʃən 'nʌmbə]

Definition: In chemistry and crystallography, the coordination number of a central atom in a molecule or crystal is the number of its nearest neighbours. This number is determined somewhat differently for molecules and for crystals. In chemistry the emphasis is on bonding structures in molecules or ions and the coordination number of an atom is determined by simply counting the other atoms to which it is bonded (by either single or multiple bonds). For example, $[Cr(NH_3)_2Cl_2Br_2]^{1-}$ has Cr^{3+} as its central cation, which has a coordination number of 6. However the solid-state structures of crystals often have less clearly defined bonds, so a simpler model is used, in which the atoms are represented by touching spheres. In this model, the coordination number of an atom is the number of other atoms which it touches. For an atom in the interior of a crystal lattice, the number of atoms touching the given atom is the bulk coordination number; for an atom at a surface of a crystal, this is the surface coordination number. 配位数 pèi wèi shù

Example:

The greater the pore coordination number was, the faster the material was dried.
配位数值越大,物料干燥越快。

Extended Terms:

lattice coordination number 品格配位数,点阵配位数;晶格配位数
pore coordination number 孔隙配位数
average coordination number 平均配位数
coordination number of the lattice 点阵配位数
throat-to-pore coordination number 喉道—孔隙配位数

---

coordination polyhedron [kəuˌɔːdiˈneiʃən ˌpɔliˈhiːdrən]

Definition: the symmetrical polyhedral chemical structure of relatively simple polyatomic aggregates having coordination numbers of 4 to 8 配位多面体 pèi wèi duō miàn tǐ

coordination: the regulation of diverse elements into an integrated and harmonious operation 配位 pèi wèi

polyhedron: a solid figure bounded by plane polygons or faces 多面体 duō miàn tǐ

Example:

According to the growth unit model of anion coordination polyhedron, the effects of KOH concentrations on the phase, particle size and morphology of nanopowder were investigated.
结合负离子配位多面体生长基元模型,研究了 KOH 浓度对粉体物相、粒径和形貌的影响。

---

corresponding twins [ˌkɔːriˈspɔndiŋ twinz]

Definition: In case of twinning by merohedry, when the twin element is twofold and the crystal is centrosymmetric, the twin operation can be described either as a rotation or as a reflection (the two operations being equivalent under the action of the center). In case of twinning by pseudomerohedry, instead, the two twin operations are no longer equivalent even in centrosymmetric crystals but produce different twins, which are called reciprocal twins or corresponding twins. One of the most classical examples is that of albite vs. pericline twins in feldspars. Pairs of corresponding twins normally do not have the same frequency of occurrence, as one would be tempted to assume from the identical value of their twin obliquity. 对称双晶 duì chèn shuāng jīng

corresponding: accompanying 对应的 duì yìng de

---

coset [ˈkəuset] *n.*

Definition: (*Mathematics*) a set composed of all the products obtained by multiplying each element of a subgroup in turn by one particular element of the group containing the subgroup [数]傍系 bàng xì

Example:

Further, we can also define if a coset graph is a graph representation, we note it GR for short, that is, the relevant group is exactly equal to the full automorphism group of the coset graph.

甚至,我们也可以定义一个陪集图是一个图表示,简称 GR,即相应的群刚好等于该陪集图的全自同构群。

Extended Terms:

double coset 重倍集;双陪集;双傍集
right coset 右陪集;右傍集
coset weight 陪集权;伴集权
coset leader 陪集首
left coset 左陪集;左倍系
coset space 陪集空间;傍集空间;傍系空间

---

cotectic crystallization [kəuˈtektik ˌkristəlaiˈzeiʃən]

Definition: simultaneous crystallization of two or more solid phases from a single liquid over a finite range of falling temperature without resorption 同结晶(共结晶)tóng jié jīng(gòng jié jīng)

cotectic: referring to conditions of pressure, temperature, and composition under which two or more solid phases crystallize at the same time, with no resorption, from a single liquid over a finite range of decreasing temperature 共结的 gòng jié de

crystallization: the formation of crystals 结晶 jié jīng

C

cotype [ˈkəutaip] *n.*

Definition: a term formerly used for either syntype or paratype 共型 gòng xíng

Origin: co+type

Extended Terms:

agroe cotype 农业生态型
cotype specimen 本模标本

---

covalency [kəuˈveiləns i] *n.*

Definition: (*Chemistry*) of, relating to, or denoting chemical bonds formed by the sharing of electrons between atoms 共价 gòng jià

Origin: co+valency

Example:

The relationship between bond valence and bond covalency has been investigated by a semiempirical method.
用一种半经验方法研究了固体材料中键电荷与键共价性的关系。

Extended Terms:

normal covalency 正常共价
coordination covalency 配位共价

---

cover slip [ˈkʌvə slip]

Definition: A cover slip or cover glass is a thin flat piece of transparent material, usually square or rectangular, about 20 mm (4/5 in) wide and a fraction of a millimeter thick, that is placed over objects for viewing with a microscope. 盖玻片 gài bō piàn

Example:

Then cover it with a cover slip. Observe the preparation under the microscope.
然后盖上盖玻片,在显微镜下观察切片。

Extended Term:

cover slip culture 玻片培养,盖片培养

---

critical radius [ˈkritikəl ˈreidiəs]

Definition: Critical radius is the minimum size that must be formed by atoms or molecules clustering together (in a gas, liquid or solid matrix) before a new-phase inclusion (a bubble, a droplet, or a solid particle) is stable and begins to grow. 临界半径 lín jiè bàn jìng

critical: at or of a point at which a property or phenomenon suffers an abrupt change especially having enough mass to sustain a chain reaction 临界的 lín jiè de

radius: the length of a line segment between the center and circumference of a circle or sphere 半径 bàn jìng

Example:

Below a critical radius, the pressure excess in a gas pocket forbids dissolved carbon dioxide to diffuse into it.

若低于某个临界半径,空腔内部的气体压力过大,便会阻止液体中溶解的二氧化碳扩散到空腔里。

Extended Terms:

critical nucleus radius 临界晶核半径
critical radius of liquid drop 液滴临界半径
radius-critical graph 半径临界图

Cromer-Mann coefficients [ˈkrəumə-mæn kəuiˈfiʃəns]

Definition: The set of nine coefficients $a_i, b_i, c\,(i = 1, \ldots, 4)$ in a parameterization of the non-dispersive part of the atomic scattering factor for neutral atoms as a function of $(\sin\theta)/\lambda$: $f^0(\sin\theta/\lambda) = \sum_{i-1}^{4} a_i \exp[-b_i(\sin\theta/\lambda)^2] + c$ for $0 < (\sin\theta)/\lambda < 2.0 Å^{-1}$. 克劳默-曼恩系数 kè láo mò màn ēn xì shù

Origin: Atomic scattering factors for non-hydrogen atoms were calculated from relativistic Hartree-Fock wavefunctions by Doyle, P. A. & Turner, P. S. [(1968). Acta Cryst. A24, 390-397. Relativistic Hartree-Fock and electron scattering factors] using the wavefunctions of Coulthard, M. A. [(1967). Proc. Phys. Soc. 91, 44-49. A relativistic Hartree-Fock atomic field calculation], and in 1968 by Cromer, D. T. & Waber, J. T. using the unpublished wavefunctions of Mann J. B. [International Tables for X-ray Crystallography (1974), Vol. IV., p. 71. Birmingham: Kynoch Press]. The latter are based on a more exact treatment of potential that allows for the finite size of the nucleus. Subsequent calculations [Fox, A. G., O'Keefe, M. A. & Tabbernor, M. A. (1989). Acta Cryst. A45, 786-793. Relativistic Hartree-Fock X-ray and electron atomic scattering factors at high angles] extended the useful range to 6 $Å^{-1}$ to accommodate the increasing numbers of applications for high-angle scattering factors.

---

cross-polarized light (cpl) [krɔːs-ˈpəuləraizd lait]

Definition: Cross-polarized light is a property of anisotropic minerals resulting from differences between refractive indices. It is the result of the mineral dividing a light wave into two mutually perpendicular components. The two waves travel through the mineral with unequal velocities and unequal wavelengths. The amount of light transmitted through the upper analyzer is function of the phase difference between the two transmitted waves. This phase difference is determined by crystal thickness, mineral orientation, birefringence, and the nature of the mineral. Interference color is often a key diagnostic criteria for mineral identification. In thin section, the thickness is uniform, thus the interference color is dependent on the birefringence and orientation of the mineral. 正交偏光 zhèng jiāo piān guāng

Example:

One of her graduate students, used a cross-polarized light microscope to examine this array of the tiny switches.

她所指导的一名研究生用正交极化光显微镜检查这组细微转换。

---

crystal [ˈkristəl] *n.*

Definition: A crystal or crystalline solid is a solid material, whose constituent atoms, molecules, or ions are arranged in an orderly repeating pattern extending in all three spatial dimensions. The scientific study of crystals and crystal formation is crystallography. The process of crystal formation via mechanisms of crystal growth is called crystallization or solidification. The word once referred particularly to quartz, or "rock crystal". Most metals encountered in everyday life are polycrystals. Crystals are often symmetrically intergrown to form crystal twins. 晶体 jīng tǐ

Origin: late Old English (denoting ice or a mineral resembling it), from Old French cristal, from Latin crystallum, from Greek krustallos "ice, crystal". The chemistry sense dates from the early 17th century.

Example:

The heat and pressure of this effect solidifies the ash into a hard, dark crystal.

此效果的高热和高压将使灰烬聚合成一块坚硬的黑色晶体。

Extended Terms:

liquid crystal 液晶;液晶体;液态晶体

crystal oscillator 晶体振荡器;石英振荡器;晶振

rock crystal 石英(二氧化硅);白水晶

crystal structure 晶体结构;结晶结构;结晶组织;晶体组织

crystal habit 晶癖;晶体形态;晶体习性;晶体惯态

---

crystal chemistry [ˈkristəl ˈkemistri]

Definition: Crystal chemistry is the study of the principles of chemistry behind crystals and their use in describing structure-property relations in solids. The principles that govern the assembly of crystal and glass structures are described, models of many of the technologically important crystal structures (Zinc Blende, Alumina, Quartz, Perovskite) are studied, and the effect of crystal structure on the various fundamental mechanisms responsible for many physical properties are discussed. The objectives of the field include: 1) identifying important raw materials and minerals as well as their names and chemical formulae; 2) describing the crystal structure of important materials and determining their atomic models; 3) learning the systematics of crystal and glass chemistry; 4) understanding how physical and chemical properties are related to crystal structure and microstructure; 5) studying the engineering significance of these ideas and how they relate to industrial products: past, present, and future. 晶体化学(又称结晶化学)jīng tǐ huà xué (jié jīng huà xué)

Example:

The mechanism of hydrate formation by high gravity was studied with the theory of crystal chemistry.
结合晶体化学方法探讨了超重力方式合成天然气水合物的机理。

Extended Terms:

crystal chemistry analysis 结晶化学分析
liquid crystal chemistry 液晶化学
ice-crystal chemistry 冰晶化学
comparative crystal chemistry 比较晶体化学
liquid crystal chemistry 液晶化学
crystal chemistry for inorganic materials 无机材料结晶化学

---

crystal class [ˈkristəl klaːs]

Definition: each of thirty-two categories of crystals classified according to the possible combinations of symmetry elements possessed by the crystal lattice 晶类 jīng lèi

Extended Terms:

crystal symmetry class 晶体对称类
rhombohedral crystal class 菱形晶族
hexagonal dipyramidal crystal class 六角双棱锥晶族
ditetragonal dipyramidal crystal class 双正方双锥晶族
hexagonal trapezohedral crystal class 六方偏形四半面像晶族

---

crystal constant [ˈkristəl ˈkɔnstənt]

Definition: coefficients or parameters used to describe crystal 晶体常数 jīng tǐ cháng shù

crystal: A crystal or crystalline solid is a solid material whose constituent atoms, molecules, or ions are arranged in an orderly repeating pattern extending in all three spatial dimensions. 晶体 jīng tǐ

constant: a coefficient or other parameter in a formula; given as a number or as a variable, but not being considered one of the arguments 常数 cháng shù

Example:

The composite-photonic-crystals are formed with two different crystal constants.
两种不同晶格常数的光子晶体构成了复合型二维光子晶体。

---

crystal defect [ˈkristəl diˈfekt]

Definition: Crystal defect is imperfection in the regular geometrical arrangement of the atoms in a crystalline solid. These imperfections result from deformation of the solid, rapid cooling from high temperature, or high-energy radiation (X-rays or neutrons) striking the solid. Located at single points, along lines, or on whole surfaces in the solid, these defects influence its mechanical, electrical, and optical behaviour.
晶体缺陷 jīng tǐ quē xiàn

Example:

The result showed that crystal defect of nanometer Cupric Oxide (CuO) prepared by pressure-hydrothermal method is less, structure is more stable, the process of charge and discharge is more steady, and duration is longer.
结果表明,压力—热液法制备的纳米氧化铜晶格缺陷少,结构稳定,充放电过程平稳,且持续时间长。

Extended Terms:

crystal lattice defect 晶格缺陷
crystal vacant defect 空位缺陷
crystal defect model 晶体缺陷模式
Crystal Defect Theory 晶体缺陷理论
lattice defect (= crystal defect) 晶格缺陷

---

crystal domains [ˈkristəl dəˈmeinz]

Definition: specific categories in crystal structure separated with crystal faces 晶畴 jīng chóu

crystal: A crystal or crystalline solid is a solid material whose constituent atoms, molecules, or ions are arranged in an orderly repeating pattern extending in all three spatial dimensions. 晶体 jīng tǐ

domain: a non-sovereign geographic area which has come under the authority of another government 域 yù

Example:

Nematic liquid crystal domains formed from liquid crystal molecules were separated alone in matrix networks.
由液晶分子形成的向列液晶微粒被独立地分散在基体网络内。

crystal edge ['kristəl edʒ]

Definition: lines joining two crystal faces 晶棱 jīng léng

crystal: A crystal or crystalline solid is a solid material whose constituent atoms, molecules, or ions are arranged in an orderly repeating pattern extending in all three spatial dimensions. 晶体 jīng tǐ

edge: a one-dimensional line segment joining two vertices 边,棱,边缘 biān, léng, biān yuán

Example:

The results of calculation showed that crystallization was dependent upon diffusion and crystal grows thicker only after the crystal edge of plate-like crystals has touched each other.
计算数据说明,玻璃的析晶过程是由扩散过程控制的,晶体的生长是板片状晶体在晶棱接触后才增厚。

crystal face ['kristəl feis]

Definition: One of the relatively flat surfaces by which a crystal is bounded. Faces are produced naturally during the process of crystal growth. Cut and polished gemstones are bounded by plane faces which are often produced artificially and which, therefore, are not crystal faces. 晶面 jīng miàn

Example:

The result shows that it was the typical intergranular fracture, and there were scrappy claw and tiny pore in the fracture of crystal face.
结果表明:它是典型的沿晶断裂,在断裂的晶面上有细小的爪状纹和微孔。

Extended Terms:

crystal shine face powder 透晶闪亮散粉
natural crystal face 自然晶面
real crystal face 真晶面
crystal face index 晶面指数
crystal face symbol 晶面符号

crystal family ['kristəl 'fæmili]

Definition: In crystallography, a crystal system or crystal family or lattice system is one of several classes of space groups, lattices, point groups, or crystals. Informally, two crystals tend to be in the same crystal system if they have similar symmetries, though there are many exceptions to this. 晶族 jīng zú

Example:

The perfume bottle but also are from a crystal family, bottle inlaid with gold collar 5 karats white decorative built drilling.

而且这款香水的瓶子也出自水晶世家巴卡莱特,瓶口镶嵌的金项圈是用 5 克拉白钻装饰打造的。

Extended Terms:

crystal family of methodologies 方法论透彻派

family of crystal planes 晶面族

---

crystal form ['kristəl fɔːm]

Definition: a set of crystal faces defined according to their relationship to the crystal axes 晶形 jīng xíng

Example:

The product was identified by crystal form, melting point and IR test.

产品经晶形、熔点和红外光谱分析确定为目标产品。

Extended Terms:

crystal-form factor 晶型因子

crystal growth form 晶体长大形式

---

crystal growth ['kristəl grəuθ]

Definition: Crystal growth is a major stage of a crystallization process, and consists in the addition of new atoms, ions, or polymer strings into the characteristic arrangement of a crystalline Bravais lattice. The growth typically follows an initial stage of either homogeneous or heterogeneous (surface catalyzed) nucleation, unless a "seed" crystal, purposely added to start the growth, was already present). 晶体生长 jīng tǐ shēng zhǎng

Example:

Modeling and prediction of solubility is a key to develop polymorph crystal growth and crystallization process.

溶解度的测定与预测对于多晶型体的晶体生长和结晶过程中的多晶型控制至关重要。

Extended Terms:

crystal grain growth 晶粒长大
gas phase crystal growth 气相晶体生长
holographic crystal growth analysis 晶体成长全像分析
skeletal crystal growth 骸晶生长
crystal growth from melt 溶液生长

crystal habit [ˈkristəl ˈhæbit]

Definition: In mineralogy, shape and size give rise to descriptive terms applied to the typical appearance, or habit of crystals. 晶体习性 jīng tǐ xí xìng

Example:

High quality AlN single crystals with typical hexagonal crystal habit were obtained by optimizing growth conditions.
通过优化实验条件制备出了六角形的高质量的氮化铝单晶体。

crystal indices [ˈkristəl ˈindisiːz]

Definition: a number or formula expressing properties of crystal 晶体指数 jīng tǐ zhǐ shù

crystal: A crystal or crystalline solid is a solid material whose constituent atoms, molecules, or ions are arranged in an orderly repeating pattern extending in all three spatial dimensions. 晶体 jīng tǐ

indices: a number or formula expressing some property, ratio, etc., of something indicated 指数 zhǐ shù

Example:

The refractive crystal indices for electrooptic and gyrotropic effect by fast and slow lights in arbitrary directions of direct current field and light propagation are calculated.
计算了电光和旋光效应中，当外界电场方向和传播方向任意时晶体的折射率。

Extended Terms:

crystal axial indices 晶棱指数；晶轴指数
crystal plane indices 晶面指数
indices of crystal face 晶面指数
indices of crystal direction 晶向指数

crystal nucleus [ˈkristəl ˈnjuːkliəs]

Definition: the tiny crystal that forms at the onset of crystallization 晶核（又称晶芽）

(crystalline germ) jīng hé (jīng yá)

Example:

It can form a layer of even crystal nucleus on the surface of metal. So can it speed up the formation of phosphide membrane, and reduce treatment time.
可在金属表面形成结晶均匀的晶核,使金属表面活性均一化,加快磷化膜的形成速度和缩短处理时间。

Extended Terms:

two dimensional crystal nucleus 二维晶核
nucleus of crystal 结晶核

---

crystal optics ['kristəl 'ɔptiks] *n.*

Definition: Crystal optics is the branch of optics that describes the behaviour of light in anisotropic media, that is, media (such as crystals) in which light behaves differently depending on which direction the light is propagating. The index of refraction depends on both composition and crystal structure and can be calculated using the Gladstone-Dale relation. Crystals are often naturally anisotropic, and in some media (such as liquid crystals) it is possible to induce anisotropy by applying an external electric field. 晶体光学 jīng tǐ guāng xué

Example:

It contains optical mineralogy and crystal optics.
它包括光性矿物学和晶体光学两部分。

Extended Terms:

liquid crystal optics 液晶光学
Crystal Optics and Petrology 晶体光学与岩石学

---

crystal orientation ['kristəl ˌɔːrien'teiʃən] *n.*

Definition: direction of covalent crystal 晶体取向 jīng tǐ qǔ xiàng

crystal: A crystal or crystalline solid is a solid material whose constituent atoms, molecules, or ions are arranged in an orderly repeating pattern extending in all three spatial dimensions. 晶体 jīng tǐ

orientation: (*Geometry*) placement of an object in a rotational coordinate system with respect to a fixed point and a reference position 方向 fāng xiàng, 目标 mù biāo

Example:

The dendrite grows preferentially when the crystal orientation is consistent with the coordinate axis.

晶体取向与坐标轴方向一致时枝晶优先生长。

Extended Terms:

crystal orientation effect 晶体效应
orientation of crystal 晶体定位;晶体方位

crystal pattern [ˈkristəl ˈpætən]

Definition: An object in the n-dimensional point space E^n is called an n-dimensional crystallographic pattern or, for short, crystal pattern if among its symmetry operations: 1) there are n translations, the translation vectors $t_1, \ldots, t_n$ of which are linearly independent; 2) all translation vectors, except the zero vector 0, have a length of at least $d > 0$. When the crystal pattern consists of atoms, it takes the name of crystal structure. The crystal pattern is thus the generalization of a crystal structure to any pattern, concrete of abstract, in any dimension, which obeys the conditions of periodicity and discreteness expressed above. 晶体结构 jīng tǐ jié gòu

Example:

The effect of inorganic additives on AN's (ammonium nitrate) crystal pattern transition is studied by DSC method, and the surface treatment of AN is investigated by polymer surfactant. 采用热分析法研究了无机添加剂对硝酸铵晶体结构转变的影响,并用高分子表面处理剂对硝酸铵进行表面处理。

Extended Terms:

neutron powder crystal diffraction pattern 中子粉末晶体衍射图样
rotating crystal pattern 旋晶衍射图
single-crystal pattern 单晶图案

crystal physics [ˈkristəl ˈfiziks]

Definition: Crystal physics is the study of the tensor properties of crystals and how these properties are related to the symmetries of the crystals. 晶体物理 jīng tǐ wù lǐ

Example:

It aims to cultivate undergraduate students and MA candidates majoring in liquid crystal physics and liquid crystal device physics to get engaged in academic research in related fields. 旨在培养液晶物理与液晶器件物理专业本科及硕士生从事相关领域科研工作。

crystal structure [ˈkristəl ˈstrʌktʃə]

Definition: In mineralogy and crystallography, crystal structure is a unique arrangement of atoms or molecules in a crystalline liquid or solid. A crystal structure is composed of a pattern, a set of atoms arranged in a particular way, and a lattice exhibiting long-range order and symmetry. Patterns are located upon the points of a lattice, which is an array of points repeating periodically in three dimensions. The points can be thought of as forming identical tiny boxes, called unit cells, that fill the space of the lattice. The lengths of the edges of a unit cell and the angles between them are called the lattice parameters. The symmetry properties of the crystal are embodied in its space group. 晶体结构(又称结晶构造) jīng tǐ jié gòu (jié jīng gòu zào)

Example:

Antimony trioxide is a white fine powder with the main crystal structure as cubical shape.
三氧化二锑是一种粒度细微、色泽洁白的结晶粉末,其晶体结构主要为立方形。

Extended Terms:

crystal lattice structure 晶格结构
crystal-structure effect 晶体结构效应
crystal atomic structure 晶体原子结构
crystal structure and reactivity 结晶结构和活性
crystal structure refinement 结晶构造精确化
magnetic crystal structure 磁晶体结构
cubic crystal structure 立方晶体结构

---

crystal system [ˈkristəl ˈsistəm]

Definition: each of seven categories of crystals (cubic, tetragonal, orthorhombic, trigonal, hexagonal, monoclinic, and triclinic) classified according to the possible relations of the crystal axes 晶系 jīng xì

Example:

This paper is to investigate if lyotropic liquid crystal system can improve oil recovery after polymer flooding.
该文研究的目的在于考察溶致液晶系在聚合物驱以后能否进一步提高采收率的问题。

Extended Terms:

liquid crystal imaging system 液晶成像系统
trigonal crystal system 三角晶系;三方晶系
orthorhombic crystal system 正交晶系;正交晶格;斜方晶系
monoclinic crystal system 单斜晶系
triclinic crystal system 三斜晶系
cubic crystal system 立方晶系;简单立方晶格;面心立方晶格;立方晶格

C

crystal zone [ˈkristəl zəun]

Definition: A crystal zone consists of a series of faces, all of which lie parallel to some one crystal direction and whose intersections with each other are all parallel to this direction. 晶带 jīng dài

Example:

The micro-structure of fracture surface of heat affect zone(HAZ) is of granular bainite, with characteristics of coarse crystal zone.

热影响区断口表层剖面组织为粒状贝氏体,显示出粗晶区的特征。

Extended Terms:

crystal growth zone 晶体生长区

equiaxed crystal zone 中心等轴晶区

float zone crystal 浮区熔化晶体

floating zone crystal growing 浮融带晶体生长法

vacuum float zone crystal 真空悬浮区熔晶体

---

crystalline substance [ˈkristəlain ˈsʌbstəns]

Definition: Crystalline substances have a definite rigid shape. The shape and size of crystals (even of the same materials) differs depending upon the conditions under which they are grown. 结晶质(简称晶质)jié jīng zhì(jīng zhì)

crystalline: 1) Being, relating to, or composed of crystal or crystals. 2) Resembling crystal, as in transparency or distinctness of structure or outline. 结晶质的 jié jīng zhì de

substance: a material with a definite chemical composition 物质 wù zhì

Example:

As a crystalline substance, calcium phosphate is known to be prone to failure in shear loading.

作为一种结晶质,我们知道,磷酸钙易于被剪切负荷所破坏。

---

crystallinity [ˌkristəˈliniti] *n.*

Definition: Crystallinity refers to the degree of structural order in a solid. In a crystal, the atoms or molecules are arranged in a regular, periodic manner. The degree of crystallinity has a big influence on hardness, density, transparency and diffusion. 结晶度 jié jīng dù

Origin: Middle English: from Old French cristallin, via Latin from Greek krustallinos, from krustallos

Example:

The elastic modulus and the yield strength are affected mainly by the crystallinity.
结晶度主要影响材料的弹性模量和屈服强度。

Extended Terms:

equilibrium crystallinity 平衡结晶度
fibre crystallinity 纤维结晶度,纤维结晶性
absolute crystallinity 绝对结晶性
crystallinity toxin 结晶性毒素
volume crystallinity 容积结晶度
stress crystallinity 应力晶性
crystallinity index 结晶度指数
crystallinity factor 结晶度系数

---

crystallization [ˌkristəlaiˈzeiʃən] *n.*

Definition: Crystallization is the (natural or artificial) process of formation of solid crystals precipitating from a solution, melt or more rarely deposited directly from a gas. Crystallization is also a chemical solid-liquid separation technique, in which mass transfer of a solute from the liquid solution to a pure solid crystalline phase occurs. 结晶作用 jié jīng zuò yòng

Origin: from crystal

Example:

The non-isothermal crystallization of polytrimethylene terephthalate (PTT) was studied by differential scanning calorimetry.
采用差示扫描量热仪对聚对苯二甲酸丙二醇酯进行非等温结晶研究。

Extended Terms:

fractional crystallization 分步结晶;分离结晶;分段结晶
oriented crystallization 定向结晶化;取向结晶动作;定向结晶
normal crystallization 正常结晶
epitaxial crystallization 外延结晶,附生结晶
syntectonic crystallization 同构造期结晶作用;同生构造结晶

---

crystallizing [ˈkristlaiziŋ] *n.*

Definition: to cause to form crystals or assume crystalline form 晶化 jīng huà

Origin: first known use in 1598

Example:

With its growing, dissolving-crystallizing process is playing a leading role gradually.

随着金刚石的长大,溶解—结晶过程逐渐起主导作用。

C

Extended Terms:

crystallizing evaporator 结晶蒸发器
crystallizing basin 结晶皿
crystallizing glass 微晶玻璃
crystallizing agent 结晶剂;结晶助剂;助晶剂
crystallizing power 结晶力;结晶能力

---

crystallogeny [ˌkristəˈlɔdʒini] *n.*

Definition: the science which pertains to the production of crystals 晶体发生学(又称晶体生成学) jīng tǐ fā shēng xué (jīng tǐ shēng chéng xué)

Example:

This paper reviews the two international hot topics: to systematically investigate the mechanism and seek the inner rule to develop and perfect crystallogeny.
本文评述了该领域国际研究的热点:系统研究单分子膜诱导晶体生长的机理,找出特殊规律以发展和完善晶体学。

---

crystallographic axis [ˌkristələuˈgræfik ˈaksis]

Definition: (*Crystallography*) One of three lines (sometimes four, in the case of a hexagonal crystal), passing through a common point, that are chosen to have definite relation to the symmetry properties of a crystal, and are used as a reference in describing crystal symmetry and structure. 结晶轴(简称晶轴) jié jīng zhóu (jīng zhóu)

Example:

A quartz crystal has crystallographic axes, and crystal cut is defined according to the cutting angle against a crystallographic axis and its associated mode of vibration.
温频特性与切割角有关,每个石英晶体具有结晶轴,晶体切割是按其振动模式沿垂直于结晶轴的角度切割的。

Extended Term:

principal crystallographic axis 主要结晶轴

---

crystallographic basis [ˌkristələuˈgræfik ˈbeisis]

Definition: A basis of n vectors e_1, e_2,..., e_n of the vector space V^n is a crystallographic basis of the vector lattice L if every integral linear combination $t = u^1e_1 + u^2e_2 +, \dots, + u^ne_n$ is a lattice vector of L. It may or may not be a primitive basis. 结晶基底 jié jīng jī dǐ

crystallographic: pertaining to crystallography 晶体的 jīng tǐ de; 晶体学的 jīng tǐ xué de

basis: the bottom or base of anything; the part on which something stands or rests 基础 jī chǔ; 根据 gēn jù

---

crystallographic form [ˌkristələuˈgræfik fɔːm]

Definition: crystal structure 晶型 jīng xíng; 结晶单形 jié jīng dān xíng

form: Form refers to the shape, visual appearance, or configuration of an object. 结构 jié gòu; 形式 xíng shì

Crystallographic Information File [ˌkristələuˈgræfik ˌinfəˈmeiʃən fail]

Definition: Crystallographic Information File (CIF) is a standard text file format for representing crystallographic information, promulgated by the International Union of Crystallography (IUCr). 晶体信息档案 jīng tǐ xìn xī dàng àn

information: a collection of facts from which conclusions may be drawn 信息 xìn xī

file: a set of related records (either written or electronic) kept together 文件 wén jiàn; 档案 dàng àn

---

Crystallographic Information Framework

[ˌkristələuˈgræfik ˌinfəˈmeiʃən ˈfreimwəːk]

Definition: A system of standards and specifications for the standardised exchange and archiving of crystallographic data built upon, but not restricted to, the Crystallographic Information File. In addition to file formats, the framework includes formal relations between specific data items expressed in a machine-readable dictionary definition language, controlled vocabularies, constraints on allowable values of certain numeric data, and procedures for validating the self-consistency of a crystal structure model. Responsibility for the maintenance of the standard is vested in COMCIFS, a Committee of the International Union of Crystallography. 晶体信息框架 jīng tǐ xìn xī

framework: a structure for supporting or enclosing something else, especially a skeletal support used as the basis for something being constructed 框架 kuàng jià

---

crystallographic orbit [ˌkristələuˈgræfik ˈɔːbit]

Definition: From any point of the three-dimensional Euclidean space the symmetry operations of a given space group *G* generate an infinite set of points, called a crystallographic orbit. The space group *G* is called the generating space group of the orbit. Two crystallographic orbits are said configuration-equivalent if and only if their sets of points are identical. 结晶轨迹 jié jīng guǐ jì

orbit: In physics, an orbit is the gravitationally curved path of an object around a point in space, for example the gravitational orbit of a planet around a point in space near a star. 轨道 guǐ dào

---

crystallographic symmetry [ˌkristələuˈgræfik ˈsimitri]

Definition: In crystallography, symmetry is used to characterize crystals, identify repeating parts of molecules, and simplify both data collection and nearly all calculations. Also, the symmetry of physical properties of a crystal such as thermal conductivity and optical activity must include the symmetry of the crystal. 晶体学对称性 jīng tǐ xué duì chèn xìng

symmetry: Symmetry generally conveys two primary meanings. The first is an imprecise sense of harmonious or aesthetically pleasing proportionality and balance; such that it reflects beauty or perfection. The second meaning is a precise and well-defined concept of balance or "patterned self-similarity" that can be demonstrated or proved according to the rules of a formal system: by geometry, through physics or otherwise. 对称(性) duì chèn (xìng); 匀称 yún chèn; 整齐 zhěng qí

---

crystallography [ˌkristəˈlɔgrəfi] *n.*

Definition: Crystallography is the experimental science of determining the arrangement of atoms in solids. Before the development of X-ray diffraction crystallography the study of crystals was based on the geometry of the crystals. 结晶学(又称晶体学) jié jīng xué (jīng tǐ xué)

Origin: The word "crystallography" is derived from the Greek words crystallon "cold drop / frozen drop", with its meaning extending to all solids with some degree of transparency, and grapho "write".

Example:

One of the favored methods for protein-structure determination is X-ray crystallography.
X 射线结晶学是一种极受欢迎的蛋白质机构测定方法。

Extended Terms:

diffraction crystallography 衍射晶体学;衍射结晶学
structural crystallography 构造结晶学

---

crystallology [ˌkristəˈlɔlədʒi] *n.*

Definition: the science of the crystalline structure of inorganic bodies 晶体结构学(又称结构晶体学)jīng tǐ jié gòu xué (jié gòu jīng tǐ xué)

Origin: from crystal

Example:

The crystallology, dislocation mechanism and geometry of twinning during plastic deformation of magnesium alloys are overviewed in this paper.
本文介绍了镁合金变形过程中孪生的晶体学、位错机理以及几何位向学。

---

crystallophysics [ˌkristəˈlɔfiziks] *n.*

Definition: Crystallophysics is the study of the tensor properties of crystals and how these properties are related to the symmetries of the crystals. 晶体物理学 jīng tǐ wù lǐ xué

Origin: crystal + physics

---

cube [kjuːb] *n.*

Definition: (*Geometry*) A cube is a three-dimensional solid object bounded by six square faces, facets or sides, with three meeting at each vertex. The cube can also be called a regular hexahedron and is one of the five Platonic solids. It is a special kind of square prism, of rectangular parallelepiped and of trigonal trapezohedron. The cube is dual to the octahedron. It has cubical symmetry (also called octahedral symmetry). A cube is the three-dimensional case of the more general concept of a hypercube. It has 11 nets. If one were to colour the cube so that no two adjacent faces had the same colour, one would need 3 colours. If the original cube has edge length 1, its dual octahedron has edge length. 立方体 lì fāng tǐ

Origin: mid 16th century: from Old French, or via Latin from Greek kubos

Example:

The crystals are the same size of cube.
此晶簇晶体为大小一致的立方体。

Extended Terms:

test cube 立方体试块;混凝土试力砖;立方形试样;立方试块
cube mould 方块造型
cube role 多维数据集角色
cube spin 立体旋转;立方体旋转
Boris Cube 模拟三维立方体;模拟三维运动混合机立方体

C

cubic system [ˈkjuːbik ˈsistəm]

Definition: (*Crystallography*) The cubic (or isometric) crystal system is a crystal system where the unit cell is in the shape of a cube. This is one of the most common and simplest shapes found in crystals and minerals. There are three main varieties of these crystals, called simple cubic (sc), body-centered cubic (bcc), and face-centered cubic (fcc), plus a number of other variants listed below. Note that although the unit cell in these crystals is conventionally taken to be a cube, the primitive unit cell often is not. This is related to the fact that in most cubic crystal systems, there is more than one atom per cubic unit cell. 立方晶系 lì fāng jīng xì

cubic: having the shape of a cube 立方体的 lì fāng tǐ de，立方的 lì fāng de

system: a group of interacting, interrelated, or interdependent elements forming a complex whole 系统 xì tǒng

Example:

In present paper a new method for determining the grain orientation of the cubic system metals has been developed by means of SACP technique.

应用 SACP 技术，本文发展了一种确定立方晶系金属中晶粒位向的新方法。

Extended Terms:

cubic crystal system 立方晶系；简单立方晶格；面心立方晶格

face-centered cubic system 面心立方系统

cubic closest packing(CCP) [ˈkjuːbik ˈkləuzist ˈpækiŋ]

Definition: Suppose the first two layers of hexagonal closest packed planes are stacked in "AB" fashion but the third layer is positioned so that its atoms lie over the three grooves in the A layer which were not covered by the atoms in the B layer. Then the third layer is in a different orientation from either A or B and is labeled "C". If a fourth layer then repeats the A layer orientation, and succeeding layers repeat the pattern ABCABCA... = (ABC), the resulting unit cell is hexagonal with three host atoms ($Z = 3$), unit cell edge $c = 3 \cdot \text{CPIS} \cdot r$ and $c:a = 1.5 \cdot \text{CPIS}$. Note that for identical atoms in all layers, (ACB) is identical to (ABC). It can be shown that this is a closest packed structure because the three host atoms occupy 74% of the total hexagonal unit cell volume. Furthermore, the standard reduced cell in this array is as follows: choose the two central atoms in the top and bottom "A" layers, and connect them to the six atoms shown in the "B" and "C" layers. This unit cell is identical to the standard reduced cell chosen for the face centered cubic lattice. Thus, the (ABC) repeat structure is identical to the face centered cubic lattice (CCP = FCC), with the stacking direction along the body diagonal of the cubic unit cell. 立方最紧密堆积 lì fāng zuì jǐn mì duī jī

cubic: having three dimensions 立方的 lì fāng de

close: at or within a short distance in space or time or having elements near each other 紧密的 jǐn mì de
pack: fill to capacity 群集 qún jí

Curie laws [ˈkjuri lɔːz]

Definition: In a paramagnetic material the magnetization of the material is (approximately) directly proportional to an applied magnetic field. However, if the material is heated, this proportionality is reduced: for a fixed value of the field, the magnetization is (approximately) inversely proportional to temperature. This fact is encapsulated by Curie's law:

$$M = C \cdot \frac{B}{T},$$

Where M is the resulting magnetisation, B is the magnetic field, measured in teslas, T is absolute temperature, measured in kelvins, C is a material-specific Curie constant. This relation was discovered experimentally (by fitting the results to a correctly guessed model) by Pierre Curie. It only holds for high temperatures, or weak magnetic fields. As the derivations below show, the magnetization saturates in the opposite limit of low temperatures, or strong fields. 居里定律 jū lǐ dìng lǜ

Origin: Pierre Curie found an approximation to this law which applies to the relatively high temperatures and low magnetic fields used in his experiments.

Curie temperature [ˈkjuri ˈtempəritʃə]

Definition: The Curie temperature (Tc), or Curie point refers to a characteristic property of a ferromagnetic or piezoelectric material. The Curie temperature of a ferromagnetic or a ferrimagnetic material is the reversible point above which it becomes paramagnetic. 居里温度 jū lǐ wēn dù

Origin: a term in physics and materials science, named after Pierre Curie (1859-1906)

Example:

In heating up of the PTCRs (positive temperature coefticient resistors) with AC (alternating current) or DC (direct current), under the Curie temperature, the resistivity is very low.
当在 PTC 元件施加交流或直流电压升温时,在居里点温度以下,电阻率很低。

Extended Terms:

Curie point temperature 居里点温度
ferromagnetic Curie temperature 铁磁居里温度
Curie temperature scale 居里温标

magnetic Curie temperature 磁转变居里温度
paramagnetic Curie temperature 顺磁居里温度

cyclic twin [ˈsaiklik twin]

Definition: Repeated twinning of three or more individual crystals according to the same twin law but with the twin axes or twin planes not parallel, commonly resulting in threefold, fourfold, fivefold, sixfold, or eightfold twins, which, if equally developed, display geometrical symmetry not found in single crystals, e. g., chrysoberyl, rutile. CF: repeated twinning; polysynthetic twinning. 轮式双晶 lún shì shuāng jīng

cyclic: (*Chemistry*) of or relating to compounds having atoms arranged in a ring or closed-chain structure 环(状)的 huán(zhuàng) de

twin: (*Mineralogy*) two interwoven crystals that are mirror images of each other 双晶 shuāng jīng

cylindrical system [siˈlindrikəl ˈsistəm]

Definition: A cylindrical coordinate system is a three-dimensional coordinate system that specifies point positions by the distance from a chosen reference axis, the direction from the axis relative to a chosen reference direction, and the distance from a chosen reference plane perpendicular to the axis. The latter distance is given as a positive or negative number depending on which side of the reference plane faces the point. 圆柱坐标系 yuán zhù zuò biāo xì

Example:

By the use of a cylindrical coordinate system a numerical simulation was conducted of a single tube return flow combustor flow field.
运用圆柱坐标系对单管回流燃烧室流场进行了数值模拟。

Extended Term:

volume element in cylindrical coordinate system 柱面坐标系中的体积元素

Dauphiné twin law [ˈdɔːfin twin lɔː]

Definition: (*Crystallography*) a twin law in which the twinned parts are related by a rotation of 180° around the *c* axis 道芬双晶律 dào fēn shuāng jīng lǜ

D-centred cell [diː-ˈsentəd cell] *n.*

Definition: The D centred cell is used for the rhombohedral description of the hexagonal lattice. Six right-handed D cells with basis vectors of equal length are obtained from the hP cell by means of one of the following transformation matrices: D_1: 10-1/01-1/111 D_2: -101/0-11/111. The other four D cells are obtained by cyclic permutation of the basis vectors. The resulting D cell has centering nodes at 1/3, 1/3, 1/3 and 2/3, 2/3, 2/3. D-心晶胞 xīn jīng bāo

Debye-Scherrer method [dəˈbaɪ ˈʃerə ˈmeθəd]

Definition: also Debye-Scherrer X-ray diffraction Powder diffraction, is a scientific technique using X-ray, neutron, or electron diffraction on powder or microcrystalline samples for structural characterization of materials 德拜-谢勒法 dé bài xiè lè fǎ

Origin: the development of a method of X-ray diffraction analysis by Paul Scherrer, a Swiss physicist who collaborated with Peter Debye, a physical chemist in U. S.

Extended Terms:

Debye-Scherrer Powder Method 德拜-谢勒法粉末法
Debye-Scherrer-Hull Method 德拜-谢勒-赫尔方法

degree of order [diˈgriː əv ˈɔːdə]

Definition: a position in a continuum of size, quantity, or quality 有序度 yǒu xù dù

degree: a position on a scale of intensity or amount or quality 程度 chéng dù

order: a degree in a continuum of size or quantity 顺序 shùn xù

Example:

Finally, the different components were confused and the degree of order in basic components system was reduced.

最后,混淆了不同的部件,降低了基础部件系统的有序度。

Extended Term:

degree of internal order 内部(分子间)有序度

deltoid [ˈdeltɔid] *n.*

Definition: (*Geometry*) also known as a deltoid, a type of quadrilateral 偏方形 piān fāng xíng

Origin: mid-18th century: from French deltoïde, or via modern Latin from Greek deltoeidēs

Example:

Construct two mutually orthogonal tangents of the deltoid. Where do they meet?
三角星形线的两条互相垂直的法线的交点在哪里?

Extended Terms:

deltoid branch 三角洲汊河;三角肌支
deltoid muscles 三角肌
deltoid plate: 三棱板;三角板

---

demorphism [di'mɔːfizəm] *n.*

Definition: also known as clastatio, weathering physical disintegration and chemical decomposition of earthy and rocky materials on exposure to atmospheric agents, producing an in-place mantle of waste 岩石的分解(风化)yán shí de fēn jiě (fēng huà)

Origin: de-+ morphism

Example:

The core of Yanchang Formation Chang-7 high quality source rock (oil shale) is black, hard, light, and it looks slip-shaped after demorphism.
鄂尔多斯盆地延长组长 7 优质烃源岩(油页岩)井下岩芯外观呈黑色、质硬、手感较轻,露头风化后呈纸片状。

---

detohedron [ˌdetə'hiːdrən] *n.*

Definition: A deltahedron (plural deltahedra) is a polyhedron whose faces are all equilateral triangles. There are infinitely many deltahedra, but of these only eight are convex, having 4, 6, 8, 10, 12, 14, 16 and 20 faces. The number of faces, edges, and vertices is listed below for each of the eight convex deltahedra. The deltahedra should not be confused with the deltohedra (spelled with an "o"), polyhedra whose faces are geometric kites.偏十二面体 piān shí èr miàn tǐ

Origin: The name is taken from the Greek majuscule delta (Δ), which has the shape of an equilateral triangle.

---

devitrification [diːˌvitrifi'keiʃən] *n.*

Definition: Devitrification is the opposite of vitrification, i.e., the process of crystallization in a formerly crystal-free (amorphous) glass. 脱玻化(晶化)tuō bō huà (jīng huà)

Origin: The term is derived from the Latin vitreus, meaning glassy and transparent.

Example:

Solution, devitrification as well as forming and recrystallization of secondary mineral can increase primary porosity as a whole.
溶性、脱玻化、次生矿物的生成和重结晶等成岩作用的总效应导致孔隙增加。

Extended Terms:

devitrification opal 失透乳白
devitrification glaze 失透釉
devitrification of glass 玻璃闷光;玻璃析晶;玻璃透明消失

---

diadochy [dai'ædəki] *n.*

Definition: Replacement or ability to be replaced of one atom or ion by another in a crystal lattice. 晶格同位 jīng gé tóng wèi

Origin: early 18th century. Greek diadokhē "succession", diadekhesthai "succeed one another", dekhesthai "take, accept"

---

diamagnetism [ˌdaiə'mægnitizəm] *n.*

Definition: (*Physics*) (of a substance or body) tending to become magnetized in a direction at 180° to the applied magnetic field 逆磁性 nì cí xìng

Origin: 1846: coined by Faraday, from Greek dia "through, across" + magnetic

Example:

Langevin is noted for his work on paramagnetism and diamagnetism, and devised the modern interpretation of this phenomenon in terms of electric charges of electrons within atoms.
朗之万以他的顺磁性及反磁性的研究而闻名,他想出了用现代的原子中的电子电荷去解释这些现象。

Extended Terms:

orbital diamagnetism 轨道抗磁性
anisotropic diamagnetism 各向异性反磁性
perfect diamagnetism 理想抗磁性;完全抗磁性,迈斯纳效应
electronic diamagnetism 电子抗磁性;电子反抗性
crystal diamagnetism 晶体抗磁性

D

diaphaneity [ˌdaiəfəˈniːəti] *n.*

Definition: In the field of optics, diaphaneity (also called transparency or pellucidity) is the physical property of allowing light to pass through a material; translucency (also called translucence or translucidity) only allows light to pass through diffusely. The opposite property is opacity. Transparent materials are clear, while translucent ones cannot be seen through clearly. 透明性 tòu míng xìng

Example:

Conversely, diaphaneity, emulate spend and decorative pattern effect can be affected by removing too much the glossiness of the floor.

反之,清除过多的地板光泽度,其透明度、仿真度和花纹效果都会受到影响。

Extended Term:

diaphaneity transparence transparency 透明度

---

dichroism [ˈdaikrəuizəm] *n.*

Definition: (of a crystal) showing different colours when viewed from different directions, or (more generally) having different absorption coefficients for light polarized in different directions 二[向]色性 èr [xiàng] sè xìng

Origin: mid 19th century: from Greek dikhroos (from di-"twice" + khrōs "colour") +-ic

Example:

The infrared dichroism of a particular vibration of known origin has been used to give valuable information about molecular orientation.

已知的特定振动频带的红外二色性可以为分子定向提供有价值的资料。

Extended Terms:

infrared dichroism 红外二向色性;红外二色性
circuit dichroism 圆偏振二向色性
fibre dichroism 纤维二色性
ultraviolet dichroism 紫外线二色性
streaming dichroism 流动二向色性

---

dielectric constant [ˌdaiiˈlektrik ˈkɔnstənt]

Definition: (*Physics*) a quantity measuring the ability of a substance to store electrical energy in an electric field 介电常数 jiè diàn cháng shù

Example:

The dielectric constant decreased significantly under large DC (direct current) bias level.

在较大的直流偏压场下，介电常数大幅度降低。

Extended Terms:

dielectric absorption constant 介电吸收常数；(电)介质吸收常数
dielectric loss constant 介电损耗常数
effective dielectric constant 等效介电常数
reciprocal dielectric constant 倒介电常数
parallel dielectric constant 平行介电常数

dielectricity [ˌdaiilekˈtrisiti] *n.*

Definition: Electricity is the movement of electrons within the molecules of a material. Although electrons can move freely within conductors, they cannot do so within insulators. When a direct current voltage is applied to an insulator, electrons do not separate from molecules, and are divided by an electrical charge (positive or negative) induced at both ends of the material. 介电性 jiè diàn xìng

Example:

It has better dielectricity, automatic quench and softness than silicone fiberglass sleeving, widely used in the insulation protection of internal line cluster in H & N machinery, household electrical appliances, electroheat device and special illuminations.
它具有比硅树脂玻璃纤维套管更加良好的介电性、自熄性和柔软性，广泛应用于 H & N 级电机、家用电器、电热设备、特种灯具等绝缘保护设备上。

difference Patterson map [ˈdifərəns ˈpætəsən mæp]

Definition: An application of Patterson methods for solution of crystal structures, typically proteins with heavy-atom derivatives, where the Patterson function is calculated using structure-factor coefficients based on the difference between the heavy-atom derivative and the native molecule. 差异性帕特森地图 chā yì xìng pà tè sēn dì tú

dihexagonal prism [daiˌheksəˈgɔnəl ˈprizəm]

Definition: a prism with a hexagonal base upon which is erected six rectangular faces 复六方柱 fù liù fāng zhù

dihexagonal: Consisting of two hexagonal parts united; thus, a dihexagonal pyramid is composed of two hexagonal pyramids placed base to base. Having twelve similar faces; as, a dihexagonal prism. 双六角的 shuāng liù jiǎo de

prism: In optics, a prism is a transparent optical element with flat, polished surfaces that refract light. 棱柱体 léng zhù tǐ

Origin: di-+ hexagonal

dihexagonal pyramid [daiˌheksəˈgɔnəl ˈpirəmid]

Definition: In geometry, a hexagonal pyramid is a pyramid with a hexagonal base upon which is erected six triangular faces that meet at a point (the vertex). Like any pyramid, it is self-dual. A right hexagonal pyramid with a regular hexagon base has C6v symmetry. 复六方单锥 fù liù fāng dān zhuī

pyramid: A pyramid (from Greek "πυραμις"-pyramis) is a structure where the outer surfaces are triangular and converge at a point. 棱锥(体) léng zhuī (tǐ); 角锥(体)jiǎo zhuī (tǐ)

dimmer [ˈdimə] *n.*

Definition: Dimmers are devices used to vary the brightness of a light. By decreasing or increasing the RMS voltage and hence the mean power to the lamp it is possible to vary the intensity of the light output. Although variable-voltage devices are used for various purposes, the term dimmer is generally reserved for those intended to control resistive incandescent, halogen and more recently compact fluorescent (CFL) lighting. More specialized pulse-width modulation equipment is needed to dim fluorescent, mercury vapor, solid state and other arc lighting. 遮光器 zhē guāng qì

Example:

A soft-touch dimmer lets you adjust the lights and when you place the leaf-head in a closed position, it gears up for some sun soaking recharge.

一个容易操作的调光器,能让你调整光度,当你将头部的叶片调至关闭状态时,它就会快速换挡吸收少许阳光进行再充电。

Extended Terms:

dimmer camera 减光器眼底照像机
dimmer device 调光设备
headlamp dimmer 转向信号灯及前照灯变光拨杆开关
dimmer switch 前照灯变光开关;减光器开关;变光开关;微调开关
stage dimmer 舞台调光机
exposure dimmer 照射衰减器;曝露暗光室

dipyramid (bipyramid) [daiˈpirəmid] *n.*

Definition: Two pyramids symmetrically placed base-to-base, also called a bipyramid. The dipyramids are duals of the regular prisms. 双锥 shuāng zhuī

Origin: bi- + pyramid

Example:

Diamond crystals are intermediate between dihydrate and monohydrate calcium oxalate, and have two crystal forms: pyramids and bipyramid, therefore, they are not dihydrate calcium oxalate.

本研究发现菱形晶属于一水草酸钙和二水草酸钙的中间晶体,并有两种晶体形式,单锥和双锥的结晶形式,因而菱形晶不是二水草酸钙。

Extended Terms:

triangular dipyramid 双三角锥
hexagonal dipyramid 六方双锥
rhombic dipyramid 斜方双锥
pentagonal dipyramid 双五角锥
tetragonal-dipyramid 正方双锥体

---

direct lattice [di'rekt 'lætis]

Definition: The direct lattice represents the triple periodicity of the ideal infinite perfect periodic structure that can be associated to the structure of a finite real crystal. 正格子 zhèng gé zi

Extended Term:

direct product lattice 直积格

---

direct methods [di'rekt 'meθədz]

Definition: (*Crystallography*) estimating the phases of the Fourier transform of the scattering density from the corresponding magnitudes 直接法 zhí jiē fǎ

Example:

This paper discusses parallel direct methods for large scale sparse linear equations which come from Computational Fluid Dynamics(CFD) problems.

该论文探讨了由 CFD 问题引出的大型稀疏线性方程组的并行直接求解法。

Extended Terms:

direct search methods 直接搜寻方法
parallel direct methods for implicit scheme 隐格式并行直接求解方法研究

D

direct product [di'rekt 'prɔdʌkt]

Definition: In mathematics, one can often define a direct product of objects already known, giving a new one. This is generally the Cartesian product of the underlying sets, together with a suitably defined structure on the product set. More abstractly, one talks about the product in category theory, which formalizes these notions. Examples are the product of sets, groups, the product of rings and of other algebraic structures. The product of topological spaces is another instance. There is also the direct sum — in some areas this is used interchangeably, in others it is a different concept. 直积 zhí jī

Example:

This article proves that a finite cyclic group can be isomorphically represented by direct product of some cyclic groups.

本文运用基础代数中有关循环群、直积、同构等理论,证明了一个循环群可以用另一组循环群的直积形式来同构表示。

Extended Terms:

direct-product code 直积码
right direct product 右直积
left direct product 左直积
weak direct product 弱直积
infinite direct product 无限直积

---

direct space [di'rekt speis]

Definition: The direct space (or crystal space) is the point space, E^n, in which the structures of finite real crystals are idealized as infinite perfect three-dimensional structures. 正空间 zhèng kōng jiān

Example:

In this paper, general expressions for coordinate transformation in crystal direct space or reciprocal space are derived by using matrix calculus.

本文用矩阵运算简明地推导了晶体正空间和倒易空间中的坐标变换关系式。

Extended Terms:

direct product space 直积空间
direct space management 直接存取存储区的管理
vector space direct sum 向量空间直和

D

dislocation [ˌdisləuˈkeiʃən] *n.*

Definition: In materials science, a dislocation is a crystallographic defect or irregularity, within a crystal structure. 位错 wèi cuò

Origin: late Middle English: from Old French, or from medieval Latin dislocatio(n-), from the verb dislocare, based on Latin locare "to place".

Example:

It was found that there exist prismatic dislocations, stacking faults, array of dislocations and dislocation network.

研究发现金刚石中存在层错、棱柱位错、位错列和位错网络等晶体缺陷。

Extended Terms:

dislocation line 位错线;错位线;断层线;差排线
perfect dislocation 全位错;完全位错;完整位错
dislocation structure 位错结构;错位构造;位错组织
single dislocation 单个位错;单一位错
dislocation step 位错阶梯

---

disordered structure [disˈɔːdəd ˈstrʌktʃə]

Definition: In contrast to the well-ordered but nonrepetitive coil structures, there are also genuinely disordered regions in proteins, which are either entirely absent on electron density maps or which appear with a much lower and more spread out density than the rest of the protein. The disorder could either be caused by actual motion, on a time scale of anything shorter than about a day, or it could be caused by having multiple alternative conformations taken up by the different molecules in the crystal. Well-ordered side chains also move, often very rapidly, but the movements are brief departures from a single stable conformation. 无序结构 wú xù jié gòu

disordered: lacking orderly continuity 无序的 wú xù de

structure: the manner of construction of something and the arrangement of its parts 结构 jié gòu

Example:

As the degree of hydration increases, the content of disordered structure decreases and those of ordered secondary structure increase.

随水合度增加,无序结构含量减少,有序二级结构含量增加。

Extended Term:

ordered and disordered structure 有序无序结构

D

dispersion [dis'pəːʃən] *n.*

Definition: In optics, dispersion is the phenomenon in which the phase velocity of a wave depends on its frequency, or alternatively when the group velocity depends on the frequency. 色散 sè sàn

Origin: late Middle English: from late Latin dispersio(n-), from Latin dispergere

Example:

Much data can be obtained by measuring the dispersion of cosmic radio waves passing near the sun.

可以通过测定穿过太阳附近的宇宙射电波的色散度来获得更多的数据。

Extended Terms:

dispersion relation 频散关系;色散关系;分散关系

linear dispersion 线色散;线性色散;曲线;线色散率

dispersion measure 频散量度;分散性测度

dispersion coefficient 分散系数;扩散系数;分散率;扩散系统磁漏系数

horizontal dispersion 水平分散;水平色散

---

dispersion of refringence [dis'pəːʃən əv ri'frindʒəns]

Definition: the phenomenon in which the wave length of heterogeneous crystal changes as a result of the change in refractive index dispersion 双折射率色散 shuāng zhé shè lǜ sè sàn

dispersion: In optics, dispersion is the phenomenon in which the phase velocity of a wave depends on its frequency or alternatively when the group velocity depends on the frequency. 色散 sè sàn

refringence: the power to refract 折射率 zhé shè lǜ

---

displacive modulation [dis'pleisiv mɔdju'leiʃən]

Definition: For a displacively modulated crystal phase, the positions of the atoms are displaced from those of a basis structure with space group symmetry (an ordinary crystal). 位移调节 wèi yí tiáo jié

displacive: not of something's or someone's usual or original position 位移性的 wèi yí xìng de

modulation: In electronics, modulation is the process of varying one or more properties of a high frequency periodic waveform, called the carrier signal, with respect to a modulating signal. 调整 tiáo zhěng;调节(对波幅、频率的) tiáo jié(duì bō fú、pín lǜ de);调制 tiáo zhì

D

displacive transformation [dis'pleisiv ˌtrænsfə'meiʃən]

Definition: A change in crystal summetry as a result of changes in bond length or bond angles (as contrasted to reconstructive transformations). The short-range order is unchanged; the long-range order is changed.位移型转变 wèi yí xíng zhuǎn biàn

transformation: complete change: a complete change, usually into something with an improved appearance or usefulness 转变 zhuǎn biàn

---

distorted crystal [dis'tɔːtid 'kristəl]

Definition: A crystal whose faces have developed unequally, some being larger than others. Some distorted crystal forms are drawn out or shortened, but the angle between the faces remains the same. 歪晶 wāi jīng

distorted:to twist out of a natural, normal, or original shape or condition 歪曲的 wāi qū de

crystal:A crystal or crystalline solid is a solid material whose constituent atoms, molecules, or ions are arranged in an orderly repeating pattern extending in all three spatial dimensions.晶体 jīng tǐ

---

distortion [dis'tɔːʃən] *n.*

Definition: A distortion is the alteration of the original shape (or other characteristic) of an object, image, sound, waveform or other form of information or representation.畸变 jī biàn

Origin: late 15th century (in the sense "twist to one side"): from Latin distort-"twisted apart", from the verb distorquere, from dis-"apart" + torquere "to twist"

Example:

A new linear approach with radial distortion model for camera calibration was designed and realized.

设计并实施了一种考虑径向畸变的逐步线性摄像机标定方法。

Extended Terms:

angular distortion 角变形;角度变形;角度畸变;角失真

linear distortion 线性失真;线性畸变

bias distortion 偏移失真;偏置失真;偏差失真

geometric distortion 几何畸变

quantizing distortion 量化失真;量化畸变

D

ditetragonal bipyramid [daiˈtetrəgənəl baiˈpirəmid]

Definition: also ditetragonal pyramid: 16-faced form with faces related by a 4-fold axis with a perpendicular mirror plane 复四方单锥 fù sì fāng dān zhuī

Extended Term: ditetragonal pyramidal class 复正方锥类

---

ditetragonal prism [daiˈtetrəgənəl ˈprizm]

Definition: The ditetragonal prism is a form consisting of eight rectangular vertical faces, each of which intersects the two horizontal crystallographic axes unequally. There are various ditetragonal prisms, depending upon their differing relations to the horizontal axes. 复四方柱 fù sì fāng zhù

---

ditrigonal prism [daiˈtraigənəl ˈprizm]

Definition: in crystallography, twice-three-sided. A ditrigonal prism is a six-sided prism, the hemihedral form of a twelve-sided or dihexagonal prism.复三方柱 fù sān fāng zhù

ditrigonal: in crystallography, twice-three-sided 双三角形的 shuāng sān jiǎo xíng de

prism: In optics, a prism is a transparent optical element with flat, polished surfaces that refract light. 棱柱体 léng zhù tǐ

---

dodecahedron [ˌdəudekəˈhiːdrən] *n.*

Definition: (*Geometry*) Also rhombic dodecahedron, a dodecahedron is any polyhedron with twelve flat faces, but usually a regular dodecahedron is meant: a Platonic solid. It is composed of 12 regular pentagonal faces, with three meeting at each vertex, and is represented by the Schläfli symbol {5,3}. It has 20 vertices and 30 edges. Its dual polyhedron is the icosahedron, with Schläfli symbol {3,5}. A large number of other (nonregular) polyhedra also have 12 sides, but are given other names. The most frequently named other dodecahedron is the rhombic dodecahedron. 菱形十二面体 líng xíng shí èr miàn tǐ

Origin: late 16th century: from Greek dōdekaedron, neuter (used as a noun) of dōdekaedros "twelve-faced"

Example:

It shows that the diamond seed crystal is also dodecahedron, coming from a crystal grain first, then growing as an dodecahedron diamond.

经分析发现,该金刚石晶体首先由晶核发育成菱形十二面体金刚石籽晶,再由此籽晶生长成菱形十二面体金刚石晶体。

Extended Terms:

regular dodecahedron 正十二面体
great dodecahedron 大十二面体
rhombic dodecahedron 菱形十二面体;斜方十二面体
pentagonal dodecahedron 五角十二面体
deltoid dodecahedron 扁方三四面体;偏菱形十二面体;三角斜方十二面体

---

domain [dəu'mein] *n.*

Definition: (*physics*) a discrete region of magnetism in ferromagnetic material 晶畴 jīng chóu

Origin: late Middle English (denoting heritable or landed property): from French domaine, alteration (by association with Latin dominus "lord") of Old French demeine "belonging to a lord"

Example:

Based on the domain analysis of the mine project management system, the data tree table objects model, database management object model, and general reports making object model are abstracted.
在对矿山项目管理系统进行领域分析的基础上,抽象出数据树表对象模型、数据库管理对象模型和通用报表对象模型。

Extended Terms:

active domain 有效域
target domain 目标域;目的域;目标领域;喻体
functional domain 功能域
domain assembly 结构域装配
cognitive domain 认知领域;认知域;认知范畴

---

domain of influence [dəu'mein əv 'influəns]

Definition: The domain of influence of a lattice point *P*, or Dirichlet domain or Voronoi domain, consists of all points *Q* in space that are closer to this lattice point than to any other lattice point or at most equidistant to it, namely such that $OP \leqslant |t-OP|$ for a any vector *t* belonging to the vector lattice *L*. It is the inside of the Wigner-Seitz cell. 影响域 yǐng xiǎng yù

D

dome [dəum] *n.*

Definition: A dome is a structural element of architecture that resembles the hollow upper half of a sphere. 坡面 pō miàn

Origin: early 16th century: from French dôme, from Italian duomo "cathedral, dome", from Latin domus; "house"

Example:

The Temples were covered in a dome of crystalline amplified light somewhat like a glowing force field.

神殿被一个闪闪发光的水晶穹顶笼罩着,(这些光)在某种程度上就像炽热的力场。

Extended Terms:

lava dome 熔岩穹丘;熔岩穹;熔岩丘;圆顶火山

dome structure 穹窿构造;穹状构造

volcanic dome 火山穹丘;火山丘

---

double coset [ˈdʌbl ˈkəuset]

Definition: (*Mathematics*) An (H,K) double coset in G, where G is a group and H and K are subgroups of G, is an equivalence class for the equivalence relation defined on G by $x \sim y$ if there are h in H and k in K with $hxk = y$. Then each double coset is of form HxK, and G is partitioned into its (H,K) double cosets; each of them is a union of ordinary right cosets Hy of H in G and left cosets zK of K in G. In another aspect, these are in fact orbits for the group action of $H\times K$ on G with H acting by left multiplication and K by inverse right multiplication. The space of double cosets can be written $H/G/K$. 双陪集 shuāng péi jí

coset: a subset of mathematical group that consists of all the products that obtained by multiplying either on the right or the left a fixed element of the group by each of the elements of a given subgroup 陪集 péi jí

---

dual basis [ˈdjuːəl ˈbeisis]

Definition: In linear algebra, a dual basis is a set of vectors that forms a basis for the dual space of a vector space. For a finite dimensional vector space V, the dual space V^* is isomorphic to V, and for any given set of basis vectors $\{e_1, \dots, e_n\}$ of V, there is an associated dual basis $\{e^1, \dots, e^n\}$ of V^* with the relation

$$e^i(e_j)=\begin{cases}1, & \text{if } i=j\\ 0, & \text{if } i\neq j.\end{cases}$$

Concretely, we can write vectors in an n-dimensional vector space V as $n\times 1$ column matrices and elements of the dual space V^* as $1 \times n$ row matrices that act as linear functionals by left matrix multiplication.

For example, the standard basis vectors of R^2(the Cartesian plane) are

$$\{e_1, e_2\} = \left\{ \begin{pmatrix} 1 \\ 0 \end{pmatrix}, \begin{pmatrix} 0 \\ 1 \end{pmatrix} \right\}$$

and the standard basis vectors of its dual space R^{2*} are

$$\{e'_1, e'_2\} = \{(1\ 0), (0\ 1)\}$$

In 3-dimensional space, for a given basis e you can find the biorthogonal (dual) basis by this formulas:

$$e_1^* = \frac{[e_2; e_3]}{(e_1; e_2; e_3)}, e_2^* = \frac{[e_3; e_1]}{(e_1; e_2; e_3)}, e_3^* = \frac{[e_1; e_2]}{(e_1; e_2; e_3)}$$

对偶基 duì ǒu jī

Example:

In order to make maximal use of generalized Ball bases in computer aided geometric design (CAGD) systems, a type of generalized Ball bases defined as β basis was investigated, which could generate the parametric curve located between the Bézier curve and the Said-Ball curve. The dual basis of β basis was provided with the combinational calculation method.

为进一步发挥广义 Ball 基在计算机辅助几何设计(CAGD)中的优越性,对能生成几何位置介于 Bézier 曲线与 Said Ball 曲线之间的参数曲线的一类广义 Ball 基即 β 基作了深入研究。

Extended Term:

dual price basis 双重价格

dynamical theory [dai'næmikəl 'θiəri]

Definition: The dynamical theory of diffraction describes the interaction of waves with a regular lattice. The wave fields traditionally described are X-rays, neutrons or electrons and the regular lattice, atomic crystal structures or nanometer scaled multi-layers or self-arranged systems. In a wider sense, similar treatment is related to the interaction of light with optical band-gap materials or related wave problems in acoustics. 动力学理论 dòng lì xué lǐ lùn

Example:

The dynamical theory foundation for the automatic weapon research can be provided by the simulation and analysis of the firing dynamics of the rapid burst rifle.

通过对发射动力学问题的仿真分析,可为该武器的研究提供动力学理论依据。

Extended Terms:

dynamical response theory 动力学响应理论

dynamical theory of tides 潮汐动力论;潮汐动力理论

dynamical theory of diffraction 动力学衍射理论

continuum theory and dynamical systems 连续统理论与动力系统
ergodic theory and dynamical systems 遍历理论和动力系统

eigensymmetry [ˈaigənˌsimitri] *n.*

Definition: The eigensymmetry, or inherent symmetry, of a crystal is the point group or space group of a crystal, irrespective of its orientation and location in space. 本征对称 běn zhēng duì chèn

Example:

Double-Fourier series type solutions are obtained by utilizing the superposition method and by taking advantage of eigensymmetry inherent in the problem.
利用叠加法和本问题所固有的本征对称性优点得到了双重傅立叶级数解。

---

eigenvector [ˈaigənvektə] *n.*

Definition: (*Mathematics & Physics*) a vector which when operated on by a given operator gives a scalar multiple of itself 本征矢量 běn zhēng shǐ liàng

Example:

A judging criterion of closely spaced modes by eigenvector turning angles is put forward in this paper.
用特征向量转角给出了一种密集模态的判别准则。

Extended Terms:

generalized eigenvector 广义本崭量
right eigenvector 右本镇量
empirical eigenvector 经验特征向量;经验特征向量
eigenvector analysis 特征向量分析
eigenvector centrality 特征向量中心度;特征向量中心性

---

elasticity [ˌelæsˈtisəti] *n.*

Definition: cord, tape, or fabric, typically woven with strips of rubber, which returns to its original length or shape after being stretched 弹性 tán xìng

Origin: mid 17th century (originally describing a gas in the sense "expanding spontaneously to fill the available space"): from modern Latin elasticus, from Greek elastikos "propulsive", from elaunein "to drive"

Example:

Although the polyester possesses great hardness, it still shows good elasticity and good oil resistance, but shows poor water resistance.

结果表明:弹性体聚酯在具有高硬度情况下,仍具有弹性,且耐油性好,但耐水性差。

Extended Terms:

retarded elasticity 延迟弹性;弹性后效;推迟弹性

perfect elasticity 完全有弹性;完全弹性,理想弹性

unit elasticity 单位弹性;恒一弹性;单一弹性

linear elasticity 线性弹性;线弹性

Anisotropic elasticity 各向异性弹性;异向弹性力学;各向异性弹性理论

---

electro negativity [i'lektrəu ˌnegə'tivəti]

Definition: Also electronegativity, symbol χ (the Greek letter chi), is a chemical property that describes the ability of an atom (or, more rarely, a functional group) to attract electrons (or electron density) towards itself. 电负性 diàn fù xìng

Example:

According to the equilibration principle of electronegativity, a new scale for acid-base capacity of hydroxides is established in the peper.

应用电负性均衡原理,建立一种新的标度氢氧化物(包括含氧酸)酸碱性强弱的定量方法。

Extended Terms:

electronegativity difference 电负度差;阴电性差

electronegativity scale 电负度标;阴电性标

electronegativity value 电负值;阴电性值

Pauling electronegativity scale 鲍林电负性标度

---

electrocaloric effect [ɪ'lektrəukəˌlɒrɪk i'fect]

Definition: The electrocaloric effect is phenomenon in which a material shows a reversible temperature change under an applied electric field. It is often considered to be the physical inverse of the pyroelectric effect. 电热效应 diàn rè xiào yìng

---

E

electroluminescence [iˈlektrəuˌljuːmiˈnesəns] *n.*

Definition: Electroluminescence (EL) is an optical phenomenon and electrical phenomenon in which a material emits light in response to an electric current passed through it, or to a strong electric field. This is distinct from light emission resulting from heat (incandescence), chemical reaction (chemiluminescence), sound (sonoluminescence), or other mechanical action (mechanoluminescence). 电致发光 diàn zhì fā guāng

Example:

But quantum dots can also be made to glow using direct electrical stimulation — a phenomenon called electroluminescence.

但同样可以使用直接的电力刺激量子点来发光——这种现象称之为“电致发光”。

Extended Terms:

electroluminescence memory 电发光存储器
electroluminescence dyes 电致发光染料
electroluminescence sensor 电荧光传感器
organic electroluminescence 有机 EL;有机电致发光;有机电激发光显示技术
intrinsic electroluminescence 本征电致发光

---

electromagnetism [iˌlektrəuˈmægnitizəm] *n.*

Definition: Electromagnetism is one of the four fundamental interactions of nature, along with strong interaction, weak interaction and gravitation. It is the force that causes the interaction between electrically charged particles; the areas in which this happens are called electromagnetic fields. 电磁性 diàn cí xìng

Example:

The biological effect of the laser is its light effect, thermal effect, electromagnetic effect, pressure effect and blast effect.

激光的生物学效应主要是光效应、热效应、电磁场效应、压力与冲击波效应。

Extended Terms:

marine electromagnetism 海洋电磁学
computational electromagnetism 计算电磁
electromagnetism protection 电磁防护贴
electromagnetism induction phenomenon 电磁感应现象
electromagnetism disturbance emission 电磁干扰的发射

E

electron affinity [iˈlektrɔn əˈfinəti]

Definition: The electron affinity of a molecule or atom is the energy change when an electron is added to the neutral species to form a negative ion. This property can only be measured in an atom in gaseous state.

$$X + e^- \rightarrow X^-$$

The electron affinity, E_{ea}, is defined as positive when the resulting ion has a lower energy, i.e. it is an exothermic process that releases energy:

$$E_{ea} = E_{initial} - E_{final}$$

Alternately, electron affinity is often described as the amount of energy required to detach an electron from asingly charged negative ion, i.e. the energy change for the process:

$$X^- \rightarrow X + e^-$$

A molecule or atom that has a positive electron affinity is often called anelectron acceptor and may undergo charge-transfer reactions.电子亲和性 diàn zǐ qīn hé xìng

electron: an elementary particle with negative charge 电子 diàn zǐ

affinity: the force attracting atoms to each other and binding them together in a molecule 亲和力 qīn hé lì

Example:

This paper proposed the quantitative relationship between electron affinity and induced effect.
本文提出了电子亲和力与诱导效应之间的定量关系。

Extended Terms:

low electron affinity 低电子亲和力
electron affinity detector 电子亲和检测器
negative electron affinity 负电子亲和力
positive electron affinity 正电子亲和力
electron affinity substance 电子亲和性物质

electron configuration [iˈlektrɔn kənˌfigjuˈreiʃən]

Definition: In atomic physics and quantum chemistry, electron configuration is the arrangement of electrons of an atom, a molecule, or other physical structure. It concerns the way electrons can be distributed in the orbitals of the given system (atomic or molecular for instance). 电子构型 diàn zǐ gòu xíng

configuration: an arrangement of parts or elements 构型 gòu xíng

Extended Term:

valence-electron configuration 价电子组态

E

electron density map [iˈlektrɔn ˈdensəti mæp]

Definition: A three-dimensional description of the electron density in a crystal structure, determined from X-ray diffraction experiments. X-rays scatter from the electron clouds of atoms in the crystal lattice; the diffracted waves from scattering planes h,k,l are described by structure factors F_{hkl}. The electron density as a function of position x,y,z is the Fourier transform of the structure factors:

$$\rho(xyz) = \frac{1}{V}\sum_{hkl} F(hkl)\exp[-2\pi i(hx + ky + lz)].$$

The electron density map describes the contents of the unit cells averaged over the whole crystal and not the contents of a single unit cell (a distinction that is important where structural disorder is present). Three-dimensional maps are often evaluated as parallel two-dimensional contoured sections at different heights in the unit cell.电子密度图 diàn zǐ mì dù tú

density: the amount per unit size 密度 mì dù

map: a diagrammatic representation of the earth's surface (or part of it) 图 tú

electrostriction [iˌlektrəˈstrikʃən] *n.*

Definition: Electrostriction is a property of all dielectric materials, and is caused by the presence of randomly-aligned electrical domains within the material. When an electric field is applied to the dielectric, the opposite sides of the domains become differently charged and attract each other, reducing material thickness in the direction of the applied field (and increasing thickness in the orthogonal directions due to Poisson's ratio). The resulting strain (ratio of deformation to the original dimension) is proportional to the square of the polarization. Reversal of the electric field does not reverse the direction of the deformation.

More formally, the electrostriction coefficient is afourth rank tensor (Q_{ijkl}), relating second order strain (x_{ij}) and first order polarization tensors (P_k, P_l).

$$x_{ij} = Q_{ijkl} \times P_k \times P_l$$

The related piezoelectric effect occurs only in a particular class of dielectrics. Electrostriction applies to all crystal symmetries, while the piezoelectric effect only applies to the 20 piezoelectric point groups. Electrostriction is a quadratic effect, unlike piezoelectricity, which is a linear effect. In addition, unlike piezoelectricity, electrostriction cannot be reversed: deformation will not induce an electric field. 电致伸缩(反压电性)diàn zhì shēn suō(fǎn yā diàn xìng)

Example:

A new method for measuring the characteristic of electrostriction by a digital speckle correlation method (DSCM) is presented.

(本文)提出了一种利用数字散斑相关方法测量材料的电致伸缩性能(DSCM)的新方法。

Extended Terms:

electrostriction phenomena 电致伸缩现象
Bragg electrostriction 布喇格电致伸缩
electrostriction material 电伸缩材料
electrostriction effect 电致伸缩效应

---

elliptical polarization [iˈliptikəl ˌpəuləraiˈzeiʃən]

Definition: In electrodynamics, elliptical polarization is the polarization of electromagnetic radiation such that the tip of the electric field vector describes an ellipse in any fixed plane intersecting, and normal to, the direction of propagation. An elliptically polarized wave may be resolved into two linearly polarized waves in phase quadrature, with their polarization planes at right angles to each other. 椭圆偏振 tuǒ yuán piān zhèn

Example:

After all this, the rules of the polarization of an elliptical polarized incident wave are concluded when it is incident to the interface vertically or at the Bruster angle.
并就椭圆极化电磁波在界面垂直入射,以布儒斯特角入射及在界面发生全反射时的极化规律进行了分析。

Extended Terms:

elliptical polarization instrument 椭圆极化仪
right-handed elliptical polarization 右旋椭圆偏振
elliptical-jacket polarization-maintaining optical fiber 椭圆夹层保偏光纤
wave polarization 椭圆波极化

---

elongation [ˌiːlɔŋˈgeiʃən] *n.*

Definition: the lengthening of something 延长 yán cháng

Origin: late Middle English: from Late Latin elongatio (n-), from elongare "place at a distance"

Example:

The breaking strength and elongation decreased as the degree of oxidation of cotton fiber increased.
氧化棉纤维的断裂强度和断裂伸长率随氧化程度的提高不断降低。

Extended Terms:

ultimate elongation 极限伸长;极限伸长(率);极限伸度
chain elongation 链延长;链延伸;链伸长
elongation viscosity 伸长黏度;拉伸黏度

E

elongation rate 延伸率;伸长率
relative elongation 相对伸长;延伸率,伸长率;相对延伸

elongation sign [ˌiːlɔŋˈgeiʃən sain]

Definition: Referring to the elongation of a substance in relation to refractive indices. If it is elongated in the direction of the high refractive index, it is said to have a positive sign of elongation. If it is elongated in the direction of the low refractive index, it has a negative sign of elongation, not to be confused with the sign of double refraction (i.e., optic sign). 延长符号 yán cháng fú hào

enantiomorphic form [enˌæntiəˈmɔːfik fɔːm]

Definition: the property of crystals that either of a pair is mirror images of each other but is not identical, and that rotate the plane of polarized light equally, but in opposite directions.左右对映形 zuǒ yòu duì yìng xíng

enantiomorphic: Either of a pair of crystals, molecules, or compounds that is mirror images of each other but is not identical, and that rotate the plane of polarized light equally, but in opposite directions. 镜像体的 jìng xiàng tǐ de

form:the phonological or orthographic sound or appearance of a word that can be used to describe or identify something 形式 xíng shì

enantiomorphic relationship [enˌæntiəˈmɔːfik riˈleiʃənʃip]

Definition: the relationship that either of a pair of crystals, molecules, or compounds that is mirror images of each other but is not identical, and that rotate the plane of polarized light equally, but in opposite directions.左右形关系 zuǒ yòu xíng guān xì

relationship:connection or association; the condition of being related 关系 guān xì

enthalpy [enˈθælpi] *n.*

Definition: Enthalpy (denoted as *H*) is a measure of the energy associated with a system. It can be thought of as the amount of energy required to create a system plus the amount of energy required to make room for it by displacing its environment.(H)焓(热函)hán(rè hán)

Origin: 1920s:from Greek enthalpein "warm in", from en-"within" + thalpein "to heat"

Example:

For the ECE method, the gas temperature is calculated based on gas components and their

enthalpy.
焓值守恒法依据的是燃气成分和焓值计算燃气温度。

Extended Terms:

free enthalpy 自由焓
specific enthalpy 比焓;比热焓
sensible enthalpy 显焓;可感热焓
initial enthalpy 初焓
bonding enthalpy 键焓

---

entropy [ˈentrəpi] *n.*

Definition: Entropy is a macroscopic property of a system that is a measure of the microscopic disorder within the system. It is an important part of the second law of thermodynamics. Thermodynamic systems are made up of microscopic objects, e.g. atoms or molecules, which "carry" energy. 熵 shāng

Origin: mid 19th century: from en-"inside" + Greek tropē "transformation"

Example:

A method based on maximum entropy principle was presented to optimize processing parameters in plane polishing.
对此提出一种基于最大熵原理的平面研抛工艺参数优化方法。

Extended Terms:

entropy increase 熵增加
entropy diagram 熵图;熵线圈
entropy filter 滤熵器
vibrational entropy 振动熵
entropy flow 熵流

---

epitaxial growth [ˌepiˈtæksiəl ɡrəuθ]

Definition: The growth on a crystalline substrate of a crystalline substance that mimics the orientation of the substrate. 外延生长(浮生) wài yán shēng zhǎng (fú shēng)

epitaxy: the growth on a crystalline substrate of a crystalline substance that mimics the orientation of the substrate(晶体)取向附生 qǔ xiàng fù shēng; 外延附生 wài yán fù shēng

growth: the process of growing 生长 shēng zhǎng

Example:

The applications of the photo assisted epitaxy in the epitaxial growth of semiconductors are introduced,

such as temperature decrease, doping control, atomic layer epitaxy and the selective epitaxy.
图片说明光辅外延在半导体材料外延生长中的具体应用,包括降低温度、掺杂控制、原子层外延和选择性外延。

Extended Terms:

epitaxial growth process 外延生长过程;外延晶膜生长
selective area epitaxial growth 选区外延生长
liquid phase epitaxial growth 液相外延生长;液相取向附生;液相晶膜生长
low pressure epitaxial growth 低压外延生长
vapor phase epitaxial growth 汽相外延生长

---

epitaxy [ˌepiˈtæksi] *n.*

Definition: Epitaxy refers to the method of depositing a monocrystalline film on a monocrystalline substrate. The deposited film is denoted as epitaxial film or epitaxial layer. It can be translated "to arrange upon". 面衍生 miàn yǎn shēng

Origin: 1930s: from French épitaxie, from Greek epi "upon" + taxis "arrangement"

Example:

The surface phase of diamond and graphite have the character of template, and they will dominate epitaxy pattern of themselves.
金刚石和石墨表面具有模板的特征,它们将主导自身外延层的生长方式。

Extended Terms:

microwave epitaxy 微波磊晶片
crystal epitaxy 晶体外延生长
solid epitaxy 固相外延
thin epitaxy 薄膜外延
epitaxy junction 外延结

---

equivalent isomorphism [iˈkwivələnt ˌaisəuˈmɔːfizm]

Definition: an isomorphism formed by equivalent ions or atoms 等价类质同像 děng jià lèi zhì tóng xiàng

equivalent: equal in force, amount, or value; also: equal in area or volume but not superposable 相等的 xiāng děng de

isomorphism: In abstract algebra, an isomorphism is a bijective map f such that both f and its inverse f^{-1} are homomorphisms, i.e., structure-preserving mappings. In the more general setting of category theory, an isomorphism is a morphism $f: X \rightarrow Y$ in a category for which there exists an "inverse" $f^{-1}: Y \rightarrow X$, with the property that both $f^{-1}f = id_X$ and $f f^{-1} = id_Y$. Informally, an isomorphism is a kind of mapping between objects, which shows a relationship between two properties or operations. If there exists an isomorphism between two structures, we call the two structures isomorphic. In a certain sense, isomorphic structures are structurally identical, if you choose to ignore finer-grained differences that may arise from how they are defined. 同形 tóng xíng;类质同像 lèi zhì tóng xiàng

equivalent point [iˈkwivələnt pɔint]

Definition: The equivalence point, or stoichiometric point, of a chemical reaction occurs during a chemical titration when the amount of titrant added is stoichiometrically equal to the amount of analyte present in the sample: the smallest amount of titrant that is sufficient to fully neutralize or react with the analyte. In some cases there are multiple equivalence points which are multiples of the first equivalent point, such as in the titration of a diprotic acid. 等效点系 děng xiào diǎn xì

Example:

Compared with other methods, the method mentioned above has advantages of the easy judgment of equivalent point, acute color transformation, rapidity and high accuracy.
与其他方法相比,此方法具有等效点易判断、变色敏锐、快速、准确度及精密度高等优点。

Extended Terms:

equivalent generic point 等价通性点
singing point equivalent 振鸣点当量

equivalent position [iˈkwivələnt pəˈziʃən]

Definition: The points, which are equivalent, are related via a symmetry operation. Start at x, y, z; a counterclockwise rotation of 90° generates a point with coordinates y, x, z; then x, y, z; and finally to y, x, z. 等效位置 děng xiào wèi zhì

Example:

We first observe that every positive ion is in a equivalent position to that of every other ion.
我们首先注意到,每一正离子的位置与另一正离子完全相当。

etch figure [etʃ ˈfigə]

Definition: a regular-faceted pit or, more rarely, surface that forms on crystal faces during etching. Etch figures are arranged in regular order in relation to the crystallographic axes. They reflect the symmetry of the crystal faces and the defects of the crystal structure. 蚀象 shí xiàng

E

Euclidean mapping [juːˈklidiən ˈmæpiŋ]

Definition: The Euclidean mapping is a special case of affine mapping that, besides collinearity and ratios of distances, keeps also distances and angles. Because of this, a Euclidean mapping is also called a rigid motion. 欧几里得映像 ōu jī lǐ dé yìng xiàng

Origin: attributed to the Alexandrian Greek mathematician Euclid

---

Ewald sphere [ˈjuːəld sfɪə]

Definition: The Ewald sphere is a geometric construct used in electron, neutron, and X-ray crystallography which demonstrate the relationship between: 1) the wavelength of the incident and diffracted X-ray beams, 2) the diffraction angle for a given reflection, 3) the reciprocal lattice of the crystal. Ewald's sphere can be used to find the maximum resolution available for a given X-ray wavelength and the unit cell dimensions. It is often simplified to the two-dimensional "Ewald's circle" model or may be referred to as the Ewald sphere. 埃瓦耳德球 āi wǎ ěr dé qiú

Origin: It was conceived by Paul Peter Ewald, a German physicist and crystallographer. Ewald himself spoke of the sphere of reflection.

Example:

Some relations of the spatial spectrum of target on the Ewald sphere with its far and near field scattering data on a measurement surface are discussed with the aid of the equiva lent source concept.
本文通过目标等效源概念,讨论远场与近场条件下测量面上散射场与目标 Fourier 域中 Ewald 球面上空间谱值之间的关系。

Extended Term:

Ewald's diffraction sphere 埃瓦尔德衍射球

---

exsolution [ˌeksəˈluːʃən] *n.*

Definition: (*Geology*) (of a mineral or other substance) separate out from solution, especially from solid solution in a rock 出溶(又称离溶、脱溶、解溶)chū róng

Origin: mid 20th century: (originally as exsolution) from ex-+ solution

Example:

Both clinopyroxene and orthopyroxene show exsolution texture. The exsolution rods generally are parallel to each other no matter they are straight or curved.
单斜辉石和斜方辉石均呈现出溶结构,出溶条纹一般相互平行,不管它们本身平直还是发生舒缓的弯曲变形。

Extended Terms:

exsolution lamellae 出溶纹层;出溶纹层,出溶条纹
exsolution structure 固溶体分解结构
noncoherent exsolution 不连贯出溶作用
subsolidus exsolution 凝析

extinction [ikˈstiŋkʃən] *n.*

Definition: (*Physics*) Reduction in the intensity of light or other radiation as it passes through a medium or object, due to absorption, reflection, and scattering. Extinction is a term used in optical mineralogy and petrology, which describes when cross-polarized light dims, as viewed through a thin section of a mineral in a petrographic microscope. 消光 xiāo guāng

Origin: late Middle English: from Latin exstinctio (n-), from exstinguere "quench"

Example:

The errors caused by prism extinction ratio variation are investigated.
本文研究棱镜消光比变化引起的误差。

Extended Terms:

oblique extinction 斜消光
anomalous extinction 异常消光
extinction position 消光位置
dast extinction 尘消
latent extinction 潜在消退
extinction agent 灭火剂

extinction angle [ikˈstiŋkʃən ˈæŋgl]

Definition: The extinction angle is the measure between the cleavage direction or habit of a mineral and the extinction. To find this, simply line up the cleavage lines/long direction with one of the cross hairs in the microscope, and turn the mineral until the extinction occurs. 消光角 xiāo guāng jiǎo

Example:

The extinction angle of the direct current (DC) converter, which is usually used to study the commutation failure of the DC inverter, is an important parameter in the DC transmittal system.
直流换流器熄弧角是直流输电系统的重要电气量,常用来显示直流输电系统逆变器的换相失败。

Extended Terms:

constant extinction angle control 恒熄弧角控制

symmetrical extinction angle 对称消光角
critical extinction angle 临界熄弧，最小熄弧角
angle of extinction 消弧角

extraordinary ray [ik'strɔːdinəri rei]

Definition: (*Optics*) (in double refraction) the light ray that does not obey the ordinary laws of refraction 非(寻)常光 fēi (xún) cháng guāng

Example:

Due to double refraction effect, the ordinary ray and the extraordinary ray have different reflections.
由于双折射效应，寻常光和非常光有不同的全内反射。

Extended Terms:

extraordinary ray spectrum 非常光线光谱
principal index for extraordinary ray 非常光线舟射率
extraordinary ray (extraordinary light) 非常光

face-centered lattice (F-lattice) [feɪs -'sɛntɚd 'lætɪs]

Definition: The face-centered cubic system (cF) has lattice points on the faces of the cube, that each gives exactly one half contribution, in addition to the corner lattice points, giving a total of 4 lattice points per unit cell ($1/8 \times 8$ from the corners plus $1/2 \times 6$ from the faces). Each sphere in an fcc lattice has coordination number 12. The face-centered cubic system is closely related to the hexagonal close packed system, and the two systems differ only in the relative placements of their hexagonal layers. The plane of a face-centered cubic system is a hexagonal grid. 面心格子(F-格子)miàn xīn gé zi

Example:

Thecrystal lattice is a face-centered cubic lattice.
晶胞类型为面心立方格子。

Extended Terms:

face-centered cubic lattice 面心立方格子；面心立方晶格
face-centered orthorhombic lattice 面心正交晶格；面心正交格子；面心斜方晶格；面心长方

晶格

factor group ['fæktə gruːp]

Definition: In mathematics, specifically group theory, a quotient group (or factor group) is a group obtained by identifying together elements of a larger group using an equivalence relation. 剩余群 shèng yú qún

Example:

Through the analysis of five factor group, the research method of topic relativity is put forward based on the negotiation history data.
同时通过对模型五元组的分析,提出了基于协商历史数据的议题相关性研究方法。

Extended Terms:

ordered factor group 有序商群
proper factor group 真剩余群;真商群
hierarchical group factor theory 等级群体因素说

Fedorov stage [fedə'rɔv steidʒ]

Definition: A stage attached to the rotating stage of a polarizing microscope that has three, four, or five axes and thin sections of low-symmetry minerals to be tilted about two mutually perpendicular horizontal axes. Also known as Fedorov stage; U stage. 弗氏旋转台 fú shì xuán zhuǎn tái

Origin: named after Fedorov

Example:

The figure shows a typical three-axis universal stage.
图显示一个典型的三轴万能旋转台。

Extended Terms:

universal adjustable stage 万能可调工作台
universal rotating stage 万能旋转台

ferrimagnetism [ˌferi'mægnitizəm] *n.*

Definition: (*Physics*) (of a substance) displaying a weak form of ferromagnetism associated with parallel but opposite alignment of neighbouring atoms. In contrast with antiferromagnetic materials, these alignments do not cancel out and there is a net magnetic moment 亚铁磁性 yà tiě cí xìng

Example:

Ferrimagnetism occurs mainly in magnetic oxides known as ferrites.
亚铁磁性主要存在于磁性氧化物,如铁氧体之中。

Extended Term:

flasher ferrimagnetism 铁氧体磁性,铁磁共振

---

ferromagnetism [ferəu'mægnitizəm] *n.*

Definition: (*Physics*) (of a body or substance) having a high susceptibility to magnetization, the strength of which depends on that of the applied magnetizing field, and which may persist after removal of the applied field. This is the kind of magnetism displayed by iron, and is associated with parallel magnetic alignment of neighbouring atoms 铁磁性 tiě cí xìng

Example:

The material with both metallic conductivity and ferromagnetism is a very kind of functional material, which is attracting more and more attentions.
导电性与铁磁性并存的功能材料是其中尤为重要的一种,并得到(分子固体研究领域)越来越多的关注。

Extended Terms:

parasitic ferromagnetism 寄生铁磁性
weak ferromagnetism 弱铁磁性
band theory of ferromagnetism 铁磁性能带理论

Flack parameter [flæk pə'ramitə]

Definition: In X-ray crystallography, the Flack parameter is a factor used to estimate the absolute configuration of a structural model determined by single-crystal structure analysis. 弗莱克参数 fú lái kè cān shù

---

forbidden band [fə'bidn bænd]

Definition: The ranges of allowed energies of electrons in a solid are called allowed bands. Certain ranges of energies between two such allowed bands are called forbidden band. 禁带 jìn dài

---

form [fɔːm] *n.*

Definition: Form refers to the shape, visual appearance, or configuration of an object. 形式 xíng shì

Origin: Middle English: from Old French forme (noun), fo(u)rmer (verb, from Latin formare "to form"), both based on Latin forma "a mould or form"

Example:

Traditional wavelet extraction methods usually have the hypothesis that the wavelet is in a certain form, such as the minimum phase.

传统的子波提取方法一般需要假设子波(相位)处于某种形式,例如假设地震子波处于最小相位。

Extended Terms:

angle form 角型

normal form 法线式;标准形;正规形式;标准型

dextrorotatory form 右旋体;右旋型

form factor 齿形系数;形状系数;形状因子;形状因数

Fourier synthesis [ˈfuriːə ˈsinθisis]

Definition: In mathematics, Fourier analysis is a subject area which grew from the study of Fourier series. The subject began with the study of the way general functions may be represented by sums of simpler trigonometric functions. Fourier analysis is named after Joseph Fourier, who showed that representing a function by a trigonometric series greatly simplifies the study of heat propagation. Today, the subject of Fourier analysis encompasses a vast spectrum of mathematics. In the sciences and engineering, the process of decomposing a function into simpler pieces is often called Fourier analysis, while the operation of rebuilding the function from these pieces is known as Fourier synthesis. In mathematics, the term Fourier analysis often refers to the study of both operations. 傅立叶合成 fù lì yè hé chéng

Example:

Fourier synthesis is the process of building the sound back up again.

傅立叶合成是将声音重新构成的过程。

Extended Terms:

Fourier series synthesis 傅式级数(天线数组)合成法

difference Fourier synthesis 差分傅立叶合成

F

free R-factor [friː r ˈfæktə]

Definition: In crystallography, the R-factor (sometimes called residual factor or reliability factor or the R-value) is a measure of the agreement between the crystallographic model and the experimental X-ray diffraction data. In other words, it is a measure of how well the refined structure predicts the observed data. It is defined by the following equation:

$$R = \frac{\sum || F_{obs} | - | F_{calc} ||}{\sum | F_{obs} |}$$

where F is the so called structure factor and the sum extends over all the reflections measured and their calculated counterparts respectively. The structure factor F is closely related to the intensity of the reflection it describes:

$$I_{hkl} \propto | F(hkl) |^2$$

For large molecules, R-factor usually ranges between 0.6 (when comparing a random set of reflections with a given model) and 0.2 (for example for a well-refined macro-molecular model at a resolution of 2.5 Ångström). Small molecules (up to 300 atoms) usually form more ordered crystals than large molecules, it is possible to attain lower R-factors. In the Cambridge Structural Database more than 95% of the 500,000 + crystals have an R-factor lower than 0.15 and 9.5% have an R-factor lower than 0.03. Crystallographers also use the Free R-Factor (R_{Free}) and the symmetric R-Factor to describe the quality of a model. 自由R系数 zì yóu R xì shù

Friedel pair [ˈfriːdl peə]

Definition: The couple of reflections h, k, l and is called a Friedel pair, or Bijvoet pair. Their intensities are equal if there is no absorption, but differ otherwise. Friedel's law then does not hold. Generally speaking, when absorption is present, equivalent reflections generated by the symmetry elements in the crystal have intensities different from those of equivalent reflections generated by the introduction of an additional inversion centre in normal scattering. Friedel, or Bijvoet pairs are used in the resolution of the phase problem for the solution of crystal structures. 夫里德耳对 fū lǐ dé ěr duì

Friedel's law [ˈfriːdlz lɔː]

Definition: Friedel's law, named after Georges Friedel, is a property of Fourier transforms of real functions.

Given a real function $f(x)$, its Fourier transform

$$F(k)=\int_{-\infty}^{+\infty} f(x)e^{ik-x}dx$$

has the following properties.

$$F(k)=F^*(-k)$$

where F^* is the complex conjugate of F.

Centrosymmetric points $(k, -k)$ are called Friedel's pairs.

The squared amplitude $(|F|^2)$ is centrosymmetric:

$$|F(k)|^2=|F(-k)|^2$$

The phase φ of F is antisymmetric:

$$\phi(k)=-\phi(-k).$$

Friedel's law is used in X-ray diffraction, crystallography and scattering from real potential within the Born approximation. Note that a twin operation (aka Opération de maclage) is equivalent to an inversion centre and the intensities from the individuals are equivalent under Friedel's law 夫里德耳定律 fū lǐ dé ěr dìng lǜ

general form [ˈdʒenərəl fɔːm]

Definition: the usual form applying to most crystal 一般(单)形 yì bān (dān) xíng

Example:

The shock wave is similar in general form to the blast wave in air, although it differs in detail.
这种冲击波的一般形式与空中爆震波相似,然而在细节方面却又不相同。

Extended Terms:

surface of general form 旋转面

general form of matrix 矩阵通式

general form equation of a plane 平面的一般方程

general property form 一般财产表

---

geniculate twin [dʒiˈnikjulit twin]

Definition: Geniculate twin (also elbow twin or knee twin), is a special kind of twinned crystal in which the twin-plane has markedly changed the shape of the crystal, rather like a knee or elbow joint. The twin-plane is a reflection plane.膝状双晶 xī zhuàng shuāng jīng

---

G

geometric crystal class [dʒiəu'metrik 'kristəl klaːs]

Definition: Geometric crystal classes (or simply "crystal classes") classify the symmetry groups of the external shape of macroscopic crystals, namely according to the morphological symmetry. There are 10 two-dimensional geometric crystal classes and 32 three-dimensional geometric crystal classes, in one to one correspondence with the 10 and 32 types of point groups in E^2 and E^3, respectively 几何晶体分类 jǐ hé jīng tǐ fēn lèi

---

geometric form [dʒiəu'metrik fɔːm]

Definition: a one-dimensional geometric object such as a pencil or line segment range 几何单形 jǐ hé dān xíng

geometric: characterized by simple geometric forms in design and decoration 几何的 jǐ hé de

Example:

The light-encoded filaments are like rays of light that hold a geometric form of language.
光编码细丝像光的射线,它拥有几何学语言形态。

Extended Term:

ideal geometric form 理想几何形状

---

glide plane [glaid plein]

Definition: (*Crystallography*) Also glide reflection plane, a glide plane is symmetry operation describing how a reflection in a plane, followed by a translation parallel with that plane, may leave the crystal unchanged. Glide planes are noted by a, b or c, depending on which axis the glide is along. There is also the n glide, which is a glide along the half of a diagonal of a face, and the d glide, which is along a fourth of either a face or space diagonal of the unit cell. The latter is often called the diamond glide plane as it features in the diamond structure. 滑移面 huá yí miàn;滑移反映面 huá yí fǎn yìng miàn

Example:

The dispersoid phase $Al_{20}Cu_2Mn_3$ in a 2024 Al alloy is commonly composed of twins. An observation of corresponding high resolution image shows that the twin boundary plane is a glide plane other than mirrors one.
2024 铝合金中的弥散相 $Al_{20}Cu_2Mn_3$ 普遍由孪晶组成,通过高分辨像的观察分析,发现其孪晶间的界面不是镜面,而是滑移面。

Extended Terms:

glide reflection plane 滑移反映面

glide plane fracture 滑移面断口
easy glide plane 易滑面
glide plane of symmetry 滑移对称面
axial glide plane 轴向滑移面

---

gnomonic projection [nəu'mɔnik prəu'dʒekʃən]

Definition: A gnomonic map projection displays all great circles as straight lines. Thus the shortest route between two locations in reality corresponds to that on the map. 心射极平投影 xīn shè jí píng tóu yǐng

Example:

An approach is first proposed by using gnomonic projection, based on the problem remaining unresolved, which is about the coating joint modeling and the joint overlap of spherical equilateral polygon.
针对球面等边多边形包覆拼接建模及出现拼接重叠等问题,提出了共球面等边多面体作心射极平投影的方法。

Extended Terms:

gnomonic map projection 日晷投影
gnomonic azimuthal projection 球心方位投影

Goldschmidt's rule ['gəuldʃmitz ruːl]

Definition: Goldschmidt's rules, also first law of crystallochemistry. In a system where the two variables, temperature and pressure, are controlled externally, the number of phases will not usually exceed the number of components. In geology the maximum number of naturally occurring minerals in a rock is equal to the number of components where a given mineral assemblage is stable over a range of temperatures and pressures.戈尔德施密特定律(晶体化学第一定律)gē ěr dé shī mì tè dìng lǜ (jīng tǐ huà xué dì yī dìng lǜ)

---

goniometer [ˌgəuni'ɔmitə] *n.*

Definition: an instrument for the precise measurement of angles, especially one used to measure the angles between the faces of crystals 晶体测角仪 jīng tǐ cè jiǎo yí

Origin: mid 18th century: from French goniomètre, from Greek gōnia "angle" + French-mètre "(instrument) measuring"

Example:

The high accuracy goniometer is generally used as standard of goniometer measurement.

高准确度的晶体测角仪一般作为角度计量标准器使用。

Extended Terms:

radio goniometer 无线电测角计;无线电测向仪;无线电方位测定器;无线电测角器;无线陀螺罗经
electrical goniometer 电测角计
autocollimating goniometer 自准直测角仪
plane goniometer 平面测向器;平面测向计
goniometer protractor 测角仪

goniometry [ˌgəuniˈɔmitri] *n.*

Definition: an instrument for the precise measurement of angles, especially one used to measure the angles between the faces of crystals 晶体测量(又称晶体测角)jīng tǐ cè liáng

Origin: mid 18th century: from French goniomètre, from Greek gōnia "angle" + French-mètre "(instrument) measuring"

Example:

The obtained self-assembled monolayers were characterized with water contact angle goniometry, auger electron spectroscopy and X-ray photoelectron spectroscopy.
使用接触角、俄歇电子能谱及 X 射线光电子能谱等方法进行表征。

Extended Terms:

radio goniometry 无线电测向
crystal goniometry 晶体测角
pole-figure goniometry 极图测角术

grind [graind] *n.*

Definition: The grind of a blade refers to the shape of the cross-section of the blade. It is distinct from the type of blade (e.g., clip point or drop point knife, sabre or cutlass, axe or chisel, etc.), though different tools and blades may have lent their name to a particular grind. 磨(片) mó (piàn)

Origin: Old English grindan, probably of Germanic origin. Although no cognates are known, it may be distantly related to Latin frendere "rub away, gnash"

Example:

The grind-hardening is a new technology that integrates grinding and surface hardening and enhances material capability of workpiece surface.
磨削强化技术是一种集磨削加工与表面淬火于一体的新技术,可对钢件表层进行强化处理。

Extended Terms:

grind stone 磨石;磨石砂轮;砥石
medium grind 中研磨
grind verb 磨碎,碾碎
flat grind 平面式打磨
concave grind 凹入式打磨

---

group [gruːp] *n.*

Definition: a number of people or things that are located close together or are considered or classed together 群 qún

Origin: late 17th century: from French groupe, from Italian gruppo, of Germanic origin; related to crop

Example:

Message type module is a group of components that together provide message handling for a particular protocol.
消息类型模式是为一个特定的协议共同提供消息操作的一组组元。

Extended Terms:

reference group 参照群体;相关群体;参照组
group technology 成组技术;成组建造技术;群工艺学;组合工艺学
strategic group 战略组;策略群组

---

group homomorphism [gruːp ˌhɔməˈmɔːfizəm]

Definition: In mathematics, given two groups $(G, *)$ and $(H,)$, a group homomorphism from $(G, *)$ to $(H,)$ is a function $h: G \to H$ such that for all u and v in G it holds that

$$h(u * v) = h(u) - h(v)$$

where the group operation on the left hand side of the equation is that of G and on the right hand side that of H.

From this property, one can deduce that h maps the identity element e_G of G to the identity element e_H of H, and it also maps inverses to inverses in the sense that

$$h(u^{-1}) = h(u)^{-1}.$$

Hence one can say that h "is compatible with the group structure".

Older notations for the homomorphism $h(x)$ may be x_h, though this may be confused as an index or a general subscript. A more recent trend is to write group homomorphisms on the right of their arguments, omitting brackets, so that $h(x)$ becomes simply $x\ h$. This approach is especially prevalent in areas of group theory where automata play a role, since it accords better

G

with the convention that automata read words from left to right.

In areas of mathematics where one considers groups endowed with additional structure, a *homomorphism* sometimes means a map which respects not only the group structure (as above) but also the extra structure. For example, a homomorphism of topological groups is often required to be continuous. 群同态 qún tóng tài

homomorphism: similarity of form

Example:

According to the definition of group homomorphism, we can predict the nature of its homomorphism group from the known group nature.

利用群同态的定义,我们可以从已知群的性质推测同态群的性质。

group isomorphism [gruːp ˌaisəuˈmɔːfizəm]

Definition: In abstract algebra, a group isomorphism is a function between two groups that sets up a one-to-one correspondence between the elements of the groups in a way that respects the given group operations. If there exists an isomorphism between two groups, then the groups are called isomorphic. From the standpoint of group theory, isomorphic groups have the same properties and need not be distinguished. 群同构 qún tóng gòu

groupoid [ˈgruːpɔid] *n.*

Definition: In mathematics, especially in category theory and homotopy theory, a groupoid (less often Brandt groupoid or virtual group) generalises the notion of group and of category in several equivalent ways. A groupoid can be seen as 1) a *Group* with a partial function replacing the binary operation; 2) a *Category* in which every morphism is invertible. A category of this sort can be viewed as augmented with a unary operation, called *inverse* by analogy with group theory. Special cases include: *Setoids*, that is: sets which come with an equivalence relation; *G-sets*, sets equipped with an action of a group *G*. Groupoids are often used to reason about geometrical objects such as manifolds. Heinrich Brandt introduced groupoids implicitly via Brandt semigroups in 1926. 广群 guǎng qún

Example:

Through these studies, the study of rough set's algebraic properties are spreaded to right involution groupoid.

这些结果将粗糙设备的集代数性质的研究扩展到右对合广群这个代数系统中。

Extended Terms:

space groupoid 空间亚群

isotopic groupoid 合痕广群

fundamental groupoid 基本广群
abelian groupoid 阿贝耳广群
free groupoid 自由广群

gypsum plate [ˈdʒipsəm pleit]

Definition: In polarized-light microscopy, an accessory plate of clear gypsum (replaced by quartz of the appropriate thickness in modern instruments) that gives a first-order red (approx. 1 lambda out of phase for 560 nm) interference color with crossed polars when inserted in the tube with its permitted electric vectors at 45 degrees to those of the polarizer and analyzer. It is used to determine fast and slow directions (electric vectors) of light polarization in crystals under view on the microscope stage by increasing or decreasing retardation of the light. Also called a sensitive-tint plate.石膏试板 shí gāo shì bǎn

Example:

The invention further provides a production process for pre-casting glass fiber reinforced gypsum plate.
本发明还提供了预铸式玻璃纤维加强石膏试板的制造方法。

Extended Term:

gypsum test plate 石膏试板

h-centred cell [eɪtʃ-ˈsentəd sel]

Definition: The H-centred cell (triple hexagonal cell) is an alternative description of the hexagonal Bravais lattice. From the conventional *hP* cell one obtains the *hH* cell by taking the new basis vectors by means of one of the following transformation matrices, which give three possible orientations of the *hH* cell with respect to the *hP* cell: H_1: 110/-120/001 H_2: 2-10/110/001 H_3: 1-20/2-10/001. The resulting *hH* cell has centring nodes at 1/3, 2/3, 0 and 2/3, 1/3, 0. H 型晶胞 H xíng jīng bāo

hand lens [hænd lenz]

Definition: A magnifying glass (called a hand lens in laboratory contexts) is a convex lens

H

which is used to produce a magnified image of an object. The lens is usually mounted in a frame with a handle. A magnifying glass works by creating a magnified virtual image of an object behind the lens. The distance between the lens and the object must be shorter than the focal length of the lens for this to occur. Otherwise, the image appears smaller and inverted, and can be used to project images onto surfaces. 放大镜 fàng dà jìng

Example:

Reporter was not willing to fight to remove the camera and continue filming, but the host staff was using hand lens, the two sides began the war of words.

记者不甘心地重举相机继续拍摄,却又被主办方的工作人员用手挡镜头,双方又起骂战。

Extended Term:

hand-lens magnifying glass 放大镜

Harker section [ˈhɑːkə ˈsekʃən]

Definition: In Patterson methods of structure solution, relationships between symmetrically related atoms produce peaks in the Patterson function on certain planes or along certain lines determined by the known crystallographic symmetries. Harker sections are portions of the Patterson map that contain a large proportion of the readily interpretable information because they contain many such Harker peaks (vectors between space-group equivalent atoms). 哈克切面 hā kè qiē miàn

---

heavy-atom method [ˈhevi-ˈætəm ˈmeθəd]

Definition: An application of Patterson method in crystal structure determination. For a compound containing a heavy atom (i.e. one with a significantly higher atomic scattering factor than the others present) the diffraction phases calculated from the position of the heavy atom are used to compute a first approximate electron density map. Further portions of the structure are recognisable as additional peaks in the map. Successive approximate electron density maps may then be calculated to solve the entire structure. 重原子法 zhòng yuán zǐ fǎ

---

hemidome [ˈhemidəum] *n.*

Definition: The upper or lower two faces of a dome resulting from symmetry lower than that required for a dome in orthorhombic or monoclinic crystal systems.半坡面 bàn pō miàn

Extended Term:

monoclinic hemidome 单斜半坡面

H

hemihedron [ˈhemihedr(ə)n] *n.*

Definition: (of a crystal) exhibiting only half the number of planes necessary for complete symmetry 半面体 bàn miàn tǐ

Extended Term:

triclinic hemihedron 三斜半面体

hemimorphism [ˌhemiˈmɔːrfizm] *n.*

Definition: Hemimorphism is the property of certain crystals in which there is no element of symmetry presents to cause the repetition of upper-hemisphere faces in the lower hemisphere. A good example is tourmaline, where there are no horizontal axes of symmetry or centre of symmetry, but only one vertical axis of three-fold symmetry and three vertical planes of symmetry. 异极像 yì jí xiàng

Origin: first-known use in International Scientific Vocabulary: about 1859

Extended Terms:

pyramidal hemimorphism 锥形异极像
holohedral hemimorphism 全面异极像
trigonal hemimorphism 三方半面像
trigonal tetartohedral hemimorphism 三方四分面异极像
axis of hemimorphism 异极轴

heterogeneous nucleation [ˌhetərəuˈdʒiːnjəs ˌnjuːkliˈeiʃən]

Definition: Heterogeneous nucleation occurs much more often than homogeneous nucleation. It forms at preferential sites such as phase boundaries or impurities like dust and requires less energy than homogeneous nucleation. At such preferential sites, the effective surface energy is lower, thus diminished the free energy barrier and facilitating nucleation. Surfaces promote nucleation because of wetting-contact angles greater than zero between phases encourage particles to nucleate. The free energy needed for heterogeneous nucleation is equal to the product of homogeneous nucleation and a function of the contact angle:

$$\Delta G_{heterogeneous} = \Delta G_{homogeneous} * f(\theta)$$

where $f(\theta) = \frac{1}{2} + \frac{3}{4}\cos\theta - \frac{1}{4}\cos^3\theta$

The barrier energy needed for heterogeneous nucleation is reduced, and less supercooling is needed. The wetting angle determines the ease of nucleation by reducing the energy needed. 非均匀成核 fēi jūn yún chéng hé

H

Example:

In heterogeneous nucleation, small particles present in the solution act as nuclei to start the crystal formation.
在非均匀成核中,溶液中的小颗粒就充当晶核,形成晶体。

heteropolar [ˌhetərəˈpəulə] *n.*

Definition: (*Chiefly Physics*) characterized by opposite or alternating polarity 异极 yì jí

Origin: late 19th century: from hetero-+ polar

Example:

All the objects in the universe are heteropolar objects and the structures of all objects in the universe are formed by the overlapping of the Taiji structure.
宇宙中的一切事物是有极事物,且所有事物结构由太极结构重叠而成。

Extended Terms:

heteropolar crystal 异极结晶;有极结晶
heteropolar switching 异极转换
heteropolar binding 异极键联
heteropolar colloid 有极胶体
heteropolar linkage 异极键

hexagonal bipyramid [hekˈsægənəl baiˈpirəmid]

Definition: A hexagonal bipyramid is a polyhedron formed from two hexagonal pyramids joined at their bases. The resulting solid has 12 triangular faces, 8 vertices and 18 edges. The 12 faces are identical isosceles triangles.六方双锥 liù fāng shuāng zhuī

hexagonal close packing (HCP) [hekˈsægənəl kləuz ˈpækiŋ]

Definition: In hexagonal close packing, layers of spheres are packed so that spheres in alternating layers overlie one another. As in cubic close packing, each sphere is surrounded by 12 other spheres.六方紧密堆积 liù fāng jǐn mì duī jī

hexagonal prism [hekˈsægənəl ˈpriz(ə)m]

Definition: (*Geometry*) The hexagonal prism is a prism with hexagonal base. 六方柱 liù fāng zhù

Extended Terms:

right hexagonal prism 直六角柱
hexagonal prism of the first order 第一六方柱
hexagonal prism of the third order 第三六方柱
hexagonal unit prism of the first order 第一六方单位柱

hexagonal pyramid [hekˈsægənəl ˈpirəmid]

Definition: (*Geometry*) A hexagonal pyramid is a pyramid with a hexagonal base upon which are erected six triangular faces that meet at a point (the vertex). Like any pyramid, it is self-dual. 六方单锥 liù fāng dān zhuī

Example:

However, the pure edge dislocation is easily etched along the dislocation line, inducing an etch pit of inverted hexagonal pyramid aligned with the surface step.
然而,纯刃位错易于沿位错线被腐蚀,从而使得倒六棱椎的腐蚀坑沿表面阶梯分布。

Extended Terms:

double hexagonal pyramid 六角双锥
hexagonal pyramid of the first order 第一六方锥

hexagonal system [hekˈsægənəl ˈsistəm]

Definition: (*Crystallography*) The hexagonal crystal system is one of the 7 crystal systems, the hexagonal lattice system is one of the 7 lattice systems, and the hexagonal crystal family is one of the 6 crystal families. They are closely related and often confused with each other, but they are not the same. The hexagonal lattice system consists of just one Bravais lattice type: the hexagonal one. The hexagonal crystal system consists of the 7 point groups such that all their space groups have the hexagonal lattice as underlying lattice. The hexagonal crystal family consists of the 12 point groups such that at least one of their space groups has the hexagonal lattice as underlying lattice, and is the union of the hexagonal crystal system and the trigonal crystal system. 六方晶系 liù fāng jīng xì

Example:

Mimetite group minerals are a group of isomorphic series minerals with hexagonal system, which include mimetite, vanadinite and pyromorphite.
砷铅矿族矿物是一族六方晶系的类质同象系列矿物,其中包括砷铅矿、钒铅矿和磷氯铅矿。

Extended Term:

hexagonal crystal system 六角晶系;六方晶系;六方最密堆积;六方密排晶格

H

hexagonal trapezohedron [hek'sægənəl trəˌpiːzəu'hiːdrən]

Definition: The hexagonal trapezohedron or deltohedron is the fourth in an infinite series of face-uniform polyhedra which are dual polyhedron to the antiprisms. It has twelve faces which are congruent kites. 六方偏方面体 liù fāng piān fāng miàn tǐ

hexahedron [ˌheksə'hedrən] *n.*

Definition: A hexahedron (plural: hexahedra) is a polyhedron with six faces. A regular hexahedron, with all its faces square, is a cube. 六面体 liù miàn tǐ

Origin: late 16th century: from Greek hexaedron, neuter (used as a noun) of hexaedros "six-faced"

Example:

When using this technique in the finite element simulation of rectangular solid upsetting, the hexahedron remeshing problems are solved well.

如将该技术应用于矩形块体镦粗过程的有限元模拟中,能较好地解决六面体网格重划问题。

Extended Terms:

rectangular hexahedron 矩形六面体
regular hexahedron 正六面体
stellated truncated hexahedron 星状截顶六面体
regular hexahedron group 正六面体群

hexoctahedron ['heksəuktəˌhedrɔn] *n.*

Definition: A crystalline form belonging to the isometric system and contained under forty-eight equal triangular faces. Also called adamantoid, because it is a common form of the diamond. 六八面体 liù bā miàn tǐ

Origin: hex- + octahedron

high relief [hai ri'liːf]

Definition: High relief is where the most prominent elements of the composition are undercut and rendered at more than 50% in the round against the background. 高突起 gāo tū qǐ

Origin: alto relievo, from Italian

Example:

Analysis of the topography across the front also shows unusually high relief there.

分析前缘的横剖面地形，也可看出地势不寻常的增高。

Extended Terms:

high pressure relief valve 高压释放阀；高压安全阀
high pressure relief system 高压回收系统
high-relief area 起伏大的地区
high pressure relief setting 高压安全调定
high mountain relief 高山地形

holohadron [ˌhɔlə'hædrɔn] *n.*

Definition: a crystal form of the holohedral class, having all the faces needed for complete symmetry 全面体 quán miàn tǐ

holohedral form [ˌhɔlə'hiːdrəl fɔːm]

Definition: the form of a crystal in which the full number of faces required for the symmetry are present 全形 quán xíng

holohedral: Having as many planes as required for complete symmetry in a given crystal system. 全对称的 quán duì chèn de

form: the shape and structure of an object 形式 xíng shì

holohedrism [ˌhɔlə'hiːdrizəm] *n.*

Definition: In crystallography, the property of having all the similar parts similarly modified, as a crystal, or of having all the planes of each form present that are crystallographically possible—that is, all that have the same position with reference to the axes. The law of holohedrism is one of the fundamental principles of crystallography, but there are certain exceptions to it, which are noted under hemihedrism. Also holosymmetry. 全对称性 quán duì chèn xìng

holohedry [hɔlə'hidri] *n.*

Definition: The point group of a crystal is called holohedry if it is identical to the point group of its lattice. In the three-dimensional space, there are seven holohedral geometric crystal classes: $\bar{1}$, $2/m$, mmm, $\bar{3}m$, $4/mmm$, $6/mmm$, $m\bar{3}m$ 全面体 quán miàn tǐ；全对称晶形 quán duì chèn jīng xíng

H

holotype [ˈhɔlətaip] *n.*

Definition: a single type specimen upon which the description and name of a new species is based 全型 quán xíng

Origin: holo-+ -type

Example:

In order to name a new species, one must have a type specimen—a holotype—from which a detailed description can be made, photographs taken, models cast and a professional scientific analysis prepared.
在研究一个新物种时,就要规定正型标本,这个正型标本上给出了该新物种的细节描述、图片、模型和专业的科学分析。

Extended Term:

holotype specimen 正基准标本

---

homogeneity [ˌhɔməudʒeˈniːəti] *n.*

Definition: translational invariance or compatibility of units in equations 均一性 jūn yī xìng

Origin: early 17th century (as homogeneity): from medieval Latin homogeneus, from Greek homogenēs, from homos "same" + genos "race, kind"

Example:

Wavelet scales can be confirmed by region homogeneity measure in this paper.
文中指出利用区域一致性测度能确定小波滤波尺度。

Extended Terms:

linear homogeneity 线性齐次性
homogeneity test 齐性测验;同质性检验;结构均一性测定;同构型检定
genetic homogeneity 成因均匀性;遗传纯合度
temperature homogeneity 温度均匀性
homogeneity assumption 同质假设

---

homogeneous nucleation [ˌhɔməˈdʒiːniəs ˌnjuːkliˈeiʃən] *n.*

Definition: The spontaneous freezing of water droplets at around −40℃ as clusters of water molecules within a droplet settle by chance into the lattice formation of ice, causing the entire droplet to freeze.均匀成核 jūn yún chéng hé

Example:

Potentially, two types of nucleation can occur, namely, homogeneous and heterogeneous

nucleation.
晶核的产生有两种可能的类型,即均匀成核和非均匀成核。

homopolar [ˌhɔmə'pəulə] *n.*

Definition: having equal or constant electrical polarity 同极 tóng jí

Origin: homo- + -polar

Example:

The (disturbance) PID controller has perfect performances in the homopolar magnetic suspension system.
PID 控制器在单磁体控制磁悬浮系统中表现出色。

Extended Terms:

homopolar compound 同无极化合物
homopolar crystal 共价结晶
Homopolar geneator 单极发电机
homopolar doublet 同极双体
homopolar binding 同极键联

homotype ['həumətaip] *n.*

Definition: a part or organ that has the same structure or function as another, especially to a corresponding one on the opposite side of the body 同型 tóng xíng

Origin: homo- + -type

Example:

The laboratory diagnosis on the Mycos that are the same homotype but not the same species can help not only to trace the infection sources of the tuberculosis, but also to eliminate them.
通过辨析结核病菌感染及诊断防治中的各种复杂因素,对同型不同物种来源的分枝杆菌进行鉴别,不仅可以追踪结核病的传染源,而且可以从源头上防治结核病。

icsohedron ['aikəsə'hedrən] *n.*

Definition: In geometry, an icosahedron is a regular polyhedron with 20 identical equilateral triangular faces, 30 edges and 12 vertices. It is one of the five Platonic solids.二十面体 èr shí miàn tǐ

Origin: Greek:εικοσαεδρον, from eikosi "twenty" + hedron "seat"

image ['imidʒ] *n.*

Definition: an optical appearance or counterpart produced by light or other radiation from an object reflected in a mirror or refracted through a lens 镜像 jìng xiàng

Origin: Middle English: from Old French, from Latin imago; related to imitate

Example:

In fact we will use a simple image to create a fake reflection.
事实上,我们将使用一个简单的图像创建一个虚假的镜像。

Extended Terms:

virtual image 虚像;虚影像;虚拟图像
image classification 图像分类;影像分类;画像分类
image filtering 图像滤光;像滤波;滤像;图像过滤
sound image 声像;声成像;音响形象;音像
image control 图像控件;图像控制

immersion method [i'mɑːʃən 'meθəd]

Definition: sinking until covered completely with oil 油浸法 yóu jìn fǎ

immersion: sinking until covered completely with liquid 沉浸 chén jìn

Example:

The result shows that inspecting cabtyre sheathing by ultrasonic is feasible, and the immersion method would be the most reasonable one.
研究结果表明采用水浸法及聚焦探头是比较合理的探伤方法,其中油浸法最为可行。

Extended Terms:

oil-immersion method 油浸法
liquid immersion method 液浸法
full immersion method 整体液浸法,全浸没法
dark-field color immersion method 暗视野彩色油浸法

impact mark ['impækt mɑːk]

Definition: Also percussion mark, it is a small, crescent-shaped scar produced on a hard, dense pebble by a blow. 击痕 jī hén

incommensurate composite crystal [ˈinkəˌmenʃərət ˈkɔmpəzit ˈkristəl]

Definition: An incommensurate composite crystal is a compound with two or more (N) subsystems that are themselves modulated structures, with basis structures that are mutually incommensurate. Each subsystem (numbered by ν) has a reciprocal lattice for its basic structure with three basis vectors a_i^{*v}. There is a basis of the vector module of diffraction spots that has at most $3N$ basis vectors A_j^* such that

$$a_i^{*v} \sum_{j-1}^{n} Z_{ij}^{v} A_j^{*} \ (i = 1,2,3),$$

where Z_{ij}^{v} are integer coefficients. If n is larger than the dimension of space (three), the composite crystal is an aperiodic crystal. n is the rank of the vector module.

incommensurate: not proportionate; not adequate 不成比例 bù chéng bǐ lì

composite: something composite 复合材料 fù hé cái liào

incommensurate magnetic structure

[ˈinkəˌmenʃərət mægˈnetik ˈstrʌktʃə]

Definition: An incommensurate magnetic structure is a structure in which the magnetic moments are ordered, but without periodicity that is commensurate with that of the nuclear structure of the crystal. In particular, the magnetic moments have a spin density with wave vectors that have at least one irrational component with respect to the reciprocal lattice of the atoms. Or, in the case of localized moments, the spin function $S(n+r_j)$ (in the unit cell) has Fourier components with irrational indices with respect to the reciprocal lattice of the crystal. 不对称磁性结构 bú duì chèn cí xìng jié gòu

incommensurate modulated structure

[ˈinkəˌmenʃərət ˈmɔdjuleitid ˈstrʌktʃə]

Definition: An incommensurate modulated crystal structure is a modulated crystal structure, for which the modulation function has a Fourier transform of sharp peaks at wave vectors that cannot all be expressed by rational coefficients in a basis of the reciprocal lattice of the basic structure. At least one of the components of the wave vectors of the modulation with respect to the basis structure should be irrational.不相称结构 bù xiāng chèn jié gòu

indicatrix [ˌindiˈkeitriks] *n.*

Definition: (*Crystallography*) an imaginary ellipsoidal surface whose axes represent the refractive indices of a crystal for light following different directions with respect to the crystal axes 光率体 guāng lǜ tǐ

Origin: late 19th century: modern Latin, feminine of Latin indicator "something that points out"

Example:

The crystal axes and the principal stress axes are obtained from the optical indicatrix axes manually by turning the projection diagram as usual.
用普通方法通过手工转图,可以从光率体主轴中获取晶轴和应力主轴。

Extended Terms:

spherical indicatrix 球面指标
biaxial indicatrix 双轴晶光率体
optic indicatrix 光率体
plane indicatrix 平面指标机构
curvature indicatrix 曲率指标线

---

interfacial angle [intə'feiʃəl 'æŋgl]

Definition: (*Crystallography*) the angle between two crystal faces 面角 miàn jiǎo

Extended Term:

law of constancy of interfacial angle 面角守衡定律;面间角守恒定律

---

interference colors [intə'fiərəns 'kʌləs]

Definition: Interference colors are combinations of different wavelengths. Anisotropic minerals, unless viewed down an optic axis, cause polarized light to be split into two rays as it travels through a grain. The rays may not travel at the same velocity or follow the exact same path. When the rays emerge from the grain, they combine to produce interference colors. 干涉色 gān shè sè

Example:

The optically variable pigments were produced by wet chemical method, and Ti-coated mica with interference colors acted as host materials.
采用具有干涉色的云母钛作为基质材料,通过湿化学法制备随角异色功能颜料。

Extended Terms:

disymmetric of interference colors 对称色散;干涉色变
dispersion of interference colors 干涉;色不对称色散
monosymmetric dispersion of interference colors 干涉色单斜对称分散

inclined dispersion of interference colors 干涉色倾斜色散
horizontal dispersion of interference colors 干涉色之水平色散

intergrowth [ˈintəgrəuθ] *n.*

Definition: a thing produced by intergrowing, especially of mineral crystals in rock 交生 jiāo shēng

Example:

SEM images of the diamond layer of the sample show that there is an extensive intergrowth between diamond grains, which is one of the characteristics PDC synthesized by the growing technique.

对样品进行扫描电镜观察,可看出金刚石层中广泛地存在金刚石颗粒的交互生长,这是生长法合成的 PDC 的结构特征之一。

Extended Terms:

lamellar intergrowth 片晶连生;片流性
micrographic intergrowth 共生;微文像共生
parallel intergrowth 平行交互生长;平行互生
dactylotype intergrowth 指状交生
micropegmatic intergrowth 共生

internal reflection [inˈtəːnəl riˈflekʃən]

Definition: Total internal reflection is an optical phenomenon that occurs when a ray of light strikes a medium boundary at an angle larger than a particular critical angle with respect to the normal to the surface. 内反射 nèi fǎn shè

Example:

Total internal reflection is responsible for rainbows, atmospheric halos, the sparkle of a diamond, and the path of light through optical fibres.

虹、大气晕圈、钻石的闪光以及光线通过光纤路径等现象都是由全内反射引起的。

Extended Terms:

internal primary reflection 内初反射
frustrated total (internal) reflection 受抑(内)全反射
total internal reflection prism 全内反射棱镜
frustrated total internal reflection 衰减全内反射
internal reflection element 内反射组件

interpenetration twin [ˈintəˌpeniˈtreiʃən twin]

Definition: (*Ccrystallography*) Two or more individual crystals so twinned that they appear to have grown through one another, also known as penetration twin. 穿插双晶（贯穿双晶、透入双晶）chuān chā shuāng jīng(guàn chuān shuāng jīng, tòu rù shuāng jīng)

interpenetration: the act of penetrating between or within other substances; mutual penetration 渗透 shèn tòu

---

interplanar spacing [intə(ː)ˈpleinə ˈspeisiŋ]

Definition: the perpendicular distance between successive parallel planes of atoms in a crystal 面网间距 miàn wǎng jiān jù

Example:

The results show that the wax crystals separated from crude oil are orthorhombic crystals whose interplanar spacing is 0.

结果表明从原油中分离出的蜡结晶为正交晶型的晶体,晶面距为0。

Extended Terms:

interplanar crystal spacing 晶面距离
crystallographic interplanar spacing 晶间距离

---

interstice [inˈtəːstis] *n.*

Definition: A small opening or space between objects, especially adjacent objects or objects set closely together, as between cords in a rope or components of a multiconductor electrical cable or between atoms in a crystal. 空隙 kòng xì

Origin: late Middle English: from Latin interstitium, from intersistere "stand between", from inter-"between" + sistere "to stand"

Example:

And in forebulge belt mixed aperture of primary pore and dissolved pore was mainly interstice of the reservoir.

前隆带储层类型为孔隙型,主要以原生孔及溶蚀扩大孔组成的混合孔为主。

Extended Terms:

discontinuous interstice 不连续间隙;(岩石的)不连续间隙
primary interstice 初生孔隙
secondary interstice 次生孔隙;次生裂隙;次生空隙;次生间隙
isolated interstice 隔离空隙

tetrahedral interstice 四面体空隙;四面体间隙

interstitial [ˌintəˈstiʃəl] *adj.*

Definition: of, forming, or occupying interstices 空隙的 kòng xì de

Example:

The boulders are quite widely separated and much interstitial mud is visible.
这些漂砾彼此相隔较远并有许多填隙泥。

Extended Terms:

interstitial cell 间质细胞;间细胞
interstitial deletion 中间缺失
interstitial impurity 填隙式杂质;间隙杂质;(晶)格隙杂质
interstitial ion 填隙离子;填隙式离子
holocrystalline interstitial 全晶间片状的

invariant subgroup [inˈvɛəriənt ˈsʌbgruːp]

Definition: subgroup with unchanged structure 不变子群 bù biàn zǐ qún

invariant: quantity which remains unchanged under certain classes of transformations. Invariants are extremely useful for classifying mathematical objects because they usually reflect intrinsic properties of the object of study. 不变的 bù biàn de

subgroup: (*Mathematics*) a mathematical group whose members are members of another group, both groups being subject to the same rule of combination 小群 xiǎo qún

Example:

This paper mainly researches the relationship among the solution coset of linear equations from the angle of the coset of invariant subgroup, in the course of which the base and the dimension of quotient space have been found out.
本文主要从不变子群的陪集的角度研究线性方程组的解陪集之间的关系,并找到了商空间的基与维数。

Extended Term:

fully invariant subgroup 全不变子群

ion exchange [ˈaiən iksˈtʃeindʒ]

Definition: the exchange of ions of the same charge between an insoluble solid and a solution in contact with it, used in water-softening and other purification and separation processes 离子交换 lí zǐ jiāo huàn

Example:

Such ion exchange is what moves charge from one electrode to the other.
透过离子交换,电荷才能从一个电极传导到另一个电极。

---

ionicity [ˌaiəˈnisəti] *n.*

Definition: (*Chemistry*) the ionic character of a solid 电离度 diàn lí dù,离子性

Origin: first known use:1890

Example:

With approximate treatment by molecular orbital method, a new method for calculation of ionicity percentage of chemical bonds for inorganic compounds and organic compounds.
用分子轨道法近似处理,得到了一个新的估算无机物和有机物化学键离子性百分数的计算公式。

Extended Term:

ionicity parameter 离子性参数

---

ionisation potential [ˌaiənaiˈzeiʃən pəuˈtenʃəl]

Definition: (*Physics / General Physics*) the energy usually required to remove an electron from an atom, molecule, or radical, usually measured in electronvolts. 电离电势 diàn lí diàn shì

ionization: the condition of being dissociated into ions (as by heat or radiation or chemical reaction or electrical discharge. 离子化 lí zǐ huà

potential: (*Physics*) The work required to move a unit of positive charge, a magnetic pole, or an amount of mass from a reference point to a designated point in a static electric, magnetic, or gravitational field; potential energy. [物]势的 shì de

---

isodesmic structure [ˌaisəuˈdezmik ˈstrʌktʃə]

Definition: an ionic crystal structure in which all bonds are of the same strength, so that no distinct groups of atoms are formed 均键结构 jūn jiàn jié gòu

isodesmic: describing a compound or crystal in which all bonds have the same strength (and normally the same length) 各向同点阵的 gè xiàng tóng diǎn zhèn de

---

isometric system [ˌaisəuˈmetrik ˈsistəm]

Definition: In crystallography, the cubic (or isometric) crystal system is a crystal system where the unit cell is in the shape of a cube. This is one of the most common and simplest shapes found in crystals and minerals.等轴晶系 děng zhóu jīng xì

isometric: of, relating to, or characterized by equality of measure; especially relating to or being a crystallographic system characterized by three equal axes at right angles 等轴的 děng zhóu de

---

isomorphism [ˌaisəuˈmɔːfizm] *n.*

Definition: In abstract algebra, an isomorphism is a bijective map f such that both f and its inverse f^{-1} are homomorphisms, i.e., *structure-preserving* mappings. In the more general setting of category theory, an isomorphism is a morphism $f: X \to Y$ in a category for which there exists an "inverse" $f^{-1}: Y \to X$, with the property that both $f^{-1}f = \mathrm{id}_X$ and $ff^{-1} = \mathrm{id}_Y$.

Informally, an isomorphism is a kind ofmapping between objects, which shows a relationship between two properties or operations. If there exists an isomorphism between two structures, we call the two structures isomorphic. In a certain sense, isomorphic structures are structurally identical, if you choose to ignore finer-grained differences that may arise from how they are defined. 类质同像 lèi zhì tóng xiàng

Origin: (Greek: σος isos "equal", and μορφη morphe "shape")

Example:

Au and Ag in ore occur mainly in more than 11 Au Ag minerals; a few in lattice of chalcopyrite, bornite, chalcocite, pyrite and galena in isomorphism.

矿石中的金银主要以金银的独立矿物形式存在，少数以类质同像赋存于黄铜矿、斑铜矿、辉铜矿、黄铁矿和方铅矿的晶格中。

Extended Terms:

anti isomorphism 反同构性；反同构
equivalent isomorphism 等价类同像
isomorphism class 同构类；同构类别
linear isomorphism 线性同构
weak isomorphism 弱同构

---

isomorphous crystals [ˌaisəuˈmɔːfəs ˈkristəlz]

Definition: Two crystals are said to be isomorphous if (a) both have the same space group and unit-cell dimensions and (b) the types and the positions of atoms in both are the same except for a replacement of one or more atoms in one structure with different types of atoms in the other (isomorphous replacement), such as heavy atoms, or the presence of one or more additional atoms in one of them (isomorphous addition). Isomorphous crystals can form solid solutions.

Origin: The notion of isomorphism was discovered by Mitscherlich who found that the crystal forms of salts such as the hydrated potassium phosphates and arsenates or the hydrated potassium copper and iron sulfates were identical (1819, 1820).

Example:

The structure analyses at higher resolution of several isomorphous crystals of this enzyme, especially that with the fluorescent NAD+ derivative, are in progress.
此酶的几种同晶型晶体,特别是荧光 NAD 衍生物晶体的较高分辨率的结构分析工作正在进行中。

---

isomorphous replacement [ˌaisəuˈmɔːfəs riˈpleismənt]

Definition: method of determining diffraction phases from the differences in intensity between corresponding reflections from two or more isomorphous crystals. Most commonly used in the determination of protein structures, where it is possible to derive isomorphous crystals of native protein and of heavy-atom derivatives. 同晶替换 tóng jīng tì huàn

Example:

Among them there exists the widespread phenomenon of isomorphism-phase replacement, thus forming a series of isomorphous mixed crystals.
它们彼此间广泛存在类质同相置换现象,从而形成一系列类质同像混合晶体。

Extended Terms:

isomorphous replacement method 同晶置换法
single isomorphous replacement 单个同形性置换;单个同构物取代
isomorphous ionic replacement 同晶型离子置换
multiple isomorphous replacement method 多对同晶置
double isomorphous replacement 双同态置换

---

isostructural crystal [ˌaisəˈstrʌktʃərəl ˈkristəl]

Definition: crystal with same type of crystalline structure in terms of minaerals and other crystalline substances 等结构的晶体 děng jié gòu de jīng tǐ

isostructural: having same type of crystalline structure: used to describe minerals and other crystalline substances that have the same type of crystalline structure 同构的 tóng gòu de;同型的 tóng xíng de

crystal: A crystal or crystalline solid is a solid material whose constituent atoms, molecules, or ions are arranged in an orderly repeating pattern extending in all three spatial dimensions. 晶体 jīng tǐ

Example:

The isostructural crystal SSP with NZP structure was obtained by the above three synthesis methods.
上述三种方法都能合成等结构的晶体 SSP 和 NZP。

---

isotherm ['aisəuθəːm] *n.*

Definition: a curve on a diagram joining points representing states or conditions of equal temperature 等热 děng rè

Origin: mid 19th century: from French isotherme, from Greek isos "equal" + thermē "heat"

Example:

An isotherm is a line on a map that joins locations having the same mean temperatures.
等温线是在地图上把具有相同平均温度的地方连接起来的线。

---

isotype ['aisəutaip] *n.*

Definition: In crystallography, an "isotype" is a synonym for isomorph. It is the replacement or ability to be replaced of one atom or ion by another in a crystal lattice 等型 děng xíng

Origin: iso- + -type

Example:

The oxygen isotype method was used to test the hydrothermal chimney and sulfide samples collected from the Mariana Island Arc, the Mariana Trough, the Okinawa Trough and the Galapagos Rift.
使用氧同位素方法,测试了西太平洋马里亚纳岛弧、马里亚纳海槽、冲绳海槽和东太平洋加拉帕戈斯裂谷的海底热液烟囱和硫化物全岩样品。

Extended Terms:

isotype method 同型法
isotype control 同型对照;阴性对照
isotype exclusion 同种(型)排斥
isotype switch 同种型转换
light chain isotype suppression(LCIS) 轻链同种型抑制

K

kinematical theory [kiniˈmætikl ˈθiəri]

Definition: The kinematical or geometrical theory, the amplitudes diffracted by a three-dimensional periodic assembly of atoms (Laue) or by a stack of planes (Darwin) is derived by adding the amplitudes of the waves diffracted by each atom or by each plane, simply taking into account the optical path differences between them, but neglecting the interaction of the propagating waves and matter. This approximation is not compatible with the law of conservation of energy and is only valid for very small or highly imperfect crystals. The purpose of the dynamical theory is to take this interaction into account. 运动学理论 yùn dòng xué lǐ lùn

kinematics: a branch of dynamics that deals with aspects of motion apart from considerations of mass and force 运动学 yùn dòng xué

Extended Term:

kinematical diffraction theory 运动学衍射理论

---

kink [kiŋk] *n.*

Definition: a sharp twist or curve in something that is otherwise straight 扭折 niǔ zhé

Origin: late 17th century: from Middle Low German kinke, probably from Dutch kinken "to kink"

Example:

The initial stage of compression, kink bands are formed on the surface of WAPET fiber at an angle of ca. 60°~65°to fiber axis.

在变形初期,因局部晶面滑移而形成与纤维轴约呈 60°~65°角的变形带。

Extended Terms:

accommodation kink 缓和节

kink point 转折点

kink waves 扭结波

kink fold 膝折褶皱;折带褶曲

L

lattice [ˈlætis] *n.*

Definition: (*Physics*) a regular repeated three-dimensional arrangement of atoms, ions, or molecules in a metal or other crystalline solid 点阵 diǎn zhèn

Origin: Middle English: from Old French lattis, from latte "lath", of Germanic origin

Example:

Lattice parameters are the important crystal structural parameters of polycrystalline materials. 晶胞参数是多晶材料晶体结构的重要参数。

Extended Terms:

crystal lattice 晶格;晶体点阵;晶体格构

lattice binding 晶格畸变;晶胞;晶格缺陷

space lattice 空间点阵;空间晶格;立体格子;晶体格子

lattice energy 晶格能;晶格能量;点阵能

layer lattice 层形点阵;层格;层形晶格;层状格子

---

lattice complex [ˈlætis ˈkɔmpleks]

Definition: A lattice complex is the set of all point configurations that may be generated within one type of Wyckoff set. All Wyckoff positions, Wyckoff sets and types of Wyckoff sets that generate the same set of point configurations are assigned to the same lattice complex. 点阵复容 diǎn zhèn fù róng

Origin: The name lattice complex comes from the fact that an assemblage of points that are equivalent with respect to a group of symmetry operations including lattice translations can be visualized as a set of equivalent lattices.

Extended Term:

quasi-complex lattice 拟复格

---

lattice constant [ˈlætis ˈkɔnstənt]

Definition: The lattice constant (or lattice parameter) refers to the constant distance between unit cells in a crystal lattice. 晶体几何常数 jīng tǐ jǐ hé cháng shù

Example:

The results show that the lattice constants of $GdCo_2$, $GdCo_5$ and Gd_2Co_{17} alloys enlarge almost linearly with the Al addition, while they all retain their respective structure.
研究结果表明:在一定范围内掺杂 Al 导致 $GdCo_2$、$GdCo_5$ 和 Gd_2Co_{17} 晶格几何常数增大,但仍保持各自的晶体结构。

---

lattice defect [ˈlætis diˈfekt]

Definition: It is also called crystal defect, lattice defect. It is a discontinuity in the lattice of a crystal caused by missing or extra atoms or ions, or by dislocations. 晶格缺陷 jīng gé quē xiàn

Origin: 1935-1940

Example: The relationships between the defect lattice and the acidity (based on acetic acid) of HMX crystals were studied.
研究了奥克托今(HMX)的晶格缺陷和酸值(以醋酸计)之间的关系。

Extended Terms:

crystal lattice defect 晶体晶格缺陷
lattice defect scattering 晶格缺陷散射

---

lattice energy [ˈlætis ˈenədʒi]

Definition: The lattice energy (LE) of an ionic compound is the energy required to break apart the ions in their lattice arrangement into the ions in the gas phase. 晶格能 jīng gé néng

Example:

They have found that there existed linear relationships between lattice energy and debye temperature for Ⅰ-Ⅶ, Ⅱ-Ⅵ, Ⅲ-Ⅴ crystal families in this research, the linearity decreases with increasing convalency.
他们发现Ⅰ-Ⅶ,Ⅱ-Ⅵ,Ⅲ-Ⅴ三族晶体的晶格能与德拜温度均呈线性关系,线性随着共价性的增加而降低。

Extended Terms:

lattice defect scattering 晶格缺陷散射
lattice defect (= crystal defect) 晶格缺陷

---

lattice point [ˈlætis pɔint]

Definition: a point at the intersection of two or more grid lines in a point lattice 格点 gé diǎn

Example: The condition for a fuzzy neural network to realize memory is discussed, and the concept of lattice point is given.
讨论了一种模糊神经网络实现记忆的条件和记忆的特点,并给出了样本组格子点分布的概念。

Extended Term:

lattice transformation point 晶格转变点

---

lattice system [ˈlætis ˈsistəm]

Definition: A lattice system of space groups contains complete Bravais flocks. All those Bravais flocks which intersect exactly the same set of geometric crystal classes belong to the same lattice system. 格系统 gé xì tǒng

Example:

Calculating the reliability of a 3-dimensional lattice system is very difficult. 计算出线性相连的三维格点系统的可靠度的精确公式是相当困难的。

Extended Term:

normalized lattice system 归一化格系统

Laue classes [ˈlauə ˈklæsiz]

Definition: The Laue classes correspond to the eleven centrosymmetric crystallographic point groups. When absorption is negligible and Friedel's law applies, it is impossible to distinguish by diffraction between a centrosymmetric point group and one of its non-centrosymmetric subgroups. 劳厄类 láo è lèi

Origin: after Max Theodor Felix von Laue (9 October 1879-1924 April 1960), a German physicist

---

Laue equations [ˈlauə iˈkweiʒəns] *n.*

Definition: In crystallography, the Laue equations give three conditions for incident waves to be diffracted by a crystal lattice. They are named after physicist Max von Laue (1879-1960). 劳厄方程 láo è fāng chéng

Origin: after Max Theodor Felix von Laue (9 October 1879-1924 April 1960), a German physicist

Extended Term:

Laue diffraction equation 劳厄衍射方程

Laue group [ˈlauə gruːp]

Definition: A Laue group is the point group symmetry of the diffraction pattern, i.e. the crystal point group plus a centre of inversion at the origin of the reciprocal lattice. 劳厄组 láo è zǔ

---

law of Bravais [lɔː ɔv brəˈvei]

Definition: The frequency with which a given face is observed is roughly proportional to the number of nodes it intersects in the lattice per unit length. 布拉维定律 bù lā wéi dìng lǜ

---

law of constancy of interfacial angles

[lɔː ɔv ˈkɔnstənsi ɔv intəˈfeiʃəl ˈæŋglz]

Definition: In all crystals of the same substance, the angles between corresponding faces have the same value when measured at the same temperature. This concept was first proposed by Steno in 1669 and was formulated as a law by Romé de l'Isle in 1772. 面角恒等定律 miàn jiǎo héng děng dìng lǜ

Origin: This concept was first proposed by Steno in 1669 and was formulated as a law by Romé de l'Isle in 1772.

---

law of whole numbers [lɔː ɔv həul ˈnʌmbəz]

Definition: one of the primary laws of crystallography and also one of the first quantitative laws of the atomic-molecular structure of solids, also law of rational indices. 整数定律 zhěng shù dìng lǜ

Origin: established by R. J. Haüy in 1784

---

length fast [leŋθ fɑːst]

Definition: Also negative elongation. In a section of an anisotropic crystal, a sign of elongation is parallel to the faster of the two plane-polarized rays. 负延性 fù yàn xìng

Example:

It is transparent under the microscope, showing parallel extinction and length fast; it is biaxial positive, with Ng≈2.248, Nm≈2.212, Np≈2.194, Ng-Np≈ 0.054 and 2V=70°.

在显微镜下明显显示出平行消失和负延性,还有二轴正光性,其中 Ng=2.248、Nm=2.212、

Np = 2.194、Ng-Np = 0.054、2V = 70°。

L

length slow [leŋθ sləu]

Definition: Also positive elongation. In a section of an anisotropic crystal, a sign of elongation is parallel to the slower of the two plane-polarized rays. 正延性 zhèng yán xìng

limiting complex [ˈlimitiŋ ˈkɔmpleks]

Definition: A limiting complex is a lattice complex L1 which forms a true subset of a second lattice complex L2. Each point configuration of L1 also belongs to L2.

limiting: restricting the scope or freedom of action 限制的 xiàn zhì de

complex: a whole structure (as a building) made up of interconnected or related structures 复合体 fù hé tǐ

local symmetry [ˈləukəl ˈsimitri]

Definition: A local symmetry is symmetry of some physical quantity, which smoothly depends on the point of the base manifold. Such quantities can be for example an observable, a tensor or the Lagrangian of a theory. If a symmetry is local in this sense, then one can apply a local transformation (resp. local gauge transformation), which means that the representation of the symmetry group is a function of the manifold and can thus be taken to act different on different points of spacetime. 部分对称 bù fen duì chèn

Example:

The d-d transition absorption spectra of the complexes Co (ad) (adH) Cl · H_2O is analysed. It belong to Cs local symmetry. The ligand field calculation is applied to molecules with Cs point group symmetry.

分析了 Co(ad)(adH)Cl · H_2O 分子的 d-d 跃迁吸收光谱,指出了它具有 Cs 局部对称。

long range order [lɔŋ reindʒ ˈɔːdə]

Definition: A solid is crystalline if it has long-range order. Once the positions of an atom and its neighbours are known at one point, the place of each atom is known precisely throughout the crystal. 长程有序 cháng chéng yǒu xù

Example:

During parent phase aging, DO_3 atomic reordering occurs, leading to an increase in the degree

of long range order.
在马氏体状态时效,产生了 DO_3 原子短程,导致长程有序度增加。

Extended Terms:

long range order parameter 长程(有)序参数
long-range order in copper-gold alloy 铜金合金的远程有序

longitude [ˈlɔndʒitjuːd] *n.*

Definition: angular distance east or west on the earth's surface, measured by the angle contained between the meridian of a particular place and some prime meridian, as that of Greenwich, England, and expressed either in degrees or by some corresponding difference in time 经度 jīng dù

Origin: 1350-1400; ME < L longitūdō length

Example:

Prolong the sponge along the longitude and latitude at an altitude.
沿着经度和纬度在一个高度延长海绵。

Extended Terms:

astronomical longitude 天文经度;黄经
geodetic longitude 大地经度

Lorentz-polarization correction [ˈlɔːrənts ˌpəulərai'zeiʃən kəˈrekʃən]

Definition: a multiplicative factor involved in converting diffracted radiation intensities to structure factors during the process of structure determination for X-ray diffraction experiments involving moving crystals 洛伦兹极更正 luò lún zī jí gēng zhèng

Lorentz:洛伦兹(姓氏)luò lún zī

polarization:The phenomenon in which waves of light or other radiation are restricted in direction of vibration.极化 jí huà

correction:a quantity that is added or subtracted in order to increase the accuracy of a scientific measure 更正 gēng zhèng

low relief [ləu riˈliːf]

Definition: Sculptural relief that projects very little from the background. Also called bas-relief, basso-relievo. 低突起 dī tū qǐ

Example:

Transfer zones are divided into 5 types in this paper:divergent transfer zone,convergent transfer

zones of high relief and low relief, synthetic transfer zones of high relief and low relief.
本文中转换带可划分为 5 种类型：离散型转换带、汇聚型高凸起、低凸起转换带、同向型高凸起与低凸起转换带。

Extended Terms:

low-relief area 起伏小的地区
low mountain relief 低山地形

---

luminescence [ˌljuːmiˈnesəns] *n.*

Definition: Luminescence is light that usually occurs at low temperatures, and is thus a form of cold body radiation. It can be caused by chemical reactions, electrical energy, subatomic motions, or stress on a crystal. 发光性 fā guāng xìng

Origin: 1885-1890; < L lūmin- （See lumen） +-escence

Example:

Auxiliary pressure and temperature measurements make the MultiFrequency Phase Fluorometer an ideal choice for luminescence sensor design, testing and calibration.
辅助压力和温度测量让多频相位荧光计是发光探测传感器设计、测试和校准的理想选择。

Extended Term:

cathode luminescence 阴极发光；阴极射致发光；电子激发光

magnetism [ˈmægnitizəm] *n.*

Definition: the properties of attraction possessed by magnets; the molecular properties common to magnets. 磁性 cí xìng

Origin: 1610-1620; < NL magnētismus. See magnet, -ism

Example:

Nickel is the most fundamental component in every nickle catalyst and has the great magnetism, so every nickel catalyst has magnetism.
做为镍催化剂主要成分的镍是一种较强的磁性物质，因此所有的镍催化剂都带有磁性。

Extended Terms:

terrestrial magnetism 地磁学；地磁场

nuclear magnetism 核磁性

M

Mallard's law ['mæləd lɔː]

Definition: The law of Mallard states that twin elements are always rational (i.e. direct lattice elements): therefore, a twin plane is a lattice plane, and a twin axis is a lattice row. These twin elements are pseudo symmetry elements for the lattice of the individual. The twin operations produce now slightly different orientations of the lattice of the individual, which are only quasi-equivalent, and no longer equivalent, as in the case of twinning by merohedry. 马拉德法则 mǎ lā dé fǎ zé

Origin: The law of Mallard was introduced by Georges Friedel (Leçons de Cristallographie 1926, page 436) to explain, on reticular basis, twinning by pseudomerohedry.

Mapping ['mæpiŋ] *n.*

Definition: Mapping may refer to The making of maps, as in cartography, surveying, and photogrammetry。映像 yìng xiàng

Origin: 1765-1775; map +-ing

Example:

Since computer mapping are more popularization, there is an urgent need for the automation of map design.

由于电脑绘图的普及化,在制图方面更迫切地需要地图设计的自动化。

Extended Terms:

Mapping Coordinates 贴图坐标

Reflection mapping 反射贴图

matrix ['meitriks] *n.*

Definition: fine material, as cement, in which lumps of coarser material, as of an aggregate, are embedded. 矩阵 jǔ zhèn

Origin: 1325-1375; ME matris, matrix < L mātrix female animal kept for breeding (LL: register, orig. of such beasts), parent stem (of plants), deriv. of māter mother

Example:

Embedded in the matrix of the cytoplasm are found variable numbers of mitochondria.

可变数目的线粒体被发现嵌入在细胞质的矩阵中。

Extended Terms:

Matrix Recognition 模式识别;模式辨认

transposed matrix 转置矩阵;转置阵;位调矩阵

maximum likelihood [ˈmæksiməm ˈlaiklihud]

Definition: Maximum likelihood, also called the maximum likelihood method, is the procedure of finding the value of one or more parameters for a given statistic which makes the *known* likelihood distribution a maximum. 最大似然率 zuì dà sì rán lǜ

Origin: Maximum-likelihood estimation was recommended, analyzed and vastly popularized by R. A. Fisher between 1912 and 1922 (although it had been used earlier by Gauss, Laplace, Thiele, and F. Y. Edgeworth). Reviews of the development of maximum likelihood have been provided by a number of authors.

Example:

According to the mechanism of speech Signal, an effective pitch detection algorithm by combined liner predictive coding with maximum likelihood was proposed.
根据语音信号机理,结合常用的线性预测和最大似然法,提出了一种有效的基音检测算法。

Extended Terms:

maximum-likelihood estimate 最大可能预估;最大可能预估
maximum-likelihood demodulator 最大相似度(判定)解调器;最大相似度(判定)解调器

mechanical twin [miˈkænikəl twin]

Definition: A twin formed in a metal crystal by plastic deformation, involving shear of the lattice. 机械双晶 jī xiè shuāng jīng

Example:

Explorations of X-ray diffraction illustrated, during the ECAE processing, that the mechanical twin ofαphase didn,t occur and thatαphase of the alloys prepared by route Bc had uniform distributions of plane (1 011) and (1 010), but they didn,t by route A.
X 射线分析表明,ECAE 过程中合金的 α 相没有发生机械双晶; 经 Bc 路径后合金 α 相的(1 011)和(1 010)晶面取向分布较分散,而经 A 路径后,其取向分布差异较大。

megacryst [ˈmegəkrist] *n.*

Definition: Any crystal or grain in an igneous or metamorphic rock that is significantly larger than the surrounding matrix. 巨晶 jù jīng

Example: In the course of formation of the labrador megacryst lapilli, the temperature and the cooling rate are the major factors controling their crystallization and growth.

托勒巴契克火山拉长石巨晶火山砾形成过程中,温度和冷却速率乃是控制其结晶生长的主导因素。

Extended Terms:

pyroxene megacryst 辉石巨晶
cotundum megacryst 刚玉巨晶
zircon megacryst 锆石巨晶
augite megacryst 普通辉石巨晶

---

merohedral [ˌmerəˈhedrəl] *adj.*

Definition: (*Crystallography*) of a crystal class in a system, having a general form with only one-half, one-fourth, or one-eighth the number of equivalent faces of the corresponding form in the holohedral class of the same system, also known as merosymmetric 缺面形 quē miàn xíng

Extended Term:

merohedral form 缺面形

---

merohedry [ˌmerəˈhedri] *n.*

Definition: The point group of a crystal is called merohedry if it is a subgroup of the point group of its lattice. 缺面象 quē miàn xiàng

---

mesh [meʃ] *n.*

Definition: any arrangement of interlocking metal links or wires with evenly spaced, uniform small openings between, as used in jewelry or sieves 网眼 wǎng yǎn

Origin: 1375-1425; late ME mesch, appar. continuing OE masc, max; akin to OHG māsca, MD maesche

Example:

Shiny and glitzy-this little purse in Silver Shade Crystal Mesh combined with silver calfskin is great to store away your lose change.
这款小牛皮钱包,极有光泽,银色光面,水晶网格,最适合那些丢三落四的人存放钱款。

Extended Terms:

mesh fabric 网眼织物;网眼布
Mesh Select 网格选择修改器

---

mesodesmic structure [ˌmiːsəuˈdesmik ˈstrʌktʃə]

Definition: An ionic crystal structure in which all bonds are of the same strength, so that no distinct groups of atoms are formed. 中键结构 zhōng jiàn jié gòu

metastable [ˌmetəˈsteibl] *adj.*

Definition: (*Metallurgy*) chemically unstable in the absence of certain conditions that would induce stability, but not liable to spontaneous transformation 准稳的 zhǔn wěn de

Origin: 1895-1900; meta-+ stable

Example:

Through this synthesis tree, the scientist may answer many and the metastable state liquid related question.

通过这棵合成树,科学家可以回答很多与亚稳态液体有关的问题。

Extended Terms:

metastable atom 亚稳原子;亚稳离子;准稳原子

metastable equilibrium 准稳平衡

metric tensor [ˈmetrik ˈtensə]

Definition: A metric tensor is defined to be a nondegenerate symmetric bilinear form on each tangent space that varies smoothly from point to point. It is an example of a tensor field. Relative to a local coordinate system, a metric tensor takes on the form of a symmetric matrix whose entries transform covariantly under changes to the coordinate system, which is to say that the metric tensor is a covariant symmetric tensor. 度规张量 dù guī zhāng liàng

Example:

The Bra Vector, ket vector and metric tensor are introduced directly from the inner produce of two rectors $A.\ B=(A,B)\ =\ (A\backslash B)$

从矢量的内积 $A\cdot B=(A,B)=(A|B)$ 可以直接引入左矢、左矢和度规张量。

Extended Terms:

contravariant metric tensor 逆变度量张量

associated metric tensor 相伴的度量张量

mica plate [ˈmaikə pleit]

Definition: a sheet mineral used in glass and electronic equipment. 云母板 yún mǔ bǎn

M

Example:

The emergence of mica plate greatly promote the development of electic-heat instrument industry.
云母板的出现极大地促进了电热电器工业的发展。

Extended Terms:

mica packing plate 衬垫云母板
cold molding mica plate 冷模制云母板

---

microphotometer [ˌmaikrəufəuˈtɔmitə] *n.*

Definition: a photometer adapted for measuring the intensity of light emitted, transmitted, or reflected by minute objects 显微光度计 xiǎn wēi guāng dù jì

Origin: 1895-1900; micro-+ photometer

Example:

Aikinite in the Qiyugou gold deposit, Henan Province has been studied by means of reflecting microscope, MPV-1 Type microphotometer, JCXA-733 Type electron microprobe and X-ray powder analysis.
利用反光显微镜、MPV-1 型显微光度计、JCXA-733 型电子探针、X 射线粉晶分析等手段对产在河南祁雨沟爆破角砾岩型金矿中的针硫铋铅矿进行了研究。

Extended Terms:

microphotometer comparator 显微光计比较器
iris microphotometer 光瞳测微光度计

---

microstructure [ˈmaikrəuˌstrʌktʃə] *n.*

Definition: the structure of a metal or alloy as observed, after etching and polishing, under a high degree of magnification 显微结构 xiǎn wēi jié gòu

Origin: 1880-1885; micro- + structure

Example:

It is proved that the technology has the advantages of convenience, multi-use and accuracy in the research of refractories microstructure.
研究表明,该技术在耐火材料显微结构研究中具有快捷、方便、多用途、精确等优点。

Extended Terms:

market microstructure 市场微观结构
transistor microstructure 晶体管的微型结构

cast microstructure 铸态组织

miller's indices [ˈmilə ˈindisiːz]

Definition: one of three integers giving the orientation and position of the face of a crystal in terms of the reciprocals, in lowest terms, of the intercepts of the face with each axis of the crystal 密勒指数 mǐ lè zhǐ shù

Origin: 1895-1900; named after W. H. Miller (1801-1880), British mineralogist

Example:

Those crystal planes should be selected such that their Miller indices are lower and the planes have high symmetry.

选择这样几组晶面,使它们的密勒指数较低且具有高度的对称性。

Extended Terms:

Miller-Bravais indices 密勒-布拉维指数

Bravais-Miller indices 布拉维-密勒指数

module [ˈmɔdjuːl ˌ-dʒuːl] *n.*

Definition: a separable component, frequently one that is interchangeable with others, for assembly into units of differing size, complexity, or function 模 mó

Origin: from Middle French module (1540s) or directly from Latin modulus

Example:

The centre module displays traffic guidance information.

中央模块显示交通引导信息。

Extended Terms:

command module 指令舱;指挥舱

lunar module 登月舱;登月小艇

molecular replacement [məuˈlekjulə riˈpleismənt]

Definition: Molecular replacement (or MR) is a method of solving the phase problem in X-ray crystallography. MR relies upon the existence of a previously solved protein structure which is homologous (similar) to our unknown structure from which the diffraction data is derived. 分子置换 fēn zǐ zhì huàn

Example:

The crystal structure was determined by the method of molecular replacement.

分子置换法的使用取决于该种晶体的结构。

Extended Terms:

molecular replacement method 分子置换法
molecular replacement technique 分子置换法

morphology [mɔːˈfɔlədʒi] *n.*

Definition: the form and structure of an organism considered as a whole 形态 xíng tài

Origin: 1820-1830; morpho- + -logy; first formed in German

Example:

The surface roughness and morphology had great influence on the transmission of the films.
表面的粗糙度和组织形态对透射率有较大的影响。

Extended Terms:

crystal morphology 晶体形态学;形态结晶学;晶体结构;晶体组织
insect morphology 昆虫形态学

mosaic crystal [məuˈzeiik ˈkristəl]

Definition: The mosaic crystal is a simplified model of real crystals proposed by C. G. Darwin. In this model, a real crystal is described as a mosaic of crystalline blocks with dimensions of 10^{-5} cm, tilted to each other by fractions of a minute of arc. Each block is separated from the surrounding blocs by faults and cracks. 镶嵌晶体 xiāng qiàn jīng tǐ

Example:

And the company also launch the mosaic of crystal with color of natural halcyon.
而公司也推出了呈现天然翡翠色彩的水晶马赛克。

Extended Term:

ideal mosaic crystal 理想镶嵌结晶

motif [məuˈtiːf] *n.*

Definition: a distinctive and recurring form, shape, figure, etc., in a design, as in a painting or on wallpaper 结构基元 jié gòu jī yuán

Example:

This content is the syzygy motif, and it expresses the fact that a masculine element is always paired with a feminine one.

这个内容正是回合主题,并且它表达了这样一个事实:阳性的元素总是伴随着一个阴性的因素。

Extended Term:

consensus motif 共有基序
dodecapeptide motif 十二肽基序

multiwavelength anomalous diffraction（MAD）

[mʌltɪˈweɪvleŋθ əˈnɒm(ə)ləs dɪˈfrækʃn]

Definition: An approach to solving the phase problem in protein structure determination by comparing structure factors collected at different wavelengths, including the absorption edge of a heavy-atom scatterer. Also known as multiple-wavelength anomalous diffraction or multiwavelength anomalous dispersion. 多波长反常散色法 duō bō cháng fǎn cháng sàn sè fǎ

Origin: This technique was introduced by W. Hendrickson (Hendrickson, W. A., 1991, Determination of macromolecular structures from anomalous diffraction of synchrotron radiation. Science, 254: 51-58.)

Example:

The crystal structure of sCD89 was solved by multiwavelength anomalous diffraction using selenomethionine derivative.
利用硒代衍生物的多波长反常散色法,我们解析得到了 sCD89 的晶体结构。

naked eye [ˈneikid ai]

Definition: the eye unaided by any optical instrument that alters the power of vision or alters the apparent size or distance of objects; "it is not safe to look directly at the sun with the naked eye" 肉眼 ròu yǎn

Origin: This expression was first recorded in 1664.

Example:

The eclipse can be viewed safely with the naked eye during totality, but experts say eye protection should be worn if even a sliver of the sun is visible.
尽管凭借肉眼观看日食全过程是很安全的,但哪怕太阳只有一丁点的光亮可见,专家还是建议人们尽可能佩戴可以护眼的用品。

Extended Terms:

naked-eye variable star 肉眼变星
naked eye appearance quality 目视检查外观品质

net [net] *n.*

Definition: the abstraction, in topology, of a sequence; a map from a directed set to a given space. 面网 miàn wǎng

Origin: Middle English net (*n.*), netten (*v.*), Old English net (t) (*n.*); German Netz

Example:

This is a study on the Petri net modeling and performance analysis of a testing system.
这是一项测试系统的 Petri 网建模和性能分析的研究。

Extended Terms:

virtual net 虚拟网;虚拟网络
polythelene net 尼龙绳网袋

Neumann's principle [ˈnjuːm(ə)n ˈprɪnsɪp(ə)l]

Definition: (*Crystallography*) The principle that the symmetry elements of the point group of a crystal are included among the symmetry elements of any property of the crystal. 诺埃曼原理 nuò āi màn yuán lǐ

Origin: Franz Neumann's (1795-1898) principle was first stated in his course at the university of Königsberg (1873/1874) and was published in the printed version of his lecture notes (Neumann F.E., 1885, Vorlesungen über die Theorie der Elastizität der festen Körper und des Lichtäthers, edited by O. E. Meyer. Leipzig, B. G. Teubner-Verlag).

node [nəud] *n.*

Definition: (*Geometry*) A node is a point on a curve or surface at which there can be more than one tangent line or tangent plane. (*Physics*) A node is a point, line, or region in a standing wave at which there is relatively little or no vibration. 结点 jié diǎn

Origin: 1565-1575; from Latin nōdus "knot"

Example:

Open registration for the client node using the root node administration feature.
通过使用根结点的管理工具(组件)为结点客户开放注册。

Extended Terms:

fiber node 光纤节点

initial node 初始结点;始节点;起始节点

non-crystalline substance [nʌn-ˈkristəlain ˈsʌbstəns] *n.*

Definition: (See amorphous substance) is something that has a random molecular formation in it's natural form (when solidified). The opposite would be semi-cristaline which has random molecular structure when heated but returns to an organised unifom state when solidified. 非晶质 fēi jīng zhì

Example:

The pathologic pictures showed eosinophilic amorphous substance with characteristic green birefringence in polarized light after staining with Congo red.

组织病理切片检查显示嗜伊红性无定形物质,以刚果红染色,在偏极光下,该物质表现出典型的苹果绿色之双折光性。

Extended Term:

amorphous ground substance 无结构基质;无定型基质

normal subgroup [ˈnɔːməl ˈsʌbgruːp]

Definition: A normal subgroup is a subgroup which is invariant under conjugation by members of the group. Normal subgroups can be used to construct quotient groups from a given group. 正规子群 zhèng guī zǐ qún

Example:

For certain finite group with a perfect normal subgroup, this paper discusses the problem of its augmentation ideals and quotient groups.

针对具有完全正规的有限群,本文研究了完全正规子群增长的理想条件和系数群。

Extended Terms:

non-normal subgroup 非正规子群

weakly normal subgroup 弱正规子群

normal twin [ˈnɔːməl twin]

Definition: a twin crystal whose twin axis is perpendicular to the composition surface 垂直双晶 chuí zhí shuāng jīng

nucleation [ˌnjuːkliˈeiʃən] *n.*

Definition: The initial process that occurs in the formation of a crystal from a solution, a liquid, or a vapour, in which a small number of ions, atoms, or molecules become arranged in a pattern characteristic of a crystalline solid, forming a site upon which additional particles are deposited as the crystal grows. 成核作用 chéng hé zuò yòng

Origin: 1860-1865; from Latin nucleātus "having a kernel or stone"

Example:

The supersaturation used is too low to produce an appreciable rate of nucleation.
所用的过饱和度太低,不能产生一个值得重视的成核率。

Extended Terms:

heterogeneous nucleation 非均匀形核；异质核化；非均质成核；非均匀形核
nucleation center 成核中心;核中心

octahedron [ˌɔktəˈhedrən] *n.*

Definition: a solid figure having eight faces 八面体 bā miàn tǐ

Origin: 1560-1570; Greek oktáedron "eight-sided" (neuter of oktáedros), equivalent. to okta-octa- + -edron-hedron

Example:

For Fe^{2+}, its chemical shift is 1.122 ~ 1.143 (mm/s), quadrupole splitting is 2.701 ~ 2.721 (mm/s), and it occupies the octahedron position.
Fe^{2+}:化学位移为 1.122mm/s,四极分裂为 2.701~2.721mm/s,占据八面体位置。

Extended Terms:

octahedron void 八面体空洞
octahedron site 八面体位

octet [ɔkˈtet] *n.*

Definition: any group of eight 八连线 bā lián xiàn

Origin: 1860-1865; oct- + -et, as in duet

Example:

What is the IP address range for the first octet in a class B address, in binary form?
以二进制表示的 B 类地址的前八位字节的 IP 地址范围是多少?

Extended Term:

octet string 八位位组串;八位字节字符串

---

opaque [əu'peik] *adj.*

Definition: a coloring matter, usually black or red, used to render part of a negative opaque 不透明 bù tòu míng

Origin: 1375-1425; late Middle English opake from Latin opācus "shaded"

Example:

Packing is white in a transparent and translucent or opaque powdery substance, which is also part of solid ink.
填料是白色透明、半透明或不透明的粉状物质,也是油墨中的固体组成部分。

Extended Terms:

opaque layer 不透明;不透明层;混浊层
opaque glaze 不透明釉

---

ogdohedry [ɔgdə'hidri] *n.*

Definition: The point group of a crystal is called ogdohedry if it is a subgroup of index 8 of the point group of its lattice. 八分区 bā fēn qū

---

optic axis ['ɔptik 'æksis]

Definition: An optical axis is a line along which there is some degree of rotational symmetry in an optical system such as a camera lens or microscope. 光轴 guāng zhóu

Example:

Crystals of this type, having only one optic axis, are called uniaxial crystals.
这种只有一根光轴的晶体称为单轴晶体。

Extended Terms:

optic elastic axis 光弹性轴
optic axis of crystal 晶体光轴

---

optic orientation [ˈɔptik ˌɔːrienˈteiʃən]

Definition: The optical orientation of paramagnetic atoms is the ordering, by means of optical radiation, of the directions of the magnetic moments and associated mechanical moments of the atoms of a gas. 光性方位 guāng xìng fāng wèi

optic: relating to or using sight 光性的 guāng xìng de

orientation: position or alignment relative to points of the compass or other specific directions 方位 fāng wèi

Origin: It was discovered in 1953 by A. Kastler.

Example:

A fatique failure criterion is proposed based on the input strain energy. This crtiterion can be used to adjust the fiber optic orientation angle and to analyse stress ratio of the composite.
在输入应变能量的基础上提出了疲劳失效准则,这一准则可用于光纤方位角的调整和应力比的分析。

Extended Term:

fibreoptic tube orientation 纤维取向

---

optical anormaly [ˈɔptikəl əˈnɔməli]

Definition: The phenomenon in which an organic compound has a molar refraction which does not agree with the value calculated from the equivalents of atoms and other structural units composing it. 光性异常 guāng xìng yì cháng

optical: of or relating to or involving light or optics 光学的 guāng xué de

anormaly: abnormal 异常的 yì cháng de

---

optical property [ˈɔptikəl ˈprɔpəti]

Definition: Optical property of a material is defined as its interaction with electro-magnetic radiation in the visible. 光学性质 guāng xué xìng zhì

Example:

This new optical property will be expected to extend the potential applications of HTlc in optical material fields.
这一光学性质的发现将有助于类水滑石在光学材料领域中的进一步应用。

Extended Terms:

optical rtansform property 光变换特性
optical property tester 光学性能测定仪

---

optic-axial angle [ˈɔptik ˈæksiəl ˈæŋgl]

Definition: the acute angle between the two optic axes of a biaxial crystal. Also known as optic angle; optic-axial angle. In air, the larger angle between the optic axes after refraction on leaving the crystal 光轴角 guāng zhóu jiǎo

optical: relating to or using sight 光性的 guāng xìng de

axial: along the same line as an axis of rotation in geometry 轴的 zhóu de

angle: the space between two lines or planes that intersect; the inclination of one line to another; measured in degrees or radians 角 jiǎo

---

order [ˈɔːdə] *n.*

Definition: a condition in which each thing is properly disposed with reference to other things and to its purpose 有序 yǒu xù

Origin: 1175-1225; Middle English ordre (*n.*), ordren (v., derivation. of the n.) from Old French ordre (*n.*) from Latin ordin- "row, rank, regular arrangement"

Example:

They work ever in harmony and order.
他们永远协调一致地工作。

Extended Terms:

purchase order 订购单;购货订单;采购订单;订货单
blanket order 总订单;总括订货单

---

order of interference color [ˈɔːdə ɔv intəˈfiərəns ˈkʌlə]

Definition: In order to distinguish the colours produced by different multiples of wavelengths, the interference colours are grouped in orders. 干扰色序 gān rǎo sè xù

interference color: colors formed by interference of a beam of light passed through a thin section of a mineral placed in a polarizing microscope 干扰色 gān rǎo sè

order: a condition of regular or proper arrangement 顺序 shùn xù

---

ordering [ˈɔːdəriŋ] *n.*

Definition: logical or comprehensible arrangement of separate elements 排序 pái xù

Origin: order- + -ing

Example:

In programming language, such as COBOL, one or more data items, the contents of which

identify the type or the location of a record, or the ordering of data.
在程序设计语言(例如 cobol)中,一个或若干个数据项,其内容用于标明某个记录的类型或所处的位置,或者标识数据的顺序。

Extended Term:
ordering relation 排序关系; 顺序关系; 次序关系

---

ordinary ray [ˈɔːdinəri rei]

Definition: The ray passes through the medium unchanged. 普通射线 pǔ tōng shè xiàn

Example:
Due to double refraction effect, the ordinary ray and the extraordinary ray have different reflections.
由于双折射效应,普通光线和异常光线有不同的反射。

Extended Terms:
ordinary X-ray 普通 X 射线

---

ore microscope [ɔː ˈmaikrəskəup]

Definition: Microscopes that are capable of both transmitted polarized light and reflected light such as is needed with thick genteel rock or a piece of mineral. Ore microscopes are often used my geologist for studying and identifying rocks and minerals, and can be used in mineralogy, petrology and geology. 矿相显微镜 kuàng xiāng xiǎn wēi jìng

ore: a mineral that contains metal that is valuable enough to be mined 矿 kuàng
microscope: magnifier of the image of small objects 显微镜 xiǎn wēi jìng

---

orientation angle [ˌɔːrienˈteiʃən ˈæŋgl]

Definition: The relative angle of the warp direction in a fabric to the chosen zero direction shown on the face of the drawing and would probably be the yarn or tow direction in a unidirectional tape. 定向角 dìng xiàng jiǎo

Example:
The orientation angle of two possibilities of this harmonic smooth site is given.
求出了该谐波光场的两个可能的取向角。

---

orthorhombic system [ˌɔːθəˈrɔmbik ˈsistəm]

Definition: In crystallography, the orthorhombic system is one of the seven lattice point groups. Orthorhombic lattices result from stretching a cubic lattice along two of its orthogonal pairs by two different factors, resulting in a rectangular prism with a rectangular base (*a* by *b*) and height (*c*) , such that *a*, *b*, and *c* are distinct. All three bases intersect at 90° angles. The three lattice vectors remain mutually orthogonal. 斜方晶系 xié fāng jīng xì

Example:

Goethite and lepidocrocite, both crystallizing in orthorhombic system, are the most common forms of iron(III) oxide-hydroxide and the most important mineral carriers of iron in soils. 针铁矿及鳞铁矿皆为斜方晶系的晶体,它们是氢氧化铁最常呈现的形式,也是铁在土壤中最重要的矿物媒介。

Extended Term:

orthorhombic crystal system 正交晶系;正交晶格;斜方晶系

---

orthorhombic(rhombic) bipyramid [ˌɔːθəˈrɔmbik baiˈpirəmid] *n.*

Definition: a polyhedron formed by joining an n-gonal pyramid and its mirror image base-to-base in a orthorhombic shape 斜方双椎 xié fāng shuāng zhuī

orthorhombic:designating or of a crystal system having three axes of unequal length, each of which intersects at right angles with the others 斜方晶系的 xié fāng jīng xì de

bipyramid:a polyhedron formed by joining an n-gonal pyramid and its mirror image base-to-base 双锥体 shuāng zhuī tǐ

---

orthorhombic(rhombic) disphenoid [ˌɔːθəˈrɔmbik daiˈsfiːnɔid]

Definition: a crystal form with four similar triangular faces combined in a wedge shape that is orthorhombic 斜方四面体 xié fāng sì miàn tǐ

orthorhombic:designating or of a crystal system having three axes of unequal length, each of which intersects at right angles with the others 斜方晶系的 xié fāng jīng xì de

disphenoid:A disphenoid is a polyhedron whose four faces are identical triangles.双半面晶形的 shuāng bàn miàn jīng xíng de

---

orthorhombic (rhombic) prism [ˌɔːθəˈrɔmbik ˈprizm] *n.*

Definition: The rhombic prism, an open form, consists of 4 faces which are parallel to 1 axis, used to display a beam laterally without changing it is direction. 斜方柱 xié fāng zhù

orthorhombic:designating or of a crystal system having three axes of unequal length, each of which intersects at right angles with the others 斜方晶系的 xié fāng jīng xì de

prism:In optics, a prism is a transparent optical element with flat, polished surfaces that refract light.棱晶 léng jīng

P

orthorhombic(rhombic)pyramid [ˌɔːθəˈrɔmbik ˈpirəmid] *n.*

Definition: an orthorhombic crystal in the form of a pyramid 斜方单锥 xié fāng dān zhuī

orthorhombic: designating or of a crystal system having three axes of unequal length, each of which intersects at right angles with the others 斜方晶系的 xié fāng jīng xì de

pyramid: a solid figure with a polygonal base and triangular faces that meet at a common point 角锥体 jiǎo zhuī tǐ

overgrowth(epitaxial growth) [ˈəuvəgrəuθ] *n.*

Definition: a growth overspreading or covering something 浮生 fú shēng

Origin: 1595-1605; over- + growth

Example:

These materials cause an overgrowth of algae and further deterioration, including oxygen depletion.

这些元素导致藻类的过快生长和进一步的恶化,包括水中氧气的耗竭。

Extended Terms:

bacterial overgrowth 细菌过度生长

overgrowth fault 超生断层

paracrystalline(poorly crystalline) [ˌpærə ˈkræstəlain] *n.*

Definition: Paracrystalline materials are defined as having short and medium range ordering in their lattice (similar to the liquid crystal phases) but lacking long-range ordering at least in one direction. 亚结晶的 yà jié jīng de

Origin: para- + -crytalline

Example:

The crystal chambers have a paracrystalline appearance connected with the crystal sheath and the plasma membrane.

结晶腔为亚结晶构造,其与晶体鞘膜及细胞膜相连接。

Extended Terms:

paracrystalline state 次晶态
paracrystalline lattice 次晶晶格

parallel hexahedron [ˈpærəlel ˌheksəˈhedrənˌ-ˈhiː-]

Definition: a solid figure having six parallel plane faces 平行六面体 píng xíng liù miàn tǐ

Extended Terms:

parallel:being everywhere equidistant and not intersecting 平行的 píng xíng de
hexahedron:a solid figure having six parallel plane faces 六面体 liù miàn tǐ

parallel intergrowth (growth) [ˈpærəlel intəgrəuθ]

Definition: Intergrowth of two or more crystals in such a way that one or more axes in each crystal are approximately parallel. Also known as parallel growth. 平行连晶 píng xíng lián jīng

Example:

The fold developed by hinge migration has the same uplift rates in its crest and limb, which cause the formation of a parallel growth strata.
以膝折带迁移为变形机制的褶皱,发展过程中翼部倾角不变,褶皱顶部与翼部的抬升速率相同,形成翼部平行状生长地层。

Extended Term:

parallel-axial growth 平行生长

parallel twin [ˈpærəlel twin]

Definition: A twinned crystal whose twin axis is parallel to the composition surface. 平行双晶 píng xíng shuāng jīng

Extended Term:

parallel-axial twin 平行轴变晶;平行轴双晶

parallel(straight)extinction [ˈpærəlel ik ˈstiŋkʃən]

Definition: (*Optics*) nearly total absorption of light that is propagating in an anisotropic crystal in a direction parallel to crystal outlines or traces of cleavage planes 直消光 zhí xiāo guāng
extinction:no longer in existence 消失 xiāo shī

P

paramagnetism [ˌpærəˈmægnetizm] *n.*

Definition: a body or substance that, placed in a magnetic field, possesses magnetization in direct proportion to the field strength; a substance in which the magnetic moments of the atoms are not aligned 顺磁性 shùn cí xìng

Origin: 1905-1910; back formation from paramagnetic

Example:

Poly (propargyl alcohol) obtained is a brown sticky-solid with lustre, and shows semi-conductivity and paramagnetism.
聚丙炔醇是有光泽的褐色黏性固体产物,为半导体,具有顺磁特性。

Extended Terms:

atomic paramagnetism 原子顺磁性
nuclear paramagnetism 核顺磁性

---

paramorphism [ˌpærəˈmɔːfizəm] *n.*

Definition: the state of being a paramorph 同质假象 tóng zhì jiǎ xiàng

Origin: 1865-1870; para- + -morphism

---

partial symmetry [ˈpɑːʃəl ˈsimitri]

Definition: The symmetry operations of a space group are isometries operating on the whole crystal pattern and are also called total operations or global operations. More generally, the crystal space can be divided in N components S_1 to S_N, and a coincidence operation $\varphi(S_i)\rightarrow S_j$ can act on just the i-th component S_i to bring it to coincide with the j-th component S_j. Such an operation is not one of the operations of the space group of the crystal because it is not a coincidence operation of the whole crystal space; it is not even defined, in general, for any component k different from i. It is called a partial operation: from the mathematical viewpoint, partial operations are space-groupoid operations. 局部对称 jú bù duì chèn

partial: being or affecting only a part; not total 局部的 jú bù de
symmetry: balance among the parts of something 对称 duì chèn

Example:

Meanwhile, the selection of the fixed communicating points on the line, and the relationship between the partial symmetry and total symmetry are also detailed.
并对线路通信的中间引下固定通信点的选择、局部对称性和全线对称性的关系作了阐述。

P

Patterson method [ˈpætəs(ə)n ˈmeθəd]

Definition: The Patterson method is a procedure for the solution of the phase problem of the Roentgen diffraction. 帕特森方法 pà tè sēn fāng fǎ

Origin: It goes back on Lindo Patterson (1902-1966), which introduced the method 1934.

Example:

This is the reason this article (and most security professionals) recommend you serve sensitive JSON data as a raw JSON file which must be passed through a patterson method.

这就是本文(和许多安全专家)推荐你对敏感 JSON 数据必须使用经过解析的原生 JSON 文件的原因。

Pauling's rules [ˈpɔːliŋ ruːl]

Definition: Pauling's rules are five rules for determining the crystal structures of complex ionic crystals. 鲍林法则 bào lín fǎ zé

Origin: published by Linus Pauling in 1929

---

pedion [ˈpediən] *n.*

Definition: a crystal form with only one face; member of the asymmetric class of the triclinic system 单面 dān miàn

Example: There are various knitting flannelette: Yaoli flannelette, pismire fabric, pedion curl fabric, towel fabric, roughing fabric, etc.

各种针织绒布类:摇粒绒、蚂蚁布、单面毛圈布、毛巾布、磨毛布等。

---

penetrate twin [ˈpenitreit twin]

Definition: Penetration twins are complete crystals that pass through one another and often share the centre of their axial systems. 贯穿双晶 guàn chuān shuāng jīng

---

pentagonal dodecahedron [penˈtægənl ˌdudekəˈhiːdrən]

Definition: A dodecahedral crystal with 12 irregular pentagonal faces; it is characteristic of pyrite. Also known as pentagonal dodecahedron; pyritoid; regular dodecahedron. 五角十二面体 wǔ jiǎo shí èr miàn tǐ

pentagonal: of or relating to or shaped like a pentagon 五角的 wǔ jiǎo de

dodecahedron: any polyhedron having twelve plane faces 十二面体 shí èr miàn tǐ

P

pentagonal icositetrahedron [penˈtæɡənl aiˌkousiˌtetrəˈhiːdrən]

Definition: In geometry, a pentagonal icositetrahedron is a Catalan solid which is the dual of the snub cube. It has two distinct forms, which are mirror images (or "enantiomorphs") of each other. 五角三八面体 wǔ jiǎo sān bā miàn tǐ

pentagonal: of or relating to or shaped like a pentagon 五角的 wǔ jiǎo de

icsotatrehedron: an icositetrahedron is a 24-faced polyhedron 三八面体 sān bā miàn tǐ

pericline twin law [ˈperiˌklain twin lɔː]

Definition: A parallel twin law in triclinic feldspars, in which the *b* axis is the twinning axis and the composition surface is a rhombic section. 肖钠长石双晶律 xiào nà cháng shí shuāng jīng lǜ

pericline: Refer to a doubly plunging anticline or syncline. It is a form of albite exhibiting elongate prismatic crystal. 肖纳长石 xiāo nà cháng shí

phase of a modulation [feiz ɔv mɔdjuˈleiʃən]

Definition: Phase modulation (PM) is a form of modulation that represents information as variations in the instantaneous phase of a carrier wave. 调制相位 tiáo zhì xiàng wèi

phase: (*Physical chemistry*) a distinct state of matter in a system; matter that is identical in chemical composition and physical state and separated from other material by the phase boundary 相位 xiàng wèi

modulation: (*Electronics*) the transmission of a signal by using it to vary a carrier wave; changing the carrier's amplitude or frequency or phase 调制 tiáo zhì

phase problem [feiz ˈprɔbləm]

Definition: (*Physics*) The phase problem is the name given to the problem of loss of information concerning the phase that can occur when making a physical measurement. The name itself comes from the field of X-ray crystallography, where the phase problem has to be solved for the determination of a structure from diffraction data. The phase problem is also met in the fields of imaging and signal processing. 相位问题 xiàng wèi wèn tí

Example:

In quantum mechanics, phase problem is an important issue, and has aroused fierce controversy

in the 20th century.
相位问题是量子力学中的重要问题,在 20 世纪曾经引起极大的争论。

Extended Term:

two phase solidation problem 二相凝固问题

---

piezoelectricity [piːˈeizəuiˌlektrisiti] *n.*

Definition: electricity, or electric polarity, produced by the piezoelectric effect 压电性 yā diàn xìng

Origin: 1890-1895; from Greek piéz (ein) "to press +-o-+ electricity"

Example:

The piezoelectricity ceramic power supply is the core components of piezoelectricity ink-jet printer.
压电陶瓷驱动电源是压电式喷墨打印机的核心部件。

Extended Terms:

piezoelectricity element 压电元件
piezoelectricity effect 压电效应

---

pinacoid [ˈpinəkɔid] *n.*

Definition: a form whose faces are parallel to two of the axes 平行双面 píng xíng shuāng miàn

Origin: 1875-1880; from Greek pinak- "slab, board + -oid"

Example:

The drawing method for the double channel impeller is studied. It is found that there is an inconformity between the dimensions of plane-view and pinacoid-view of fluid cross-section.
研究了双流道污水泵叶轮常用的图纸表达方法,发现平面图上的断面投影结果与轴面图中确定的断面形状尺寸不一致的问题。

Extended Terms:

hemibasal pinacoid 半底面
brachy pinacoid 短轴轴面

---

plane of symmetry [plein ɔv ˈsimitri]

Definition: Plane symmetry means a symmetry of a pattern in the Euclidean plane; that is, a transformation of the plane that carries any directioned lines to lines and preserves many different distances. If one has a pattern in the plane, the set of plane symmetries that preserve the pattern forms a group. 对称面 duì chèn miàn

plane: an unbounded two-dimensional shape 平面 píng miàn

symmetry:an attribute of a shape or relation; exact reflection of form on opposite sides of a dividing line or plane 对称 duì chèn

---

plane-polarization [plein ˌpəulərai'zeiʃən]

Definition: Polarization of an electromagnetic wave in which the electric vector at a fixed point in space remains pointing in a fixed direction, although varying in magnitude. Also known as plane polarization. 平面极化 píng miàn jí huà

polarization:the condition of having or giving polarity 极化 jí huà

---

plane-polarized light(ppl) [plein-'pəuləraizd 'lait]

Definition: A polarized light vibrating in a single plane perpendicular to the direction of propagation; Light in which electric field oscillates in a single plane; The superposition of equal intensities of leff and right circularly polarized light. 单偏光 dān piān guāng

Example:

These two enantiomers are nonsuperimposable mirror images that can only be distinguished on the basis of their different rotation of plane-polarized light.

这两个对映异构体是不可重叠的镜像关系,只能根据它们对平面偏振光不同的旋光加以识别。

Extended Term:

plane-polarized light and rotation 偏振光及旋光性

---

pleochroism [pli'ɔkrəuizəm] *n.*

Definition: the property of certain crystals of exhibiting different colors when viewed from different directions under transmitted light 多色性 duō sè xìng

Origin: 1855-1860; pleochro(ic) +-ism

Pleochroism: distinct in strong colored gem varieties, kunzite is violet-purple/colorless and hiddenite is green to blue-green/colorless to pale green.

晶体多色性:不同类的锂辉石具有明显的强烈颜色,紫锂辉石具有紫罗兰色到紫色或者无色,希登石是绿色到蓝绿色或者无色到淡绿色。

Extended Terms:

infrared pleochroism 红外线多色性

reflection pleochroism 反射多色性

pleochroism brown-yellow [pli'ɔkrəuizəm braun-'jeləu]

Definition: Pleochroism is an optical phenomenon in which mineral grains within a rock appear to be brwn and yellow when observed at different angles under a polarizing petrographic microscope. 褐-黄色多色性 hè huáng sè duō sè xìng

point group [pɔint gruːp]

Definition: In geometry, a point group is a group of geometric symmetries (isometries) leaving a point fixed. 点群 diǎn qún

Example:

The design and expression of computer three dimensional animation for the molecular point group were studied.

(本文)探讨了采用计算机三维动画和多媒体技术描述分子点群的表达方法。

Extended Terms:

cubic point group 立方点群
point group symmetry 点群对称
cyclic group point 循环点群

point symmetry [pɔint 'simitri]

Definition: Point symmetry is when every part has a matching part: the same distance from the central point, but in the opposite direction. 点对称 diǎn duì chèn

Example:

The result indicates that under the point symmetry transformations, only the autonomous Hamiltonian Ermakov systems possess form invariance.

结果表明,在点对称变换下,只有自治的哈密顿埃玛科夫系统才具有形式不变性。

Extended Terms:

point symmetry structure 点对称结构
point symmetry transformation 点对称转换
zero point symmetry 原点对称

polar axis ['pəulə 'æksis]

Definition: (*Mathematics*) The fixed line in a system of polar coordinates from which the

P

polar angle, θ, is measured anticlockwise. 极轴 jí zhóu

Example:

It can be used as Altazimuth mount when polar axis inclined to 90°.
当极轴倾斜 90°时,它可以被用来作为经纬仪挂载。

Extended Terms:

polar symmetry axis 极性对轴;极性对称轴
polar axis tracking system 极轴跟踪系统

---

polar lattice [ˈpəulə ˈlætis]

Definition: The polar lattice is a lattice dual of the direct lattice, which is the ancestor of the reciprocal lattice. 极格 jí gé

Origin: It was introduced by Auguste Bravais in a "mémoire" presented to the Académie de Sciences de Paris on 11 December 1848.

Example:

In this paper, the excitons in polar lattice is investigated by the perturbation method that is developed to discuse polaron problem by Haga When we neglect. The interaction of phonon of different wave vector in the recoil effect, the rest energy and effect mass of exciton is derived.
在本文中采用 Haga 研究极化子时提出的微扰法讨论极化晶体中的激子,在忽略反冲效应中不同波矢的声子之间的相互作用时,导出激子的基态能量和有效质量。

---

polarization [ˌpəuləraiˈzeiʃən] *n.*

Definition: (*optics*) A state, or the production of a state, in which rays of light or similar radiation exhibit different properties in different directions. 偏振 piān zhèn

Origin: 1805-1815; polarize +-ation

Example:

However, for light with orbital angular momentum (OAM), the energy spirals around the beam axis. Ordinary beams carry only "spin angular momentum", encoded in the polarization of light.
然而,对于具有轨道角动量(OAM)的光,其能量绕着光轴螺旋,普通的光仅仅携带"自旋角动量",用光的偏振态进行编码。

Extended Terms:

dielectric polarization 电介质极化
circular polarization 圆偏振;圆偏振光;圆偏化;圆光

---

polarized light [ˈpəulərаizd lait]

Definition: light that is reflected or transmitted through certain media so that all vibrations are restricted to a single plane 偏光 piān guāng

Example:

All tubes also remain perfectly flexible so that in polarized light applications the microscope table can still be rotated.

所有的连接管保持柔性，保证在偏光下使用时，显微镜载物台仍然可以旋转。

Extended Terms:

polarized-light optical system 振光光学系统

single-frequency polarized light 单频偏振光

---

polarizing microscope [ˈpəulərаiziŋ ˈmaikrəskəup]

Definition: a microscope that utilizes polarized light to reveal detail in an object, used especially, to study crystalline and fibrous structures 偏光显微镜 piān guāng xiǎn wēi jìng

Example:

Its heating and optical performance was studied by polarizing microscope and Differential Scanning Calorimetry (DSC).

利用差示扫描量热仪、偏光显微镜等研究了记录层材料热变化中的形态与性能。

Extended Term:

polarizing light microscope 偏光显微镜

---

polarizing prism [ˈpəulərаiz ˈprizm]

Definition: a type of optical prism used as a linear polarizer—that is, linearly polarized optical radiation can be obtained by means of it. It usually consists of two or more triangular prisms, at least one of which is cut from an optically anisotropic crystal. 偏振棱镜 piān zhèn léng jìng

Example:

A polarizing prism was designed for distributed polarization coupling analyzer (DPCA).

偏振棱镜是用来测试偏振耦合系统的。

Extended Terms:

Ahrens polarizing prism 阿伦斯偏振棱镜

Glan-Foucault polarizing prism 格兰-傅科偏振棱镜

P

polish [ˈpɔliʃ] *n.*

Definition: smoothness and gloss of surface 抛光 pāo guāng

Origin: 1250-1300; Middle English polishen; from Middle French poliss-, “long”; from Latin polīre “to polish”

Example:

High quality $Y_2O_3ZrO_2$ single crystal rectangular waveguides were fabricated from bulky Y_2O_3 stabilized ZrO_2 single crystal by precise cut and fine polish with cross section larger than 1mm× 1mm and length of 45mm~65mm.

通过精确的切割和良好的抛光，从 Y_2O_3稳定的 ZrO_2块状单晶制得可用于红外激光传输及光纤高温传感的高品质 Y_2O_3 ZrO_2单晶矩形光波导，获得的矩形波导截面大于 1mm× 1mm，长度为 45mm~ 65mm。

Extended Terms:

electrochemical polish 电化学抛光

polish machine 抛光机

---

polymorph [ˈpɔliˌmɔːf] *n.*

Definition: (*Crystallography*) any of the crystal forms assumed by a substance that exhibits polymorphism 同质多像变体 tóng zhì duō xiàng biàn tǐ

Origin: poly- + -morph

Example:

Modeling and prediction of solubility is a key to develop polymorph crystal growth and crystallization process.

溶解度的测定与预测对于多晶型体的晶体生长和结晶过程中的多晶型控制至关重要。

Extended Terms:

polymorph transformation 同质多像转变

polymorph stability 晶型稳定性

---

polymorphism [ˌpɔliˈmɔːfizm] *n.*

Definition: Polymorphism (biophysics), also referred to as lipid polymorphism, is the property of amphiphiles that gives rise to various aggregations of lipids. 同质多像 tóng zhì duō xiàng

Origin: poly- + -morphism

Example:

Explicit case analysis on the type of an object is usually an error. The designer should use polymorphism in most of these cases.
对对象的类型的精细情况分析一般是错误的。在大多数情况下,设计者应当使用多态。

Extended Terms:

chromosomal polymorphism 染色体多态性;染色体多型性
lipid polymorphism 脂多形态;脂多型性

---

polysome [ˈpɔlisəum] *n.*

Definition: Polyribosomes (or polysomes), also known as ergosomes, are a cluster of ribosomes, bound to a messenger RNA(mRNA) molecule, first discovered and characterized by Jonathan Warner, Paul Knopf, and Alex Rich in 1963. Polyribosomes read one strand of mRNA simultaneously, helping to synthesize the same protein at different spots on the mRNA, mRNA being the "messenger" in the process of protein synthesis. They may appear as clusters, linear arrays, or rosettes in routine: this is aided by the fact that mRNA is able to be twisted into a circular formation, creating a cycle of rapid ribosome recycling, and utilization of ribosomes. 多晶体现象 duō jīng tǐ xiàn xiàng

Origin: poly-+some

Example:

Several ribosomes may be actively engaged in protein synthesis along the same mRNA molecule, forming a polyribosome, or polysome.
几个核糖体可能沿着同一个信使核糖核酸(mRNA)分子参与蛋白质的合成,形成多聚核糖体。

Extended Terms:

polysomatic 多晶体的
polysomatism 多晶体性

---

polysynthetic twin [ˌpɔliːsinˈθetik twin]

Definition: A repeated twin in which all successive composition surfaces are parallel. 聚片双晶 jù piàn shuāng jīng

polysynthetic: forming derivative or compound words by putting together constituents each of which expresses a single definite meaning 聚片状的 jù piàn zhuàng de

Example:

Sylvanite occurs in columnar, tabular and grained forms and assume polysynthetic twin with

strong anisotropy and bireflectance.
针碲银金矿呈柱状、板状、粒状晶形，聚片双晶，具强非均质性及双反射和反射多色性。

polytype [ˈpɔlitaip] *n.*

Definition: a type of polymorph whose different forms are due to more than one possible mode of atomic packing 多型体 duō xíng tǐ

Origin: poly- + -type

Example:

The sericite in the porcelain stone belongs to dioctahedral typewith polytype being 2M_1 and unit cell parameters being a=5.15, b=8.97, c=20.10 and β=96.18°.
光福瓷石中绢云母为二八面体型，多型为2M1，其晶胞参数a=5.15，b=8.97，c=20.10，β=96.18°。

Extended Terms:

polytype transformation 多型转变
polytype characteristics 多型特征
mineral polytype 矿物多型

polytypism [ˌpɔliˈtaipizəm] *n.*

Definition: An element or compound is polytypic if it occurs in several structural modifications, each of which can be regarded as built up by stacking layers of (nearly) identical structure and composition, and if the modifications differ only in their stacking sequence. Polytypism is a special case of polymorphism: the two-dimensional translations within the layers are essentially preserved. 多型性 duō xíng xìng

Origin: poly- + -type +-ism

Example:

Also the kinds of polytypism with different size distributions can be synthesized by changing the reaction conditions.
改变相应条件，也能合成不同分布结构(的晶粒)的多型性。

Extended Terms:

polytypism ventricular premature beat 多型性室性早搏
polytypism premature beat 多型性早搏

positive form [ˈpɔzətiv fɔːm]

Definition: In complex geometry, the term positive form refers to several classes of real differential forms of Hodge type (p, p). 正形 zhèng xíng

Example:

P wave in leads Ⅰ, Ⅱ, Ⅲ and aVF is of positive form and that in aVR and aVL, of negative form.

P 波的形态在Ⅰ,Ⅱ,Ⅲ及 aVF 导联中呈正向,而在 aVR,aVL 导联中呈负向。

Extended Terms:

positive definite form 正定型

conditional positive form 条件正定型

---

preferred orientation [priˈfəːd ˌɔːrienˈteiʃən ˌəu-]

Definition: the nonrandom orientation of planar or linear fabric elements in structural petrology 择优取向 zé yōu qǔ xiàng

Example:

The arrangement must show a preferred orientation.

排列必须呈现特定的指向。

Extended Terms:

preferred orientation column 最适方位小柱

dimensional preferred orientation 空间选择定位

---

primary extinction [ˈpraiməri ikˈstiŋkʃən]

Definition: (*Solid-state Physics*) A weakening of the stronger beams produced in X-ray diffraction by a very perfect crystal, as compared with the weaker. 初级消光 chū jí xiāo guāng

Example:

Making use of Chen's integral table (1965) and Becher's primary extinction coeffieient(1974), we obtained the solutions of the equations.

引用 1965 年陈箎积分表及 1974 年 Becker 等人的初级消光系数,得到(这种情况的)微分动力学方程组的解。

Extended Term:

primary extinction effect 原生消灭效应

---

primitive basis [ˈprimitiv ˈbeisis]

Definition: A primitive basis is a crystallographic basis of the vector lattice L such that every lattice vector t of L may be obtained as an integral linear combination of the basis vectors, a,

P

b, *c*. In mathematics, a primitive basis is often called a *lattice basis*, whereas in crystallography the latter has a more general meaning and corresponds to a crystallographic basis. 原始基础 yuán shǐ jī chǔ

Example:

On the basis of both historical and modern documents, the geographical distribution of sports talents of nowadays was discussed followed by three conclusions: the geographical environment where people are living is the primitive basis of the distribution of sports talents.
采用文献资料研究方法,对现代体育运动人才的地理分布进行了探讨,探讨有三:其一认为人类所处的地理环境是现代体育人才分布的原始基础。

---

primitive cell ['primitiv sel]

Definition: (*Crystallography*) A parallelepiped whose edges are defined by the primitive translations of a crystal lattice; it is a unit cell of minimum volume. 原胞 yuán bāo

Example:

A matrix method to solve reciprocal lattice primitive cell and its drawing by using Matlab is offered.
采用矩阵计算倒易晶格原胞和利用 Matlab 绘制原胞图。

Extended Terms:

primitive cell nucleus 原始核
smallest primitive cell 最小素单胞

---

primitive lattice ['primitiv 'lætis]

Definition: (*Crystallography*) A crystal lattice in which there are lattice points only at its corners. Also known as simple lattice. 初级点阵 chū jí diǎn zhèn

Example:

It's about reciprocal lattice primitive cell and basic conversion.
有关倒易晶格原胞与基底变换。

Extended Terms:

primitive cubic lattice 单纯立方晶格
non-primitive lattice 非初基晶格

---

principal optic axis ['prinsəpəl 'ɔptik 'æksis]

Definition: a line extending through the center of a lens at a right angle to the lens surface

光学主轴 guāng xué zhǔ zhóu

Example:

The image of sphericity which does not located the principal optic axis is an ellipse based on pinhole model, and the image center offsets from the projection of the sphere center.
针孔成像模型下非主光轴上的球体成像为椭圆,椭圆中心偏离球心投影点。

---

principle refractive index [ˈprinsəpl riˈfræktiv ˈindeks]

Definition: The Refractive Index of a transparent isotropic medium may be loosely defined as the "bending" power of the medium for a ray of light obliquely incident on its surface. 主折射率 zhǔ zhé shè lǜ

principle: a basic truth or law or assumption 原则 yuán zé
refractive: capable of changing the direction (of a light or sound wave) 折射的 zhé shè de
index: a number or ratio (a value on a scale of measurement) derived from a series of observed facts; can reveal relative changes as a function of time 指数 zhǐ shù

---

prism [prizm] *n.*

Definition: a solid figure whose bases or ends have the same size and shape and are parallel to one another, and each of whose sides is a parallelogram 柱 zhù

Origin: 1560s, a type of solid figure, from L.L. prisma (Martianus Capella), from Gk. prisma (Euclid), lit. "something sawed", from prizein "to saw". Meaning in optics is first attested 1610s.

Example:

The device can be calibrated so that the prism-rotating dial reads directly in wavelength.
这个装置可以经过校准,使得校镜旋转度盘直接读取波长。

Extended Terms:

prism object 棱柱体
achromatic prism 消色差棱镜;消色棱镜
objective prism 物镜棱镜;物端棱镜

---

pseudo symmetry [ˈpsjuːdəu ˈsimitri]

Definition: apparent symmetry of a crystal, resembling that of another system; generally due to twinning 假对称 jiǎ duì chèn

Example:

It is concluded that the crystal structure of Dingdaohengite-(Ce) is a superstructure with space group P2_(1)/a. It possesses pseudo symmetry corresponding to the space group C2/m.
根据结构分析及衍射数据消光规律统计认为,丁道衡矿的空间群应该为 P21/a,而 C2/m 为假对称空间群,结构属于具有 C2/m 假对称的 P21/a 超结构,是一种超结构的新类型。

---

pseudomorph ['psju:dəumɔ:f] *n.*

Definition: A mineral that has the crystalline form of another mineral rather than the form normally characteristic of its own composition. 假象 jiǎ xiàng

Origin: pseudo- + -morph

Example:

It can be divided into three types as follows:granule glauconite,cemented glauconite and clastic pseudomorph glauconite.
它可分为三种类型:团粒状海绿石、胶结物海绿石和碎屑假象海绿石。

Extended Terms:

incrustation pseudomorph 结壳假象
substitution pseudomorph 交换假象;替换假象;替代假象

---

pycnometer [pik'nɔmitə] *n.*

Definition: a standard vessel used in measuring the density or specific gravity of materials 比重瓶 bǐ zhòng píng

Example:

The relative densities of flavors were determined by pycnometer (standard method) and densimeter.
目前测定香精密度的方法主要有标准传统法——比重瓶法和密度仪法。

Extended Terms:

quartz pycnometer 石英比重瓶;石英比色计
mercury pycnometer 水银比重计

---

pyramid ['pirəmid] *n.*

Definition: A pyramid is a structure where the outer surfaces are triangular and converge at a point. The base of a pyramid can be trilateral, quadrilateral, or any polygon shape, meaning that a pyramid has at least three triangular surfaces (at least four faces including the base). 单锥

dān zhuī

Origin: 1550s (earlier in Latin form piramis, late 14th century), from French pyramide (Old French piramide, 12c.), from Latin pyramides, pl. of pyramis "one of the pyramids of Egypt," from Greek. pyramis (pl. pyramides), apparently an alteration of Egyptian pimar "pyramid." Related: Pyramidal.

Example:

Next, we put a pyramid on top of the cube.
接着,我们在正立方体的顶面摆一个金字塔。

Extended Terms:

human pyramid 叠罗汉;搭人塔
contrast pyramid 对比度金字塔

---

pyroelectricity [ˌpaiərəuilekˈtrisiti] *n.*

Definition: generation of electric charge on a crystal by change of temperature 热电性 rè diàn xìng

Origin: pyro-(hot) +-electricity

Example:

In dissimilar mineralizing conditions and forming sections, the values of mineral pyroelectricity and types of thermal conductivity are different.
矿物的热电性大小或热导电类型等在不同的成矿环境、部位中表现不同。

Extended Term:

pyroelectricity of crystal 晶体的热电性

quartz wedge [kwɔːts wedʒ]

Definition: A very thin wedge of quartz cut parallel to an optic axis; used to determine the sign of double refraction of biaxial crystals, and in other applications involving polarized light and its interaction with matter. 石英楔 shí yīng xiē

Example:

The depolarization principle, design and effectiveness of depolarizer made of quartz wedge are

Q

theoretically analysed, and the optimal depolarization condition and measured datum are given in this paper.
本文对石英楔作为退偏振器的原理、设计及有效性进行了理论分析和实验研究,指出了最佳退偏条件。

Extended Term:

quartz wedge compensator 石英楔补偿器

---

quasi-crystal [ˈkweizai ˈkristəl] *n.*

Definition: a phase of solid matter that, like a crystal, exhibits long-range orientational order and translational order but whose atoms and clusters repeat in a sequence defined by a sum of periodic functions whose periods are in an irrational ratio 准晶体 zhǔn jìng tǐ

Example:

But after annealing of 10h milled sample, the fraction of quasi-crystal phase decrease with increasing time and there is almost no evidence of it after 50h of milling.
但球磨 10 小时以后,继续延长球磨时间,准晶体相的成分比例反而减少,到 50 小时已几乎看不到准晶体的迹象。

Extended Term:

photonic quasi-crystal 准光子晶体

---

quasiperiodicity [ˈkweizaiˌpiəriəˈdisiti] *n.*

Definition: Quasiperiodicity is the property of a system that displays irregular periodicity. 拟周期 nǐ zhōu qī

Origin: quasi-+period+-city

Example:

By spectrum analysis, the variation of tree-ring δ13C sequence shows a quasi periodicity of 2-3a, which is coincident with “quasi two-year tropic barometric oscillation” (QBO), and it explains that the tree-ring in Guangdong Province can record large-scale information of ENSO.
通过谱分析发现,树轮记录的 δ13C 序列变化中 2~3a 周期与热带气候的“准两年振荡”(QBO)十分一致,这从另一个侧面说明了广东樟树树轮记录了大范围的 ENSO 信息。

Extended Term:

quasiperiodicity point 准周期点

R

R-factor [aː ˈfæktə] *n.*

Definition: R-factor is an old name for a plasmid that codes for antibiotic resistance. R 型因子 R xíng yīn zǐ

Example:

In this paper, Karhunen-Loeve transformation and the R-Factor Analysis have been used in compression-processing the multi-spectrum data on the land satellite Computer Compatible Tape (CCT).

本文应用 Karhunen-Love 变换与 R 型因子分析，对陆地卫星计算机兼容磁带的多光谱数据作了压缩处理。

Extended Term:

R-factor bacterial population 因子细菌群

---

real-space correlation coefficient [ˈriəl-spies kɔːriˈleiʃən kəuiˈfiʃənt]

Definition: The real-space correlation coefficient, RSCC, is a measure of the similarity between an electron-density map calculated directly from a structural model and one calculated from experimental data. 实空间相关系数 shí kōng jiān xiāng guān xì shù

real space: Tenominates the 3-dimensional space in classical physics. The real space coordinates specify the position of an object. For instance, the trajectory in classical mechanics is described in real space. 实空间 shí kōng jiān

correlation: A statistic representing how closely two variables co-vary; it can vary from-1 (perfect negative correlation) through 0 (no correlation) to +1 (perfect positive correlation). 相关性 xiāng guān xìng

coefficient: a constant number that serves as a measure of some property or characteristic 系数 xì shù

---

real-space residual [ˈriəl-speis riˈzidjuəl]

Definition: The real-space residual, RSR, is a measure of the similarity between an electron-density map calculated directly from a structural model and one calculated from

R

experimental data. 实空间误差 shí kōng jiān wù chā

residual:something left after other parts have been taken away 剩余 shèng yú

reciprocal lattice [ri'siprəkəl 'lætis]

Definition: A lattice array of points formed by drawing perpendiculars to each plane (*hkl*) in a crystal lattice through a common point as origin; the distance from each point to the origin is inversely proportional to spacing of the specific lattice planes; the axes of the reciprocal lattice are perpendicular to those of the crystal lattice. 倒易格子 dào yì gé zi

Example:

On the base of translational symmetry, reciprocal lattice and related concepts are introduced.
基于转换对称,介绍了如倒易格子及相关概念。

Extended Terms:

reciprocal-lattice coordinates 倒易晶格坐标
Reciprocal-lattice 倒格矢

reciprocal space [ri'siprəkəl speis]

Definition: The reciprocal and direct spaces are reciprocal of one another, that is the reciprocal space associated to the reciprocal space is the direct space. They are related by a Fourier transform and the reciprocal space is also called Fourier space or phase space. 倒易空间 dǎo yì kōng jiān

Example:

The model solved the equation on reciprocal space using 2-D projection for a 3-D system.
该模型将三维问题进行二维投影,在倒易空间求解。

Extended Term:

reciprocal space method 倒易空间法

reconstructive transformation [ˌriːkən'strʌktiv ˌtrænsfə'meiʃən]

Definition: a type of crystal transformation that involves the breaking of either first-or second-order coordination bonds 重建型转变 chóng jiàn xíng zhuǎn biàn

reconstructive:helping to restore to good condition 重建的 chóng jiàn de

transformation:a function that changes the position or direction of the axes of a coordinate system 转换 zhuǎn huàn

recrystallization [riːˌkristəlaiˈzeiʃən] *n.*

Definition: In chemistry, recrystallization is a procedure for purifying compounds. 再结晶 zài jié jīng

Example:

The longer the heat treatment time is, the easier the recrystallization is.
热处理时间越长,则再结晶越容易发生。

Extended Terms:

secondary recrystallization 二次再结晶;次生重结晶;次级再结晶
primary recrystallization 一次再结晶;初级再结晶

---

refinement [riˈfainmənt] *n.*

Definition: the act or the result of refining; an improvement or elaboration 提纯 tí chún

Origin: 1610s, "act or process of refining," from refine + -ment. Meaning "fineness of feeling" is from 1708.

Example:

Knowledge is one thing, virtue is another; good sense is not conscience, refinement is not humility, nor is largeness and justness of view faith.
知识是一回事,美德则是另外一回事。好意不是良心,优雅不是谦让,广博与公正的观点也不是信仰。

Extended Term:

element refinement 元素修饰词;元素限定

---

reflection [reˈflekʃən] *n.*

Definition: (*Physics*, *Optics*) the return of light, heat, sound, etc., after striking a surface; something so reflected, as heat or especially light. 反射 fǎn shè

Origin: late 14 century, in reference to surfaces, from Late Latin reflexionem (nominative reflexio) "a reflection," literally "a bending back," noun of action from past participle stem of reflectere, from re- "back" + flectere "to bend." Meaning "remark made after turning back one's thought on some subject" is from 1650s.

Example:

Any given culture is a reflection of the politics and economics of a given society.
一定的文化是一定社会的政治和经济的反映。

Extended Terms:

reflection dimming 反射暗淡

total reflection 全反射;总反射

reflection conditions [reˈflekʃən kənˈdiʃən]

Definition: The reflection conditions describe the conditions of occurence of a reflection (structure factor not systematically zero). There are two types of systematic reflection conditions for diffraction of crystals by radiation: General conditions and Special conditions. 反射条件 fǎn shè tiǎo jiàn

Example:

Two kinds of reflection conditions of −h + k+l=3n and h−k+l=3n were observed in one crystal simultaneously. These special reflection conditions should be caused by a twinning.

发现在晶体中存在着的所有衍射点均符合−h+k+l=3n 或 h−k+l=3n 的衍射条件。

reflectivity [ˌriːflekˈtiviti] *n.*

Definition: (*Physics*) the ratio of the energy of a wave reflected from a surface to the energy possessed by the wave striking the surface 反射率

Origin: 1620-1630; reflect +-ive

Example:

The variation of the absorption, transmission and reflectivity of the amorphous and crystalline phase change film with the wavelength are studied.

同时,研究了晶态和非晶态相变薄膜的吸收率、透射率和反射率随波长的变化。

Extended Term:

radar reflectivity 雷达反射率;雷达反射系数

reflectometer [ˌriːflekˈtɔmitə] *n.*

Definition: an instrument for measuring the reflectance of a surface 反射计 fǎn shè jì

Origin: reflect +-meter

Example:

The methods employ difference traces obtained from optical time domain reflectometer (OTDR) measurements.

识别方法利用由光时域反射计(OTDR)测量获得的差分迹线。

Extended Terms:

domain reflectometer 时域反射计
pulse reflectometer 脉冲反射计

refraction [ri'frækʃən] *n.*

Definition: Refraction is the change in direction of a wave due to a change in its speed. 折射 zhé shè

Origin: 1570s, from Late Latin. refractionem (nominative refractio) "a breaking up," noun of action from past participle stem of refringere "to break up", from re- "back" + comb. form of frangere "to break"

Example:

Shallow Seismic Refraction Survey is one kind of celerity, effective and economic means in engineering exploration.
浅层折射波法是工程勘探中常用的一种经济并且快速有效的方法。

Extended Term:

wave refraction 波折射;波浪折射

refractive index [ri'fræktiv 'indeks]

Definition: the ratio of the speed of light in a vacuum to the speed of light in a medium under consideration, also called the index of refractive 折射率 zhé shè lǜ

Example:

A simple method based on total internal reflection is presented for accurately measuring the refractive index of biological tissue.
提出一种利用全反射原理,精确测量生物组织以及一般均匀介质折射率的方法。

Extended Term:

absolute refractive index 绝对折射率;绝对折射指数

relief [ri'li:f] *n.*

Definition: the variations in elevation of an area of the earth's surface 突起 tū qǐ

Origin: from Anglo-French relif, from Old French relief "assistance," literally "a raising, that which is lifted", from stressed stem of relever. Meaning "aid to impoverished persons" is attested from 14th century; that of "deliverance of a besieged town" is from 14th century.

Example:

Do not obstruct any pressure-relief device. Dirt, paint, corrosion, or other materials prevent pressure-relief devices from functioning properly.
不要阻塞任何减压装置。污垢、油漆、腐蚀物或其他材料均会阻止减压装置正常运行。

Extended Term:

bas relief 基底突现

---

resolution [ˌrezəˈluːʃən] *n.*

Definition: the act or process of separating or reducing something into its constituent parts: *the prismatic resolution of sunlight into its spectral colors* 分辨率 fēn biàn lǜ

Origin: late 14th century, "a breaking into parts," from Old French resolution or directly from Latin resolutionem (nominative resolutio) "process of reducing things into simpler forms," from past participle stem of resolvere "loosen". Originally sense of "solving" (as of mathematical problems) first recorded 1540s, that of "holding firmly" (in resolute) 1530s, and that of "decision or expression of a meeting" is from 16th century.

Example:

It is the first folio edition slide and this is actually digitized version at high resolution of that particular quotation from Hamlet and then it goes on in the normal way.
这张是第一对开本的幻灯片,这是电子版,高分辨率的是那句哈姆雷特经典台词,其他照旧。

Extended Term:

Test Resolution 测试分辨率

---

reticular density [riˈtikjulə ˈdensəti]

Definition: the number of points per unit area in a two-dimensional lattice, such as the plane of a crystal lattice 面网密度 miàn wǎng mì dù

reticular: resembling or forming a network 网状的 wǎng zhuàng de

density: the amount per unit size 密度 mì dù

---

rhombic dodecahedron [ˈrɔmbik dəudekəˈhiːdrən]

Definition: A crystal form in the cubic system that is a dodecahedron whose faces are equal rhombuses. 菱形十二面体 líng xíng shí èr miàn tǐ

rhombic: resembling a rhombus 菱形的 líng xíng de

R

dodecahedron: any polyhedron with twelve flat faces 十二面体 shí èr miàn tǐ

Example:

The coarse grains, are thought to be formed by a solution/deposition process. Most of the Nb_3Sn crystals have the appearance of rhombic dodecahedron and orthogonal parallelepiped. It is evident that the interfacial energy of the {110} and {100} planes of the Nb_3Sn crystal is lower than the others.

固态—液态界面上生长的 Nb_3Sn 分成两层,靠近 Nb 的内层晶粒细小,排列致密,外层晶粒粗大,分布零散,后者是前者经过溶解/沉积过程引起的,晶体形貌大多数呈菱形十二面体,部分呈正交平行六面体,说明 Nb_3Sn 的{110},{100}面的界面能低。

---

rhombohedral system [ˌrɔmbəuˈhiːdrəl ˈsistəm]

Definition: a division of the hexagonal system embracing the rhombohedron, scalenohedron, etc. 菱形晶系 líng xíng jīng xì

rhombohedral: having threefold symmetry 斜方六面体的 xié fāng liù miàn tǐ de

Example:

XRD patterns showed that the prepared $LaNiO_3$ was single-plase perovskite in structure, they are rhombohedral system with space group R3M.

XRD 分析表明所合成的镍酸镧($LaNiO_3$)为单相的钙钛矿型微晶,属于三方晶系,空间群为 R3M。

---

rhombohedron [ˌrɔmbəuˈhiːdrən] *n.*

Definition: a trigonal crystal form that is a parallelepiped, the six identical faces being rhombs, also known as rhomb 菱面体 líng miàn tǐ

Example:

The amount of bond types 1321, 1311 and 1301 that relate to the rhombohedron is about 14%.

与菱面体结构相关的键型 1321,1311,1301 的成键数之和约为 14%。

Extended Term:

obtuse rhombohedron 钝菱面体

---

rotation method [rəuˈteiʃən ˈmeθəd]

Definition: The rotation method is a concept used by the philosopher Søren Kierkegaard in *Either/Or* to describe the mechanism used by higher level aesthetes in order to avoid boredom. The method is an essential hedonistic aspect of the aesthetic way of life. 旋转法 xuán zhuǎn fǎ

Example:

The applications of such teaching methods as body platform demonstration, paper cut sample and prototype rotation method can greatly improve the teaching quality and effect.
运用人台演示法、纸样切展法和原型旋转法等教学方法能较好地提高教学质量和效果。

Extended Term:

rotation vector method 旋转向量法

---

rotoinversion axis [ˌrəutəuinˈvəːʃən ˈæksis]

Definition: a type of crystal symmetry element that combines a rotation of 60°, 90°, 120°, or 180° with inversion across the center. Also known as symmetry axis of rotary inversion; symmetry axis of rotoinversion. 旋转反伸轴 xuán zhuǎn fǎn shēn zhóu

axis: a straight line through a body or figure that satisfies certain conditions 轴 zhóu

---

row [rəu] *n.*

Definition: a series of objects placed next to each other, usually in a straight line 行列 háng liè

Origin: "noisy commotion," 1746, Cambridge University slang, of uncertain origin, perhaps related to rousel "drinking bout" (1602), a shortened form of carousal. Klein suggests a back-formation from rouse (n.), mistaken as a plural (cf. pea from pease).

Example:

In this way you can select the whole row normally.
用这种方法你能正常地选择全部的行列。

Extended Term:

row effects 行效应

scalenohedron [ˌskeilənɔˈhiːdrən] *n.*

Definition: a closed crystal form whose faces are scalene triangles 偏三角面体 piān sān jiǎo miàn tǐ

Origin: 1850-1855; from Greek skalēnó (s) unequal +-hedron

Example:

8, 8 m, 82, 8/m, 8/mmm, and 2 m. Single forms: octagonal prism, dioctagonal prism, octagonal pyramid, dioctagonal pyramid, octagonal dipyramid, dioctagonal dipyramid, octagonal scalenohedron, dioctagonal scalenohedron, and octagonal traperohedron. For d odecahegonal quasicrystals, the point groups are 12.
推导了八次对称准晶体的点群为 8,8m,82,8/m,8/mmm,2m,单形为八方柱,复八方柱,八方单锥,复八方单锥,八方双锥,复八方双锥,八方偏三角面体,复八方偏三角面体,八方偏方面体。

Extended Term:

tetragonal scalenohedron 正方偏三角面体

---

screw axis [skruː ˈæksis]

Definition: a symmetry element of some crystal lattices, in which the lattice is unaltered by a rotation about the axis combined with a translation parallel to the axis and equal to a fraction of the unit lattice distance in this direction 螺旋轴 luó xuán zhóu

screw: a propeller with several angled blades that rotates to push against water or air 螺旋 luó xuán

axis: a straight line through a body or figure that satisfies certain conditions 轴 zhóu

Origin: 1900-1905

Extended Term:

instantaneous screw axis 瞬时螺旋轴

---

secondary extinction [ˈsekəndəri ikˈstiŋkʃən]

Definition: increased absorption or decreased diffraction of X-rays by a crystal lattice, due to previous reflection of the X-rays from suitably placed crystal planes 次级消光 cì jí xiāo guāng

secondary: being of second rank or importance or value; not direct or immediate 次级的 cì jí de

extinction: the act of extinguishing; causing to stop burning 消光 xiāo guāng

Example:

An absorption and secondary extinction correction gave final R = 0.058. The Bi and V atoms occupy a respective set of 4e special positions and the O atoms two sets of general positions.
经吸收和次级消光校正后,R=0.058。Bi 和 V 原子各占据一套 4e 特殊点系,O 原子占据两套一般点系。

S

short range order [ʃɔːt reindʒ ˈɔːdə]

Definition: a regularity in the arrangement of atoms in a disordered solid or a liquid in which the probability of a given type of atom having neighbors of a given type is greater than one would expect on a purely random basis 短程有序 duǎn chéng yǒu xù

short: (primarily spatial sense) having little length or lacking in length 短 duǎn

range: an area in which something acts or operates or has power or control 范围 fàn wéi

order: a degree in a continuum of size or quantity 序列 xù liè

Example:

The chemical short range order and the atomic neighbouring structure feature of metallic glass Cu_(70)Ti_(30) have been discussed.

讨论了 Cu_(70)Ti_(30)金属玻璃中化学短程有序及 Cu 原子近邻结构的特征。

skeleton crystal [ˈskelitən ˈkristəl]

Definition: a crystal formed in microscopic outline with incomplete filling in of the faces 骸晶 hái jīng

Example:

The SiC crystal that was grown on the surface of heat-treated TiC fiber was observed by SEM, and itsmorphology was found to be skeletal. The formation reason and growth mechanism of their skeleton crystal of SiC was discussed in detail.

用扫描电镜对 TiC 质长纤维经高温处理后表面形成的 SiC 晶体的形貌进行观察,发现其晶形是骸晶体,文中对这种骸晶的形成原因与生长机理作详细讨论。

slide [slaid] *n.*

Definition: a small flat rectangular piece of glass on which specimens can be mounted for microscopic study 载玻片 zài bō piàn

Origin: before 950; Middle English sliden (v.), Old English slīdan; from Middle Low German slīden, Middle High German slīten; akin to sled

Example:

For the incompact and oil saturated rock samples, it is necessary to concrete and fix them on the slide glass by glue 502 or epoxy resin and grind them flat at first, then the thin sections can be made.

饱含油胶结程度低松散状砂岩样品,需先用 502 胶或环氧树脂胶在载物片上进行固定和固结,并制成平面,然后再进行制片。

Extended Terms:

slide culture 玻片培养物;悬滴培养
slide stainer 组织切片染色机

---

solid solution [ˈsɔlid səˈljuːʃən]

Definition: a solid, homogeneous mixture of substances, as glass or certain alloys, (in a crystal structure) the more or less complete substitution of one kind of atom, ion, or molecule for another that is chemically different but similar in size and shape; isomorphism 固溶体 gù róng tǐ

Example:

a new phenomenon of re-dissolution of second phases and re-formation of supersaturate solid solution found in processes of severe plastic deformation was summarized
综述了在室温强塑性变形过程中发现的新现象:第二相颗粒的回溶,合金基体重新形成过饱和固溶体。

Extended Terms:

continuous solid-solution series 连续固溶体系列
liquid-solid-solution phase transfer and separation 相转移、相分离

---

solvus [ˈsɔlvəs] *n.*

Definition: in a phase or equilibrium diagram, the locus of points representing the solid-solubility temperatures of various compositions of the solid phase 固溶线 gù róng xiàn

Example:

The results show that the near solvus pre-precipitation can be limited to grain boundary and enhance the discontinuity of grain boundary prelipitates in the sequent age.
研究结果表明:在近固溶线条件下于晶界处产生预析出,并提高了后续时效状态下晶界析出相不连续分布的程度。

Extended Terms:

solvus tempreture 溶线温度
solvus line 溶解度曲线

---

space group [speis gruːp]

Definition: a set of symmetry elements that brings a periodic arrangement of points on a

S

Bravais space lattice to its original position 空间群 kōng jiān qún

Example:

The present paper investigated the varied regularities of space group of crystal structure of helix type polymer chains in terms of coloured symmetry theory.
本文利用有色对称的理论讨论了螺旋型高分子链晶体结构所属空间群的演变规律性。

Extended Term:

space-group maximum 空间群峰

---

space lattice [spies 'lætis]

Definition: a space frame built of lattice girders 空间格子 kōng jiān gé zi

Example:

The out-plane global stability of space lattice tilted column frame of triangle section must be taken into consideration at the design stage.
截面为三角形的空间格子构式斜腿刚架的平面外稳定性是设计中必须考虑的问题。

Extended Term:

Bravais space lattice 布拉维空间晶格

---

sphenoid ['sfiːnɔid] *n.*

Definition: (*Anatomy*) the sphenoid bone 楔形 xiē xíng

Origin: 1725-1735; from New Latin sphēnoīdēs from Greek sphēnoeidés.

Example:

The objective is to investigate the approach and experience of endoscopic surgery in sphenoid sinus occupying lesions associated with a deviated septum.
目的是探讨利用鼻内镜手术治疗蝶窦良性病变伴有鼻中隔偏曲的方法和体会。

Extended Terms:

sphenoid bone 蝶骨；楔形骨
tetragonal sphenoid 正方楔；正方榍

---

spherical coordinate ['sferikəl kəu'ɔːdineit]

Definition: any of a set of coordinates in a three-dimensional system for locating points in space by means of a radius vector and two angles measured from the center of a sphere with respect to two arbitrary, fixed, perpendicular directions 球面坐标 qiú miàn zuò biāo

Example:

Moving object coordinate measurement is performed using laser tracking measurement system (LTS) that is based on spherical coordinate method.
在球坐标激光跟踪测量系统中采用球坐标法可将空间运动物体的三维坐标求出。

Extended Term:

spherical rectangular coordinate 球面直角坐标

---

spinodal decomposition [spaiˈnəudəl ˌdiːkɔmpəˈziʃən]

Definition: an unmixing process in which crystals with bulk composition in the central region of the phase diagram undergo exsolution 拐点分解 guǎi diǎn fēn jiě

Example:

The formation of the skin layer agrees with the spinodal decomposition (SD) mechanism, and the thermodynamic factor plays the leading role in this process.
膜的皮层按旋节线分相机理形成,主要由热力学因素决定。

---

stability [stəˈbiliti] *n.*

Definition: the state or quality of being stable 稳定性 wěn dìng xìng

Origin: 1400-1450; from Latin stabilitās, equivalent to stabili (s) stabile + -tās--ty; from late Middle English or Old French stablete

Example:

The objective is to investigate the stability of glipizide capsules.
目的在于考察格列吡嗪胶囊的稳定性。

Extended Terms:

structural stability 结构稳定性;构造稳定性
seismic stability 抗震稳定性

---

stabilizer [ˈsteibilaizə] *n.*

Definition: any of various substances added to foods, chemical compounds, etc., to prevent deterioration, the breaking down of an emulsion, or the loss of desirable 稳定剂 wěn dìng jì

Origin: 1905-1910; stabilize +-er

Example:

It has been found that initiator, plasticizer, stabilizer, crosslinking agent and composition of

blends influence significantly the crosslinking reaction.
发现引发剂、交联剂、稳定剂和增塑剂等对共混物中的交联反应均有较大的影响。

Extended Term:

voltage stabilizer 稳压器;电压稳定器

stacking fault ['stækiŋ fɔːlt]

Definition: a defect in a face-centered cubic or hexagonal close-packed crystal in which there is a change from the regular sequence of positions of atomic planes 堆垛层错 duī duò céng cuò

Example:

The formation of twin structure depends on the slip of stacking fault for nucleating and extending.
孪晶结构的形成依靠堆垛层错的滑移运动而形成核和扩展。

Extended Terms:

extrinsic stacking fault 非本昭垛层错;插入型层错
stacking fault energy 堆垛层错能;叠差能;层错能

stereogram ['steriəu'græm] *n.*

Definition: a picture or diagram designed to give the impression of solidity, or a stereograph 二维图形 èr wéi tú xíng

Origin: 1865-1870; stereo- + -gram

Example:

A 3-D stereogram is given showing the primary topography in AUTOCAD, utilizing the vectorized primary relief map of the slope of a mining junkyard in a certain iron ore stope.
利用矢量化的某铁矿排土场边坡原地表地形图在 AUTOCAD 中形成原地表三维立体图。

Extended Terms:

isometric stereogram 等距图
split stereogram 分裂式实体镜画

stereographic projection [ˌsteriəu'græfik prəu'dʒekʃən]

Definition: a one-to-one correspondence between the points on a sphere and the extended complex plane where the North Pole on the sphere corresponds to the point at infinity of the plane 极射赤平投影 jí shè chì píng tóu yǐng

Origin: 1695-1705

Example:

Stereographic projection mode will affect the hardware choices, the image quality and the final system cost.

立体影像生成模式将影响硬件设备的选择、立体影像的质量以及系统的最终造价。

Extended Terms:

stereographic polar projection 极射赤平投影;极球面投影

polar stereographic map projection 极球面投影

striation [strai'eiʃən] *n.*

Definition: the state of being striated or having striae 条痕 tiáo hén

Origin: 1849, from Modern Latin stria "strip, streak," in classical Latin "furrow, channel, flute of a column;" cognate with Dutch striem, Old High German strimo, German strieme "stripe, streak," from Proto-Indo-European base *streig-.

Example:

The presence of striation area is one of the main characteristic on the surface generated by abrasive jet.

波纹区域(的存在)是磨料射流切割表面的主要特征之一。

Extended Terms:

striation cast 擦痕铸型

growith striation 生长辉纹

structure amplitude ['strʌktʃə 'æmplitjuːd]

Definition: the absolute value of a structure factor 结构振幅 jié gòu zhèn fú

Example:

With further exploration and development of oil fields, the reservoirs with low amplitude structure are paid more attention to.

随着油田勘探开发的深入,低幅度构造的勘探逐渐受到重视。

structure determination ['strʌktʃə diˌtəːmi'neiʃən]

Definition: Structure determination in crystallography refers to the process of elaborating the three-dimensional positional coordinates (and also, usually, the three-dimensional anisotropic

S

displacement parameters) of the scattering centres in an ordered crystal lattice. Where a crystal is composed of a molecular compound, the term generally includes the three-dimensional description of the chemical structures of each molecular compound present. 结构测定 jié gòu cè dìng

Example:

Yale University Center for Structural Biology houses X-ray equipment and computational resources for macromolecular structure determination.
耶鲁大学结构生物学中心备有测定大分子结构的 X 光仪器和计算资源。

Extended Terms:

Spectroscopy and Structure Determination 光谱学和结构鉴定
powder structure determination 粉晶法结构测定

---

structure factor [ˈstrʌktʃə ˈfæktə]

Definition: It is a factor which determines the amplitude of the beam reflected from a given atomic plane in the diffraction of an X-ray beam by a crystal, and is equal to the sum of the atomic scattering factors of the atoms in a unit cell, each multiplied by an appropriate phase factor. 结构因子 jié gòu yīn zǐ

Example:

It is shown that the logarithm of the structure factor of crystal is proportional to the function of the change in the integral gray values of diffraction spots.
结果表明:晶体结构因数的对数值与其相应的衍射斑点灰度积分值差的函数呈线性关系。

Extended Terms:

layer structure factor 成层构造因子
geometrical structure factor 几何构造因素;几何结构因子;几何结构因素

---

subgroup [ˈsʌbgruːp] *n.*

Definition: a distinct group within a group; a subdivision of a group 子群 zǐ qún

Origin: 1835-1845; sub-+ group

Example:

The West Nile virus is an RNA flavivirus of the Japanese encephalitis subgroup.
西尼罗河病毒是日本脑炎病毒亚组的一种 RNA 黄病毒。

Extended Terms:

universal subgroup 通用子群
torsion subgroup 挠子群

S

superlattice [ˈsjuːpəˌlætis]

Definition: (*Solid-state Physics*) An ordered arrangement of atoms in a solid solution which forms a lattice superimposed on the normal solid solution lattice. Also known as artificial crystal; artificially layered structure; superstructure. 超晶格 chāo jīng gé

Example:

This paper reviews the main achievements in dielectric superlattice field. Much has been focused on the work done in our Lab.

本文评述在介电体超晶格领域中的主要成果,特别是本实验室所做的工作。

Extended Term:

superlattice structure 超点阵结构;超晶格结构

---

superspace [ˈsjuːpəspeis] *n.*

Definition: the space of all three-geometries on a three-manifold, used in discussions of quantum gravity 超空间 chāo kōng jiān

Origin: super + space

Example:

Quantum entanglement is the key to quantum information and quantum teleportation, and special quantum correlations with superspace and nonlocality.

量子纠缠是量子信息与量子隐形传态的关键。量子纠缠是一个特殊的超空间、非定域的量子关联。

Extended Term:

superspace group 超空间群

---

symmetry [ˈsimitri] *n.*

Definition: (*Physics*) A property of a physical system that is unaffected by certain mathematical transformations as, for example, the work done by gravity on an object, which is not affected by any change in the position from which the potential energy of the object is measured. 对称 duì chèn

Origin: 1535-1545; from Latin symmetria from Greek symmetría commensurateness.

Example:

The experiments show that the formulas set up in this paper can lower the deformation of non-

symmetry extruded sections efficiently.
实验证明利用本文提出的计算式进行模具设计能有效地减小挤压非对称型材的弯扭畸变。

Extended Term:

radial symmetry 辐射对称;径向对称;放射对称

symmetry class ['simitri klɑːs]

Definition: one of 32 categories of crystals according to the inversions, rotations about an axis, reflections, and combinations of these which leaves the crystal invariant 对称型 duì chèn xíng

Example:

This paper proves a combinatorial formula, which is based on an arbitrary irreducible character x of a subgroup G of S_m, and gives its application in the symmetry class of tensors corresponding to G and x.
本文证明了一个组合公式是基于 S_m 的子群 G 的任意不可约特征标 x 的,本文也给出了此公式在对应于 G 和 x 的张量对称类上的应用。

Extended Terms:

crystal symmetry class 晶体对称类
symmetry class of tensors 张量对称类

symmetry element ['simitri 'elimənt]

Definition: Symmetry element is some combination of rotations and reflections and translations which brings a crystal into a position that cannot be distinguished from its original position. It is also known as symmetry operation; symmetry transformation. The rotational axes, mirror planes, and center of symmetry characteristic of a given crystal. 对称要素 duì chèn yào sù

Example:

Rotation-reflection axis Sn, an abstract symmetry element in molecular structure, is described and the relation between rotation-reflection axis and optical activity of organic compounds is revealed in the paper.
本文描述了分子结构中较为抽象的对称要素——像转轴 Sn,并揭示了像转轴与有机物旋光性的关系。

Extended Terms:

macroscopic symmetry element 宏观对称素
space symmetry element 空间对称元

S

symmetry operation [ˈsimitri ˌɔpəˈreiʃən]

Definition: A symmetry operation may be defined as a permutation of atoms such that the molecule or crystal is transformed into a state indistinguishable from the starting state. 对称操作 duì chèn cāo zuò

Example:

By using group theory, two-dimensional quasicrystal point groups and their symmetry operation, generating operation, order of group, direct product groups or semidirect product groups were derived and drawn.

用群论的方法逐一推导了二维准晶系列(包括五角、十角、八角和十二角晶系)的各个点群,得出了每个点群的全部对称操作、生成操作、群阶、直积群或半直积群关系式。

Extended Term:

improper symmetry operation 非常规对称操作

---

symmetry plane [ˈsimitri plein]

Definition: in certain crystals, a symmetry element whereby reflection of the crystal through a certain plane leaves the crystal unchanged 对称面 duì chèn miàn

Example:

And the inclusion has the "V" shape at the narrow face and symmetry plane under the nozzle.

而在铸坯窄面和水口下方对称面处,夹杂物形成"V"形分布。

Extended Term:

axis of symmetry axis plane 轴平面

---

systematic absence [ˌsistiˈmætik ˈæbsəns]

Definition: (*Solid-state Physics*) A factor which determines the amplitude of the beam reflected from a given atomic plane in the diffraction of an X-ray beam by a crystal, and is equal to the sum of the atomic scattering factors of the atoms in a unit cell, each multiplied by an appropriate phase factor. 系统消光 xì tǒng xiāo guāng

systematic: characterized by order and planning 系统的 xì tǒng de

absence: epilepsy characterized by paroxysmal attacks of brief clouding of consciousness 消失 xiāo shī

Example:

These programs are efficient when some lines disappeared by either accident or systematic

absence
即使在因偶然或系统消光而失去某些衍射线的情况下,该程序亦能有效地确定粉末 X 射线衍射数据的指标。

tensor [ˈtensəˌ -sɔː] *n.*

Definition: an object relative to a locally euclidean space which possesses a specified system of components for every coordinate system and which changes under a transformation of coordinates; a multilinear function on the cartesian product of several copies of a vector space and the dual of the vector space to the field of scalars on the vector space. 张量 zhāng liàng

Origin: muscle that stretches a part, 1704, Modern Latin, agent noun of Latin tendere "to stretch"

Example:

This decomposition may give another definition of high order tensor's covariant derivative.
这种分解可以给出高阶张量协变导数的另一种定义。

Extended Term:

metric tensor 度规张量;度量张量;基本张量

---

tetartohedry [tiˌtaːtəuˈhiːdri] *n.*

Definition: In geometry, a tetrahedron (plural: tetrahedra) is a polyhedron composed of four triangular faces, three of which meet at each vertex. 四分面相 sì fēn miàn xiàng

Extended Terms:

trapezohedral tetartohedry 偏方四分像
sphenoidal tetartohedry 楣形四分面像

---

terrestrial meridian [tiˈrestriəl məˈridiən]

Definition: A line on the surface of the earth connecting points having the same astronomical longitude. Also known as terrestrial meridian. 地球子午圈 dì qiú zǐ wǔ quān

terrestrial: of or relating to or characteristic of the planet Earth or its inhabitants 地球的 dì qiú de

meridian: the highest level or degree attainable; the highest stage of development 子午线 zǐ wǔ xiàn

tetragonal trapezohedron [te'trægənəl trə,pi:zəu'hi:drən]

Definition: The tetragonal trapezohedron or deltohedron is the second in an infinite series of face-uniform polyhedra which are dual to the antiprisms. It has eight faces which are congruent kites. 四方偏方面体 sì fāng piān fāng miàn tǐ

tetragonal: of or relating to or shaped like a quadrilateral 正方晶系的 zhèng fāng jīng xì de

trapezohedron: a polyhedron whose faces are trapeziums 三八面体 sān bā miàn tǐ

Example:

This article is about crystallization habits of the trigonal trapezohedron simple form of synthetic quartz.

本文是关于水晶中三方偏方面体单形的结晶习性。

tetrahedron [,tetrə'hi:drət, -'he-] *n.*

Definition: (*Geometry*) a solid contained by four plane faces; a triangular pyramid or any of various objects resembling a tetrahedron in the distribution of its faces or apexes 四面体 sì miàn tǐ

Origin: 1570, from Late Greek. tetraedron, originally neuter of tetraedros (adj.) "four-sided," from tetra- "four" + hedra "seat, base, chair, face of a geometric solid," from PIE base sed- "to sit".

Example:

The Ge-O tetrahedron network in Bi_2O_3-GeO_2 glasses is gradually depolymerized with increasing Bi_2O_3 content, however, in P_2O_5-GeO_2 glasses, the Ge-O tetrahedron network is seriously disordered by addinig P_2O_5.

随着 Bi_2O_3 含量的增加，Bi_2O_3-GeO_2 玻璃中的 Ge-O 四面体网络逐步解聚，随着 P_2O_5 含量的增加，P_2O_5-GeO_2 玻璃中的 Ge-O 四面体网络更加无序。

Extended Terms:

regular tetrahedron 正四面体

supplementary tetrahedron 补充四面体

tetrahexahedron [trə,pi:zəu'hi:drən] *n.*

Definition: a solid in the isometric system, bounded by twenty-four equal triangular faces, four corresponding to each face of the cube 四六面体 sì liù miàn tǐ

T

thermal conductivity [ˈθəːməl ˌkɔndʌkˈtiviti]

Definition: The heat flow across a surface per unit area per unit time, divided by the negative of the rate of change of temperature with distance in a direction perpendicular to the surface. Also known as coefficient of conductivity; heat conductivity. 热传导率 rè chuán dǎo lǜ

Example:

In the most general case the thermal conductivity is a tensor with six components.
在最一般的情况下,热导率是一个具有六个分量的张量。

Extended Terms:

thermal semi-conductivity sensor 热半传导传感器
thermal-conductivity detector 热导侦测器

thermal expansion [ˈθəːməl ikˈspænʃən]

Definition: the dimensional changes exhibited by solids, liquids, and gases for changes in temperature while pressure is held constant 热膨胀 rè péng zhàng

Example:

By installing a backflow preventer on any residential water system, you create a closed system that won't accommodate thermal expansion.
通过在任何住宅水系统上安装一个防回流阀,可以形成一个无法适应热膨胀的封闭系统。

Extended Terms:

thermal cubic expansion coefficient 体积膨胀系数
thermal electric expansion valve 热电膨胀阀;电子热力膨胀阀

thermodynamics [ˌθəːməudaiˈnæmiks] *n.*

Definition: (used with a sing. verb) Physics that deals with the relationships and conversions between heat and other forms of energy. (used with a *pl.* verb) Thermodynamic phenomena and processes. 热动力学 rè dòng lì xué

Origin: theory of relationship between heat and mechanical energy, 1854, from adj. thermodynamic (1849), from thermo-+ dynamic.

Example:

In thermodynamics we usually deal with functions of two or more variables.
在热力学中,我们经常涉及两个或更多变量的函数。

Extended Terms:

irreversible thermodynamics 非平衡态热力学;不可逆过程热力学;不可逆热力学
statistic thermodynamics 统计热力学

thermoluminescence [ˈθəːməuˌljuːmiˈnesəns] *n.*

Definition: A phenomenon in which certain minerals release previously absorbed radiation upon being moderately heated. 热发光 rè fā guāng

Example:

The quality control of thermoluminescence dosimetry is discussed.
讨论了热释光剂量测量的质量控制问题。

Extended Terms:

thermoluminescence dating 热发光定年法
phantom thermoluminescence 体模热释光剂量仪

thin section [θin ˈsekʃən]

Definition: A piece of rock or mineral specifically prepared to study its optical properties; the sample is ground to 0.03-millimeter thickness, then polished and placed between two microscope slides. Also known as section. 薄片 báo piàn

Example:

Comparative observations were made on the morphology and anatomy of secretory cavities in stems of 14 genera, 23 species and one varieties by paraffin and thin section method.
利用石蜡切片和薄切片方法对芸香科 14 属 23 种和 1 变种植物幼茎中分泌囊的分布和结构进行了比较研究。

Extended Terms:

thin-section analysis 薄片分析
thin-section staining 薄片染色

topotaxy [ˈtɔpəutæksi, ˈtəu-] *n.*

Definition: all chemical solid state reactions that lead to a material with crystal orientations which are correlated with crystal orientations in the initial product 拓扑关系 tuò pū guān xi

Example:

In addition, combining the transverse task of DACHANGGAOFENG mine of Guangxi province, this paper discussed the full process fromcollecting data、building modeL building the

topotaxy、spacial analyst、thevisualization of model to roam.
并结合广西大厂高峰矿的一个横向课题，论述了从数据采集、模型建立、拓扑关系的建立、空间分析、模型可视化到漫游等完整过程。

tranformation twin [ˌtrænsfəˈmeiʃən twin]

Definition: a crystal twin developed by a growth transformation from a higher to a lower symmetry 转变双晶 zhuǎn biàn shuāng jīng

tranformation:the act of changing in form or shape or appearance 转变 zhuǎn biàn

transition phase [trænˈziʃən feiz]

Definition: A phase transition is the transformation of a thermodynamic system from one phase or state of matter to another. 过渡相 guò dù xiāng

Example:

Finally the igniting point was defined as the bottom of the transition phase.
根据瞬间着火的定义将过渡阶段的谷底定义为着火点。

Extended Term:

the current transition phase 当前的过渡阶段

transparent [trænsˈpærənt] *adj.*

Definition: capable of transmitting light so that objects or images can be seen as if there were no intervening material 透明 tòu míng

Origin: early 15 century, from Midieval Latin transparentem (nominative transparens), present participle of transparere "show light through," from Latin trans- "through" + parere "come in sight, appear." Figurative sense of "easily seen through" is first attested 1590s. The attempt to back-form a verb transpare (c.1600) died with the 17 century.

Example:

Realize transparent display dialog:the dialog box above the text transparent and opaque.
实现对话框透明显示:对话框透明而上面的文字不透明。

Extended Terms:

transparent paper 透明纸;玻璃纸
transparent positive 透明正片

T

trapezohedron [trəˌpiːzəuˈhiːdrən] *n.*

Definition: (*crystallography*) a crystal form having all fac trapezoid trisoctahedron 偏方面体 piān fāng miàn tǐ

Origin: 1810-1820; trapez(oid) +-o +-hedron

Example:

This reflects the change of the crystal-habit of the trigonal trapezonhedron simple form in synthetic smoky quartz.
这变化反映了人工烟水晶三方偏方面体单形结晶习性的变化。

Extended Terms:

hexagonal trapezohedron 六方偏方六面体; 六方偏方六面体
trigonal trapezohedron 三方偏方面体;三方偏方八面体;三方偏形体

---

triclinic system [traiˈklinik ˈsistəm]

Definition: The most general and least symmetric crystal system, referred to by three axes of different length which are not at right angles to one another. 三斜晶系 sān xié jīng xì

Example:

Description: Bismuth nitrate pentahydrate belongs to five-color triclinic system crystal, owns hydroscopicity and sour flavor.
性状:五水硝酸铋为五色三斜晶系结晶,有吸湿性,具有酸味。

Extended Term:

triclinic crystal system 三斜晶系

---

trigonal bipyramid [ˈtrigənəl baiˈpirəmid]

Definition: In chemistry a trigonal bipyramid formation is a molecular geometry with one atom at the center and 5 more atoms at the corners of a triangular dipyramid. 三角双锥 sān jiǎo shuāng zhuī

bipyramid: crystal having the form of two pyramids that meet at a plane of symmetry 双椎体 shuāng zhuī tǐ

Example:

The crystal structure of complex indicates that the 2-carbethoxy-6-iminopyridine is coordinated to the cobalt as a tridentate ligand using [N, N, O] atoms and the coordination geometry of the central cobalt is the distorted trigonal bipyramid, with the pyridyl nitrogen atom and the two chlorin atoms forming the equatorial plane.

晶体结构分析表明:2-乙氧甲酰基作为三齿配体以[N,N,O]原子和两个氯离子与中心钴(Ⅱ)配位,形成畸变的三角双锥配位环境,其中吡啶氮原子和两个氯原子形成赤道平面。

Extended Terms:

distorted trigonal bipyramid 畸变的三角双锥
trigonal bipyramid structure 三角双锥结构

---

trigonal pyramid [ˈtriɡənəl ˈpirəmid]

Definition: In chemistry, a trigonal pyramid is a molecular geometry with one atom at the apex and three atoms at the corners of a trigonal base.三方锥 sān fāng zhuī

pyramid: a polyhedron having a polygonal base and triangular sides with a common vertex 椎体 zhuī tǐ

---

trigonal system [ˈtriɡənəl ˈsistəm]

Definition: a crystal system which is characterized by threefold symmetry, and which is usually considered as part of the hexagonal system since the lattice may be either hexagonal or rhombohedral 三方晶系 sān fāng jīng xì

Example:

The research indicates LaCo1-2xMnxCuxO_3 (x = 0.05, 0.1, 0.2) is the pure perovskite structure, belongs to the trigonal system.

研究表明合成的钙钛矿型催化剂 LaCo1-2xMnxCuxO_3(x=0.05,0.1,0.2)为单一钙钛矿结构,属于三方晶系。

Extended Terms:

trigonal crystal system 三角晶系;三方晶系
rhombohedral system (=trigonal system)三角晶系

---

trigonal trapezohedron [ˈtriɡənəl trəˌpiːzəuˈhiːdrən]

Definition: In geometry, the trigonal trapezohedron or deltohedron is the first in an infinite series of face-uniform polyhedra which are dual to the antiprisms. It has six faces which are congruent rhombi. 三方偏方面体 sān fāng piān fāng miàn tǐ

trapezohedron: a polyhedron whose faces are trapeziums 偏方三八面体 piān fāng sān bā miàn tǐ

T

Example:

Moreover, a correlation between the crystallization habit of the trigonal trapezohedron and the crystal defects has been studied for various seed orientations, namely the Z-cut, Y-cut, X-cut and the cut which intersects the Z-axis at 55°.
从籽晶取向(Z 切,Y 棒,X 棒与 Z 轴相交 55°切型等)研究了三方偏方面体结晶习性与晶体缺陷之间的关系。

---

twin [twin] *n.*

Definition: two interwoven crystals that are mirror images of each other 双晶 shuāng jīng

Origin: Old English twinn "double thread," from Proto-Germanic *twizna- (cf. Dutch twijn, Low German twern, German zwirn "twine, thread")

Example:

A new approach based on microscopy has been developed to determine the angle of inclination (θ) for the plane-law twin composition plane (010) of albitein the plagioclase slice obliquely cut along [010] crystal zone.
本文作者研究出来的测定倾斜角的新方法,以测定斜长石斜切[010]晶带切片中钠长石面律双晶结合面(010)的倾角(θ)。

Extended Terms:

twin crystal 孪晶
twin lamellae 双芯片

---

twin axis [twin 'æksis]

Definition: the crystal axis about which one individual of a twin crystal may be rotated (usually 180°) to bring it into coincidence with the other individual 双晶轴 shuāng jīng zhóu

Example:

It is built a mathematical model of optimization design of I-beam, and analyzed an example of I-beam using ANSYS. The result of optimization is got, and it is put forward some advices of the design of a beam with twin axis symmetrical I-section.
根据优化设计原理给出梁优化设计的数学模型,利用通用有限元计算软件 ANSYS 对某平台梁进行优化实例分析,得出优化结果,并根据优化结果提出构件设计的建议。

---

twin law [twin lɔː]

Definition: a statement relating two or more individuals of a twin to one another in terms of

T

their crystallography (twin plane, twin axis, and so on) 双晶律 shuāng jīng lǜ

Example:

The spring-back law of the thick-plate in twin-curvature mould press is studied.
对双曲率模冲压后板料的回弹规律进行了模拟。

Extended Term:

Brazil twin law 巴西双晶律;巴西双晶律

---

twin plane [twin plein]

Definition: the plane common to and across which the individual crystals or components of a crystal twin are symmetrically arranged or reflected, also known as twinning plane 双晶面 shuāng jīng miàn

Example:

A special twinning of huangheite-(Ce) ($BaCe(CO_3)_2F$, R3m), in which twin plane and composition plane are parallel to (0001), was first discovered by authors by means of RASA-5RP four-circlesingle crystal diffractometer.
本文用转靶四圆单晶衍射仪首次发现黄河矿($BaCe(CO_3)_2F$,R3m)的一种特殊类型的双晶,其双晶面和c面平行于(0001)。

Extended Terms:

stepped twin plane 阶梯状双晶面
twin-motored plane 双发动机飞机

---

twinning ['twiniŋ] *n.*

Definition: (*Mineralogy*) the formation of twin crystals 双晶形成 shuāng jīng xíng chéng

Origin: 1565-1575; twin+-ing

Example:

The influences of material variables on twinning, including crystal orientation, temperature, strain rate, grain size and prestrain, are analyzed.
分析了影响双晶形成的基本变量,包括晶粒取向、变形温度、变形速度、晶粒尺寸、预变形。

Extended Terms:

deformation twinning 塑变双晶;形变孪生;变形双晶
twinning crystal 双晶

U

uniaxial crystal [ˌjuːniˈæksiəl ˈkristəl]

Definition: a doubly refracting crystal which has a single axis along which light can propagate without exhibiting double refraction 一轴晶 yī zhóu jīng

Example:

This paper provided a simple method of calculating the internal reflection direction in anisotropic uniaxial crystal, and gave the concrete example.

提供了一般情况下直接求解光在各向异性的单轴晶体内表面上反射轨迹的一种较为简便的方法,并给出了计算实例。

Extended Terms:

uniaxial negative crystal 负单轴晶体

uniaxial positive crystal 正单轴晶体

---

unit cell [ˈjuːnit sel]

Definition: the smallest building block of a crystal, consisting of atoms, ions, or molecules, whose geometric arrangement defines a crystal's characteristic symmetry and whose repetition in space produces a crystal lattice 晶胞 jīng bāo

Example:

A unit cell model was applied to study the creep damage behavior after fiber fractures in the fiber reinforced composites at high temperature.

首先利用复合材料纤维断裂单胞模型,编制蠕变损伤子程序,对单胞模型进行蠕变损伤分析。

Extended Term:

unit-cell parameters 单个电池参数;晶格单元参数

vacancy [ˈveikənsi] *n.*

Definition: a crystal defect caused by the absence of an atom, ion, or molecule in a crystal lattice 空缺 kòng quē

Origin: c.1600, "state of being vacant," from Late.Latin vacantia, from vacans. Meaning "available room at a hotel" is recorded from 1953.

Example:

Therefore, we must concern "vacancy rate" of the data.
因此,当前必须关注"空置率"这个数据。

Extended Term:

domestic vacancy 空置住宅楼宇;空置住宅单位

---

valence band [ˈveɪl(ə)ns bænd]

Definition: the highest electronic energy band in a semiconductor or insulator which can be filled with electrons 价带 jià dài

Example:

The upshift of valence band edge under mechanical strain increases the majority hole concentration at the oxide/Si interface.
价电带的向上移动会使得氧化层以及矽的接面的电洞浓度增加。

Extended Terms:

valence bond band 价带
virtual valence band 虚价带

---

vector [ˈvektə] *n.*

Definition: (*Mathematics*) quantity, such as velocity, completely specified by a magnitude and a direction 矢量 shǐ liàng

Origin: quantity having magnitude and direction," 1704, from Latin vector "one who carries or conveys, carrier," from past participle stem of vehere "carry, convey" (see vehicle).

Example:

Synthesize gene and clone the object gene into suitable vector.
合成基因,将目的基因克隆到合适的载体。

Extended Terms:

vector equation 向(相)量方程;矢量方程;矢量方程式
vector diagram 向量图;矢量图

vector space [ˈvektə speis]

Definition: a system consisting of a set of generalized vectors and a field of scalars, having the same rules for vector addition and scalar multiplication as physical vectors and scalars 向量空间 xiàng liàng kōng jiān

Example:

By means of the expression of vector space, mathematical modeling of Jakobson's theory on distinctive features of phonemes is made.
本文采用向量空间的表述形式,对雅柯布森的音位区别特征学说进行数学建模。

Extended Terms:

vector-space model 向量空间模型;向量模型
vector space direct sum 向量空间直和

Wigner-Seitz cell [ˈwignə-seits sel]

Definition: The Wigner-Seitz cell, named after Eugene Wigner and Frederick Seitz, is a geometrical construction used in the study of crystalline material in solid-state physics. The unique property of a crystal is that its atoms are arranged in a regular 3-dimensional array called a lattice 维格纳-赛茨晶胞 wéi gé nà sài cí jīng bāo

Wulff net [wulf net]

Definition: A method of displaying the positions of the poles of a crystal in which poles are

X

projected through the equatorial plane of the reference sphere by lines joining them with the south pole for poles in the upper hemisphere, and with the north pole for poles in the lower hemisphere. 吴氏网 wú shì wǎng

Wyckoff position [ˈwikɒf pəˈzɪʃ(ə)n]

Definition: In crystallography, a Wyckoff position is a point belonging to a set of points for which site symmetry groups are conjugate subgroups of the space group. 威科夫定位 wēi kē fū dìng wèi

X-ray crystallography [ˌeks-rei ˈkristəˈlɒgrəfi]

Definition: the study of crystal structure by means of X-ray diffraction. X-射线晶体学 shè xiàn jīng tǐ xué

Example:
The structures of drug-target complexes obtained by X-ray crystallography provide direct and convictive evidences for drug design.
利用晶体X-射线衍射的方法获得药物与靶标复合物的结构,为药物设计提供最直接有力的依据。

Extended Term:
X-ray crystallography X-射线结晶学

X-rays [ˈeks-reiz] *n.*

Definition: a relatively high-energy photon having a wavelength in the approximate range from 0.01 to 10 nanometers. A stream of such photons, used for their penetrating power in radiography, radiology, radiotherapy, and scientific research. Often used in the plural. Also called roentgen *ray*. 伦琴线(X-射线)lún qín xiàn(X shè xiàn)

Origin: 1896, translation of German X-strahl, from X, algebraic symbol for an unknown quantity, + Strahl (plural Strahlen) "beam, ray". Coined 1895 by German scientist Wilhelm Conrad Röntgen (1845-1923), who discovered them.

Example:
There is also concern in America about the overuse of medical X-rays, especially in emergency

rooms.
美国也考虑到了在医疗过程中,尤其是在急诊室中,X 射线的过度使用。

Extended Terms:

characteristic X-rays 标识 X 射线;特性 X 线
soft X-rays 软 X 射线;弱 X 射线;软 X 线

zone [zəun] *n.*

Definition: a ringlike or cylindrical growth or structure of crystal 晶带 jīng dài

Origin: late 14 century, from Latin zona "geographical belt, celestial zone," from Greek zone "a belt," related to zonnynai "to gird," from Proto-Indo-European base yes- "to gird, girdle" (cf. Avestan yasta- "girt," Lithuanian juosiu "to gird," Old Church Slavonic po-jasu "girdle"). Originally one of the five great divisions of the earth's surface (torrid, temperate, frigid; separated by tropics of Cancer and Capricorn and Arctic and Antarctic circles); meaning "any discrete region" is first recorded 1822. Zone defense in team sports is recorded from 1927. Zoning "land-use planning" is recorded from 1912. Zoned (adj.) in drug-use sense is attested 1960s, from ozone, which is found high in the atmosphere; the related verb to zone is from 1980s.

Example:

They throw below the strike zone more than they throw above it.
投出的坏球低于好球带的数量比高于好球带多。

Extended Terms:

transition zone 交换区域;过渡区;转变区;过渡带
weld zone 焊接区;焊缝区

---

zone axis [zəun ˈæksis]

Definition: a line through the center of a crystal which is parallel to all the faces of a zone 晶带轴 jīng dài zhóu

axis: a straight line through a body or figure that satisfies certain conditions 轴 zhóu

---

Z

zone law [zəun lɔː]

Definition: a law which states that the Miller indices (h, k, l) of any crystal plane lying in a zone with zone indices (u, v, w) satisfy the equation $hu + lv + kw = 0$ 晶带定律 jīng dài dìng lǜ

zone: any encircling or beltlike structure 带 dài

law: a generalization that describes recurring facts or events in nature 定律 dìng lǜ

Example:

Flatness error was obtained by minimum zone law which decided by triangle rule and cross rule. 最小晶带定律可求得平面度误差,而最小晶带定律又取决于三角形准则和交叉准则。

Extended Term:

Weiss zone law 韦斯晶带定律

第二部分

矿物名称词汇

acmite [ˈækmait] *n.*

Definition: (*Mineralogy*) A brown or green silicate mineral of the pyroxene group, often in long, pointed prismatic crystals; hardness is 6-6.5 on Mohs scale, and specific gravity is 3.50-3.55; found in igneous and metamorphic rocks 锥辉石 zhuī huī shí

Formula: $NaFeSi_2O_6$

Origin: Aegirine-augite is also known as acmite and the chrome analogue of jadeite where Cr replaces Al is known as kosmochlor (formerly ureyite) and is also a jade mineral.

Examples:

Based on outcrop observation, microscopic identification and EPMA, it is determined that the mineral association complexes in the Pingchau Formation of the Pingchau Island, Hong Kong, are glauberite pseudocrystals replaced by analcime, acmite and calcite, which show bipyramid shapes and scattered occurrence in mud beds.

本文通过野外地质调查、显微鉴定分析和 EPMA 测试,确定香港坪洲岛坪洲组地层中的方沸石、锥辉石和次生方解石等矿物部分是由钙芒硝假晶形成的,它们多以集合体方式构成钙芒硝假晶,具有钙芒硝的棱面体形态,并可以和区域上含钙芒硝的盐系地层进行对比。

It is believed that the Fe for forming acmite is mainly from the erosion of pipes with minor Fe from the trace Fe in water.

形成锥辉石中的铁主要来源于水管的腐蚀,当水管表面形成一层水垢后,进水中的微量铁成为另外一个重要来源。

Extended Term:

acmite-augite 霓辉石

---

actinolite [ækˈtinəˌlait] *n.*

Definition: a green mineral of the amphibole group containing calcium, magnesium, and iron and occurring chiefly in metamorphic rocks and as a form of asbestos

阳起石 yáng qǐ shí

Formula: $Ca_2(Mg,Fe)_5[Si_4O_{11}]_2(OH)_2$

Origin: late 18th century: from Greek aktis, aktin-"ray" + lithos "stone" (because of the ray-like crystals)

Example:

The results show that this type of nephrite jade mainly consists of tremolite-actinolite isomorphismseries. The mineral components are tremolite dominantly, with minor actinolite, diopside, epidote, clinozoisite, magnetite, pyrite and other impurities.
结果表明,该类碧玉主要是由透闪石、阳起石类质同像系列矿物组成,矿物成分主要为透闪石,并含有少量阳起石、透辉石、绿帘石、斜黝帘石、磁铁矿、黄铁矿等杂质矿物。

Extended Terms:

mangan-actinolite 锰阳起石
actinolite asbestos 阳起石石棉
actinolite schist 阳起片岩
actinolite rock 阳起石岩

---

adamite ['ædəmait] *n.*

Definition: (*Mineralogy*) a zinc arsenate hydroxide mineral 羟砷锌石 qiǎng shēn xīn shí

Formula: Zn_2AsO_4OH

Origin: named after French mineralogist Gilbert Joseph Adam (1795-1881) in 1866

Example:

Roughing and intermediate rolls for structure and wire mills were made of adamite steel containing 1.3%-1.5%C.
型钢和线材轧机的粗轧和中间机架的轧辊采用含 C 1.3%~1.5%。

Extended Terms:

adamite roll 高碳铬镍耐磨铸铁轧辊
forged cast iron and adamite roll 锻造白口铁和半钢轧辊

---

adularia (adular) [ˌædju'lɛəriə] *n.*

Definition: a variety of orthoclase feldspar found as colorless to white prismatic crystals in cavities in metamorphic rocks 冰长石 bīng cháng shí

Formula: $KAlSi_3O_8$

Origin: Italian, from French adulaire, after Adula, a mountain group of southeast Switzerland

Example:

The adularia associated with Au, Ag mineralization in Bitian gold deposit is identified by

Electron Microprobe, Scanning Electron Microscope, and X-ray Diffractometer.
电子探针、扫描电子显微镜、X-射线衍射等测试方法的应用,揭示了碧田金矿床冰长石的存在及其矿物学特征。

Extended Terms:

sodian adularia 钠冰长石
adularia moonstone 冰长月光石
adularia sericite type 冰长石绢云母型
adularia-calcite assemblage 冰长石方解石组合

---

aegirine [ˈiːdʒəˌriːn] *n.*

Definition: (*Mineralogy*) a mineral with monoclinic crystals with the chemical formula $NaFe_3+(Si_2O_6)$ belonging to the pyroxene group 霓石 ní shí

Formula: $NaFe[Si_2O_6]$

Origin: The name aegirine is after the Teutonic god of the sea, Aegir, and was given when the first specimens of the mineral were discovered in Norway.

Examples:

For Aegirine stone of sheet glass the author has made a petrographic analysis by optical microscope, X-ray differaction structure analysis and shape observation of chemical component distribution by scanning electron microscope.
作者对平板玻璃的析晶霓石结石进行了光学显微镜岩相分析、X-射线衍射结构分析和扫描电镜化学成分分布的形态观察。
The lkali granite are mineralogically characterized by the occurence of aegirine and arfvedsonite and chemically by high silicon and alkali, low calcium and magnesium, and enrichment of HFS elements, being typical A-type granites.
这些碱性花岗岩以出现霓石、钠铁闪石、高硅、高碱、低钙、低镁、富集高场强元素为特征,属于典型的 A 型花岗岩。

Extended Terms:

aegirine-augite 霓辉石
aegirine-felsite 霓霏细岩
aegirine-granite 霓花岗岩
aegirine-hedrumite 霓淡正长岩

---

albite [ˈælbait] *n.*

Definition: a sodium-rich mineral of the plagioclase feldspar group, typically white, occurring widely in igneous rocks 钠长石 nà cháng shí

Formula: $Na_2O \cdot Al_2O_3 \cdot 6SiO_2$

Origin: early 19th century: from Latin albus "white" +-ite

Example:

Mineral constituents consist mainly of jadeite, aegirine-augite, magnesio-riebeckite, winchite, magnesio-hornblende, actinolite, quartz, albite, rutile and sphene.
主要组成矿物为硬玉、霓辉石、镁钠闪石、蓝透闪石、镁角闪石、阳起石、石英、钠长石、金红石和榍石。

Extended Terms:

albite law 钠长(双晶)律
albite twin 钠长石双晶
high albite 高温型钠长石
oligoclase albite 奥钠长石
albite porphyrite 钠长玢岩
albite twin law 钠长石双晶律

---

alkali ['ælkəlai] *n.*

Definition: a compound with particular chemical properties including turning litmus blue and neutralizing or effervescing with acids; typically, a caustic or corrosive substance of this kind such as lime or soda 碱 jiǎn

Formula: -OH

Examples:

a natural black bitumen used in the manufacture of acid, alkali, and waterproof coatings
用来制造酸、碱和防水涂层的天然的黑色沥青
A Primary Study on the Cornea Repair after Alkali Burn
碱烧伤后角膜修复的初步研究

Extended Terms:

alkali resistance 抗碱的，耐碱的
alkali cellulose 碱纤维素
alkali chloride 碱金属氯化物

---

alkali amphibole ['ælkəlai 'æmfibəul]

Definition: (碱性角闪石 jiǎn xìng jiǎo shǎn shí) any of a class of rock-forming silicate or aluminosilicate minerals typically occurring as fibrous or columnar crystals 闪石 shǎn shí

Formula: 闪石 $(Ca,Na)2\text{-}3(Mg^{2+},Fe^{2+},Fe^{3+},Al^{3+})_5[(Al,Si)_4O_{11}](OH)_2$

Origin: 闪石 early 19th century: from French, from Latin amphibolus "ambiguous" (because of the varied structure of these minerals), from Greek amphibolos, from amphi- "both, on both sides" + ballein "to throw"

Example:

...and mineralogically an assemblage mainly of calcite and dolomite with alkali amphibole, felspar, ae-girite, apatite, fluorite, magnetite and REE minerals.
在矿物组合上,以白云石为主,方解石次之,伴生一套碱性闪石、长石、霓石、磷灰石、萤石、磁铁矿、稀土矿物组合。

Extended Terms:

alkali-fluids 碱流体
alkali-cement 碱胶凝材料
alkali-resisting 抗碱

---

alkali feldspar [ˈælkəlai ˈfeldspɑː]

Definition: any of the group of feldspars rich in sodium and/or potassium
碱性长石 jiǎn xìng cháng shí

Definition: an abundant rock-forming mineral typically occurring as colourless or pale-coloured crystals and consisting of aluminosilicates of potassium, sodium, and calcium 长石 cháng shí

Formula: 长石 $KAlSi_3O_8 \cdot NaAlSi_3O_8 \cdot CaAl_2Si_2O_8$

Origin: 长石 mid 18th century: alteration of German Feldspat, Feldspath, from Feld "field" + Spat, Spath "spar" (see spar). The form felspar is by mistaken association with German Fels "rock"

Examples:

a light-colored igneous rock consisting essentially of alkali feldspar
一种浅色火成岩,主要由碱性长石构成
The metamorphosed volcanic rock of the upper member of the Haizhou group is rich in alkali-feldspar (60% to 70%) and high potash feldspar (35% to 60%).
苏北海州群云台组上段变火山岩富含碱性长石(60%~70%)和钾长石(35%~60%)。
The outer circle consists of alkaline gabbro, quartz syenite and alkali-feldspar granite. The inner circle consists of quartz diorite-granodiorite-adamellite-moyite of calc-alkaline series.
外环为碱性辉长岩、石英正长岩、碱长花岗岩,内环为钙碱系列的石英闪长岩、花岗闪长岩、二长花岗岩、钾长花岗岩。

Extended Terms:

barium feldspar 钡长石

calcium feldspar 钙长石

alkali pyroxene [ˈælkəlai paiˈrɔksiːn]

Definition: (碱性辉石 jiǎn xìng huī shí, Mineralogy) any of a group of crystalline minerals containing silicates of iron, magnesium and calcium 辉石 huī shí

Formula: 辉石 XYT_2O_6

Origin: The name pyroxene comes from the Greek words for fire (πυρ) and stranger (ξένος). Pyroxenes were named this way because of their presence in volcanic lavas, where they are sometimes seen as crystals embedded in volcanic glass.

allanite orthite [əˈlælənait ˈɔːθait]

Definition: a brownish-black mineral of the epidote group, consisting of a silicate of rare earth metals, aluminium, and iron 褐帘石 hè lián shí

Formula: $(Ce,Ca,Y,La)_2(Al,Fe^{3+})_3(SiO_4)_3(OH)$

Origin: early 19th century: named after Thomas Allan (1777-1833), Scottish mineralogist, +-ite

Example:

After its alteration the chemical compositions of the secondary products of allanite have been undertaken great change. Besides SiO_2, the contents of Al, Ce^{4+} increase 2 times more than their original states, while Fe, Ti, Mg, Ca and some REE have nearly all lost.
褐帘石发生蚀变后,其次生物的化学成分变化很大,除 SiO_2 外,A_1、Ce^{4+} 增高一倍以上,而 Fe,Ti,Mg,Ca 及多种稀土成分几乎全部消失。

Extended Term:

mangan-orthite 锰褐帘石
nd-rich allanite 富钕褐帘石

alleghanyite [əlegˈheniːait] *n.*

Definition: (*Mineralogy*) a pink mineral consisting of basic manganese silicate 粒硅锰矿 lì guī měng kuàng

Formula: $Mn_5(SiO_4)_2(OH)_2$

Example:

Quartz, hausmannite, rhodochrosite, tephroite, ribbeite, pyroxmangite, and caryopilite are major minerals; calcite, kutnahorite, alleghanyite, spessartine, rhodonite, clinochlore, and parsettensite are second in abundance.

石英石、黑锰矿、菱锰矿、锰橄榄石、硅镁锰石、锰辉石和肾硅锰矿是主要矿物,方解石、锰白云石、粒硅锰矿、锰铝榴石、蔷薇辉石、绿泥石和红硅锰矿是第二丰富的矿石。

---

allophane [ˈæləfein] *n.*

Definition: (*Mineralogy*) an amorphous hydrous aluminium silicate clay mineral 水铝英石 shuǐ lǚ yīng shí

Formula: $mAl_2O_3 \cdot nSiO_2 \cdot pH_2O$

Example:

The clay minerals in shale are mainly hydromica, next to it is kaolinite, halloysite, montomorillonite and little allophane.
页岩中的粘土矿物主要为水云母,次为高岭石、多水高岭石、蒙脱石,少量水铝英石。

Extended Term:

humic allophane soil 腐殖质铝英土

---

almandine [ˈɔːlmənˌdain] *n.*

Definition: a deep violet-red garnet, found in metamorphic rocks and used as a gemstone, a kind of garnet with a violet tint 铁铝榴石 tiě lǚ liú shí

Formula: $FeAl_2Si_3O_{12}$

Origin: late Middle English: from obsolete French, alteration of alabandine, from medieval Latin alabandina (gemma), "jewel from Alabanda", an ancient city in Asia Minor where these stones were cut.

Example:

The garnets consist mainly of almandine and grossularite, and belong to C-type eclogite.
石榴子石以铁铝榴石和钙铝榴石为主,属于 C 类榴辉岩。

Extended Terms:

almandine spinel 红尖晶石,贵榴石尖晶石
pyrope-almandine 镁铝—铁铝榴石

---

alunite [ˈæljunait] *n.*

Definition: (*Mineralogy*) a gray, water-soluble mineral, potassium aluminium sulphate; the natural source of alum 明矾石 míng fán shí

Formula: $KAl_3(SO_4)_2(OH)_6$

Origin: from its earlier name aluminilite, named in 1824

Example:

In the experiments, natural alunite ores, alunitized wall rocks and waste materials were used after extracting alum, alum syrup, alum residue and furnace ash in the mine which could serve as starting materials. Aqueous solutions of certain concentrations of different acids of HCl, HBr, $HClO_4$, HF, $C_6H_6O_8$ and $C_6H_5CH_3$ are prepared as extraction solutions. The experiments lasted 120～168 hours in high pressure vessels at 300℃ and 500×10^5 Pa.
实验采用天然的明矾石矿石、矾化围岩和提炼明矾后的废料——矾浆、矾碴和炉灰作为试料，配制含一定浓度的HCl、HBr、$HClO_4$、HF、$C_6H_6O_8$和$C_6H_5CH_3$等不同酸类组合的水溶液作为萃取溶液，在300℃的温度和500×10^5Pa压力下的高压釜中持续了120～168小时的互相作用。

Extended Terms:

lead-alunite 铅明矾
alunite deposit 明矾石矿床
alunite high strength cement 明矾石高强水泥
alunite expansive cement 明矾石膨胀水泥
alunite expansion agent for concrete 明矾石混凝土膨胀剂

amazonite [ˈæməzənait] *n.*

Definition: a green variety of microcline, often used as a semiprecious stone. Also called amazon stone.天河石 tiān hé shí

Formula: Rb_2O

Origin: The name is taken from that of the Amazon River, from which certain green stones were formerly obtained, but it is doubtful whether green feldspar occurs in the Amazon area.

Examples:

Quartz and topaz are the earliest phases crystallized from the melt, whereas amazonite is formed from a fluid-rich residual magma by interstitial fillig and replacing earlier minerals.
石英和黄玉是最早从熔体中结晶出的矿物相，而天河石是由富含流体的残余熔体填隙结晶或与先存矿物反应而成。

In this paper, the authors analyzed the geological characteristics of the amazonite, studied the color mechanism of the amazonite, and set forth the main method of color perfecting, which designated a way to improve the value of the amazonite, furthermore, provided the theory basis for further developing and utilizing this ore bed.
在本文中，作者分析了(西榆皮)天河石的宝石学特征，研究了(西榆皮)天河石颜色的呈色机理，探讨了(西榆皮)天河石颜色优化处理的可能性，为提高(西榆皮)天河石的宝石价值指明了方向，为该矿床的进一步开发利用提供了理论依据。

amblygonite [æm'bligənait] *n.*

Definition: (*Mineralogy*) a mineral, a mixed fluoride and aluminophosphate of sodium and lithium that is an important ore of lithium 锂磷铝石 lǐ lín lǚ shí

Formula: $(Li,Na)Al(PO_4)(F,OH)$

Origin: The mineral was first discovered in Saxony by August Breithaupt in 1817, and named by him from the Greek amblus, blunt, and gouia, angle, because of the obtuse angle between the cleavages.

Examples:

Whether the amblygonite occurs or not is closely related to the Li, Rb and Cs contents of the melt. Both amblygonite and feldspars are the main hosts of whole-rock phosphorus in the Yashan topaz-lepidolite granite.
磷锂铝石是雅山黄玉锂云母花岗岩中的主要磷酸盐矿物,其产出与否同体系的 Li,Rb,Cs 含量密切相关。

If amblygonite is present in the granite, the mineral is the main contributor of whole-rock phosphorus and feldspars are the subordinate contributor; if no amblygonite crystallizes out, feldspars will be the main contributor of whole-rock phosphorus.
磷锂铝石和长石矿物都是雅山黄玉锂云母花岗岩中磷的主要贮体,并且相互之间呈互补关系,当出现磷锂铝石时,磷锂铝石为全岩磷的主要贡献者,当无磷钾铝石晶出现时,长石矿物为全岩磷的主要贡献者。

---

amethyst ['æməθist] *n.*

Definition: a precious stone consisting of a violet or purple variety of quartz
紫水晶 zǐ shuǐ jīng

Formula: Fe_2O_3

Origin: Middle English: via Old French from Latin amethystus, from Greek amethustos "not drunken" (because the stone was believed to prevent intoxication)

Example:

Amethyst crystals uniformly coloured under irradiation can be obtained using KOH and NH_4F aqueous solution as reaction medium, dissolvable iron-salt as dopant, r-cut slices as seeds and other appropriately adjusted technical conditions.
采用 KOH+NH_4F 混合水溶液为反应介质,以可溶性铁盐为掺杂物,选用 r 切籽晶,并调节其他工艺条件,水热条件下可制得经 γ 射线辐照着色均匀且较深的人工紫水晶晶体。

Extended Terms:

false amethyst 假紫晶

synthetic amethyst crystal 人工紫水晶
amethyst eryngo 紫水晶刺芹

amphibole [ˈæmfibəul] *n.*

Definition: any of a class of rock-forming silicate or aluminosilicate minerals typically occurring as fibrous or columnar crystals.闪石 shǎn shí

Formula: Amphibole compositions in the system $Mg_7Si_8O_{22}(OH)_2$ (anthophyllite)-$Fe_7Si_8O_{22}(OH)_2$(grunerite)-"$Ca_7Si_8O_{22}(OH)_2$."

Origin: early 19th century: from French, from Latin amphibolus "ambiguous" (because of the varied structure of these minerals), from Greek amphibolos, from amphi-"both, on both sides" + ballein "to throw"

Examples:

a fibrous amphibole, used for making fireproof articles; inhaling fibers can cause asbestosis or lung cancer
一种纤维状的闪石,用作防火材料,吸入纤维会导致石棉沉滞症或肺癌。
The whole-rock and amphibole separate yield a Sm-Nd isochronic age of 128 ± 12 Ma, whereas the Rb-Sr whole-rock ages is 131.8 ± 6.2 Ma and,311.1 ± 7.9 Ma, respectively.
测得 Sm-Nd 全岩及角闪石单矿物内部等时线年龄为 128±12 Ma,Rb-Sr 全岩等时线年龄为 131.8±6.2 Ma 和 311.1±7.9 Ma。

Extended Terms:

manganese amphibole 锰闪石
amphibole asbestos 闪石石棉
sodic amphibole 钠角闪石
amphibole family 闪石族

analbite [ˈænəlbait] *n.*

Definition: (*Mineralogy*) a triclinic albite which is not stable and becomes monoclinic at about 700℃高温钠长石 gāo wēn nà cháng shí

Formula: $(Na,K)AlSi_3O_8$

Example:

The free energy, entropy and enthalpy of sodium feldspar in thermal equilibrium and in metastable states are derived from investigations of the heat capacities of albite, analbite, ordered and disordered Or31.
经调查自由能量、熵和焓的钠长石在热平衡状态,导出了稳态的比热的钠长石、高温钠长石、有序和无序 Or31。

Extended Term:

albite 钠长石

analcime [ə'nælsaim] (also called analcidite, analcite [ə'nælsait])

Definition: (*Mineralogy*) a mineral, a sodium aluminosilicate with a chemical formula $NaAlSi_2O_6 \cdot H_2O$, having a zeolite structure, found in alkaline basalts 方沸石 fāng fèi shí

Formula: $NaAlSi_2O_6 \cdot H_2O$

Origin: named in 1801 after the property of attaining weak electricity when heated or subjected to friction, from the Greek word meaning "weak"

Examples:

It is proposed that hydrothermal alteration and replacement processes played a main role for the glauberite to have been changed to analcime, acmite and calcite, as the primary glauberite provided Na_2O and CaO and the hydrothermal activity produced SiO_2-bearing fluid.

本文认为,原始钙芒硝为上述矿物的形成提供了 Na_2O 和 CaO,热液活动带来了部分 SiO_2,通过交代蚀变作用形成了丰富的钙芒硝假晶。

In 20℃ and 40℃, the selectivity of Cd^{2+} to analcime is less than that of Na^+ to analcime.

在 20℃和 40℃时,方沸石对 Cd^{2+} 的选择性小于对 Na^+ 的选择性。

Extended Term:

analcime ana 方沸石

anatase [ˌænə'teiz] *n.*

Definition: one of the tetragonal forms of titanium dioxide, usually found as brown crystals, used as a pigment in paints and inks 锐钛矿 ruì tài kuàng

Origin: early 19th century: from French, from Greek anatasis "extension", with allusion to the length of the crystals.

Formula: TiO_2

Example:

The XRD spectra showed that doped-Fe^{3+} 0.1% Fe-N/TiO_2 obtained at 500℃ and 550℃ kept anatase and the relative anatase crystals had the size of 15.3 nm and 20.6 nm respectively, and the N1s XPS spectra was observed.

XRD 谱图结果表明,分别在 500℃和 550℃焙烧后的 0.1%Fe-N/TiO_2,均为锐钛矿型,晶体粒度分别为 15.3 nm 和 20.6 nm,XPS 谱图显示 Fe-N/TiO_2 有 N1s 谱峰。

Extended Terms:

anatase titanium dioxide 锐钛型二氧化钛
anatase octahedrite 锐钛矿
anatase type 锐钛矿型

A

andalusite [ˌændəˈluːsait] *n.*

Definition: a grey, green, brown, or pink aluminosilicate mineral occurring mainly in metamorphic rocks as elongated rhombic prisms, sometimes of gem quality 红柱石 hóng zhù shí

Formula: Al_2SiO_5

Origin: early 19th century: from the name of the Spanish region of Andalusia +-ite

Example:

Adding alumina fibre (5%, 10%, 15% and 20% respectively) to andalusite of 0.2-101.5 μm and 0.1-34.7 μm particle size, alumina fibre reinforced andalusite composite is prepared by traditional pressureless sintering process firing at 1350℃ and 1500℃.
以南非红柱石和多晶氧化铝纤维为原料,在纤维加入量(w)分别为5%、10%、15%和20%,烧成温度分别为1350℃和1500℃的条件下,研究了红柱石原料粒度为0.2~101.5μm和0.1~34.7μm时对传统无压烧结工艺制备的氧化铝纤维增强红柱石基复合材料烧结性能的影响。

Extended Terms:

andalusite concentrate 红柱石精矿
natural andalusite 天然红柱石

andesine [ˈændiziːn] *n.*

Definition: (*Mineralogy*) sodium calcium aluminum silicate, ($Na_xCa_xAlSi_3O_8$), a plagioclase feldspar, the third member of the albite-anorthite solid solution series, found in fine-grained andesite lavas; crystals are rarely seen 中长石 zhōng cháng shí

Formula: $Na_xCa_xAlSi_3O_8$

Origin: named for the type locality in the Andes Mountains

Examples:

The plagioclases from amphibolites are chiefly andesine, An = 42%-66%, Ab = 34%-58%.
斜长角闪岩中的斜长石主要为中长石,An为42%~66%,Ab为34%~58%,分布范围较集中。
Pyroxene is exclusively augite with high Na_2O, FeO and MnO contents, and plagioclase is low

CaO andesine, which share characteristics with their counterparts from gabbros formed in an intra-plate setting and are largely different from those from gabbros associated with island arc or active continental margin.
岩体主要矿物辉石属普通辉石,Na_2O、FeO 和 MnO 较高,长石属中长石,低 CaO,这些特征明显区别于岛弧或活动大陆边缘环境中的同类矿物,而与板内伸展环境中形成的辉长岩的矿物特征相似。

---

andradite [ˈandrədait] *n.*

Definition: A mineral of the garnet group, containing calcium and iron. It occurs as yellow, green, brown, or black crystals, sometimes of gem quality.钙铁榴石 gài tiě liú shí

Formula: $Ca_3Fe_2(SiO_4)_3$

Origin: mid 19th century: named after J. B. de Andrada e Silva (c.1763-1838), Brazilian geologist, +-ite

Example:

Garnets in the lherzolite are andradite enriched in Ca and Fe and depleted in Mg and Al (And 95-97, Pyr 0.27-5.06, Gro 0-2.62), indicating that they formed by metamorphism. Cumulates consist mainly of dunite, wehrlite, pyroxenite and gabbro.
含石榴石二辉橄榄岩中的石榴石为钙铁榴石,富 Ca 和 Fe,贫 Mg 和 Al(And95~97, Pyr 0.27~5.06,Gro0~2.62),为变质成因。

Extended Terms:

iron andradite 铁榴石
grossular-andradite 钙铝榴石—钙铁榴石

---

anglesite [ˈæŋgliˌsait] *n.*

Definition: (*Mineralogy*) a crystalline mineral form of lead sulfate, $PbSO_4$, formed by the weathering of galena 铅矾 qiān fán 又称硫酸铅矿。

Formula: $PbSO_4$

Origin: after the place Anglesey in Wales, in 1832

Example:

Based on summary and optimization of the conventional analyzing methods, new and more accurate methods were put forward for chemical phase analysis of lead in this paper. 20% NaCl took the place of 25% NaCl as the extractant of anglesite, greatly reducing the effect of background on the experiment results;
本研究在对传统分析方法进行总结、优化的基础上,提出了采用 20%NaCl 作为铅矾的浸取剂代替传统的 25%NaCl 的新方法,可大大减少背景的影响。

anhydrite [æn'haidrait] *n.*

Definition: a white mineral consisting of anhydrous calcium sulphate. It typically occurs in evaporite deposits.硬石膏 yìng shí gāo

Formula: $Ca[SO_4]$

Origin: early 19th century:from Greek anudros (See anhydrous) +-ite

Example:

The result shows, for this composite cement system, retarding action of dihydrate gypsum is more obvious than that ofanhydrite, meanwhile, anhydrite tends to cause setting time short and water needed increased.

结果表明:二水石膏对该种复合体系水泥的缓凝作用比硬石膏明显,硬石膏易引起复合体系水泥急凝和需水量增大。

Extended Terms:

maleic anhydrite 顺酐
anhydrite hydration 硬石膏水化
plastering anhydrite 粉刷石膏

ankerite ['æŋkərait] *n.*

Definition: (*Mineralogy*) any of a group of mixed carbonate minerals, principally of calcium with iron, magnesium and manganese 铁白云石 tiě bái yún shí

Formula: $Ca(Mg,Fe)[CO_3]_2$

Origin: after the Austrian mineralogist Mathias Joseph Anker (1771-1843) in 1825

Examples:

The veined ankerite is the dominant source of Mn, Cu, Ni, Pb, and Zn, which are as high as 0.09%, 74.0 μg/g, 33.6 μg/g, 185 μg/g, and 289 μg/g in this coal seam, respectively.

铁白云石是煤中 Mn、Cu、Ni、Pb 和 Zn 富集的主要原因,这 5 种微量元素的含量分别为 0. 09%、74.0μg/g、33.6μg/g、185μg/g 和 289μg/g。

Pb isotopic compositions for sulfides from the studied ores are comparable to those for feldspar Pb from the Himalayan alkaline rocks and 87 Sr/ 86 Sr ratios for calcite and ankerite are similar to or higher than those for the Himalayan alkaline rocks,indicating mantle derived Sr and Pb.

矿石中铅同位素组成与盆地中喜马拉雅期碱性岩长石中铅的同位素组成一致,方解石和铁白云石的 87Sr/86 Sr 比值接近或稍高于碱性岩的 87Sr/86 Sr 的比值,显示矿石锶、铅与碱性岩锶、铅是同源的,均来自于上地幔。

Extended Term:

ankerite alteration 铁白云石蚀变

annabergite [ˈænəˈbəːgait] *n.*

Definition: (*Mineralogy*) a mineral consisting of a hydrous nickel arsenate, chemical formula $Ni_3(AsO_4)_2 \cdot 8H_2O$, with an apple-green colour 镍华 niè huá

Formula: $Ni_3(AsO_4)_2 \cdot 8H_2O$

Origin: named after the locality, Annaberg-Buchholz in Germany

Examples:

The low solubility crystalline annabergite is synthesized by diluting under the conditions of $n(Ni)/n(As) = (1.5,)$ pH=7 and $\theta = 22$ ℃ for 2 h.

采用稀释法在 pH 值为 7,$n(Ni):n(As) = 1.5$,反应温度为 22℃,反应时间为 2 小时的条件下合成了具有较低溶解度的砷酸镍化合物。

The solubility of annabergite is characterized by the concentrations of Ni and As.

结晶型砷酸镍的溶解性用溶液中的 Ni 和 As 的浓度来表征。

Extended Term:

crystalline annabergite 结晶型砷酸镍

annite [əˈnit]

Definition: (*Mineralogy*) a phyllosilicate mineral related to biotite.羟铁云母 qiǎng tiě yún mǔ

Formula: $KFe_3^{2+}[(OH)_2|AlSi_3O_{10}]$

Origin: named after the locality at Rockport, Cape Ann, Essex Co., Massachusetts, USA

Example:

The Losevka pluton of rare-metal albite granite, which was explored as a possible source of columbite-zircon-malacon ore, is composed of quartz, sodic plagioclase, potassium feldspar, annite, protolithionite, lepidomelane, and Li-muscovite.

Losevka 的深层岩体是一种稀有的以金属覆盖的纳长花岗岩,它被发现能成为泥铁矿、锆石、水锆石等矿的资源。这种岩体含有石英、斜长石、钾长石、铁云母、黑鳞云母、铁黑云母、锂白云母。

Extended Term:

salitrite-annite jacupirangite 铁云霞霓辉岩

anorthite [ə'nɔːθait] *n.*

Definition: a calcium-rich mineral of the feldspar group, typically white, occurring in many basic igneous rocks.钙长石 gài cháng shí

Formula: $CaAlSi_3O_8$

Origin: mid 19th century: from an-+ Greek orthos "straight" +-ite

Examples:

K^+ and Ca^{2+} stem are mainly from the dissolution of K-feldspar and anorthite in plagioclase, respectively.

K^+主要来自钾长石的溶解,Ca^{2+}主要与斜长石中钙长石组分的溶解有关。

At first exothermal peak temperature of 820℃, it is found that a surface-induced crystallization is dominated and anorthite ($CaAl_2Si_2O_8$) as major phase and wollastonite ($CaSiO_3$) as minor phase are precipitated.

在第一放热峰温度(820℃)析晶时,玻璃样品为表面析晶,析出晶体为钙长石($CaAl_2Si_2O_8$)和硅灰石($CaSiO_3$)。

Extended Terms:

anorthite diabase 钙长辉绿岩
anorthite-basalt 钙长玄武岩
manganese-anorthite 锰钙长石

---

anorthoclase [æ'nɔːθəuˌkleiz] *n.*

Definition: (*Mineralogy*) any of a group of feldspars being mixed sodium and potassium aluminosilicates 歪长石 wāi cháng shí

Formula: $(Na,K)[AlSi_3O_8]$

Example:

Most of these xenoliths are spinel lherzolites with the assemblage of olivine + orthopyroxene + clinopyroxene + spinel. The pyroxene megacrysts are mostly diopside and augite in composition. Feldspar megacrysts are mainly anorthoclase and minor plagioclase.

包体主要是尖晶石二辉橄榄岩,造岩矿物为橄榄石+斜方辉石+单斜辉石+尖晶石,辉石巨晶主要为透辉石和普通辉石,长石巨晶主要是歪长石和少量斜长石。

---

anthophyllite [ˌænθə'filait] *n.*

Definition: (*Mineralogy*) a dark brown amphibole mineral; a mixed silicate of iron and magnesium 直闪石 zhí shǎn shí

Formula: $(Mg, Fe)_7Si_8O_{22}(OH)_2$

Origin: The name is derived from the Latin word anthophyllum, meaning "clove", an allusion to the most common color of the mineral.

Examples:

The metamorphism is reflected not only in the microscopic deformation of minerals and macroscopic structural deformation of the rock body, but also in the re-association and re-construction of minerals within the rock body, forming two assemblages of metamorphic minerals characterized by containing amphibolite (anthophyllite and tremolite), i.e. Ath-Ol-En and Tre-Ol-Di.
其变质作用不仅导致微观上矿物变形和宏观上岩体的构造变形,而且也引起岩体内部矿物重新组合和再造,形成以含角闪石(直闪石和透闪石)为特征的 Anth-Ol-En 和 Tre-Ol-Di 两个变质矿物共生组合。
The pre-exist amphibolite facies rocks in the fracture belt have been superimposed by retrograde green-schist facies with neoformation of chlorite, anthophyllite, sericite and cordierite.
断裂破碎带岩石在先存的五台期角闪岩相变质岩之上,叠加了绿片岩相动力退变质作用,普遍出现绿泥石、直闪石、绢云母、堇青石等新生的变质矿物。

Extended Term:

mangano-anthophyllite 锰直闪石

---

antigorite [ænti'gɔrait] *n.*

Definition: a mineral of the serpentine group, occurring typically as thin green plates 叶蛇纹石 yè shé wén shí

Formula: $Mg_3Si_2O_5(OH)_4$

Origin: mid 19th century: from Antigorio, a valley in Piedmont, Italy, +-ite

Example:

At 25℃, the coefficient of friction of chrysotile gouge is very low ($\mu \approx 0.2 \sim 0.25$), while lizardite and antigorite gouge are much stronger, with $\mu \approx 0.39$ and 0.45, respectively.
在室温下,纤蛇纹石具有很低的摩擦系数(0.2~0.25),而利蛇纹石和叶蛇纹石的摩擦系数较高,分别为 0.39 和 0.45 左右。

Extended Term:

iron antigorite 铁叶蛇纹石

---

antimonite ['æntəməˌnait] *n.*

Definition: (*Mineralogy*) a grey mineral, antimony sulfide (Sb_2S_3) that is the main source of antimony; stibnite 辉锑矿 huī tī kuàng

Formula: Sb_2S_3

Examples:

We used potassium antimonate to precipitate "exchangeable cellular Ca (superscript 2+)"-calcium that is sufficiently loosely bound to combine with antimonite, to investigate the feature of calcium distribution during anther development of Lycium barbarurn L.
在枸杞花药发育过程中,用焦锑酸钾沉淀的钙颗粒显示出了一个与花药发育事件有关的分布特征:在孢原细胞时期的花药中钙颗粒很少。
Gold is mainly found in pyrite and antimonite.
金主要存于黄铁矿和辉锑矿等硫化物中。

Extended Terms:

meta-antimonite 偏亚锑酸盐
sodium antimonite 焦锑酸钠
sodium antimonite containing glycerin 甘油亚锑酸钠

---

antimonselite [ˌænti'mɔnsəlait] *n.*

Definition: (*Mineralogy*) an orthorhombic sulfosalt mineral containing antimony and selenium 硒锑矿 xī tī kuàng

Formula: Sb_2Se_3

Origin: Contraction of antimony triselenide, with -ite

---

antimony ['æntiməni] *n.*

Definition: The chemical element of atomic number 51, a brittle silvery-white semimetal. Antimony was known from ancient times; the naturally occurring black sulphide was used as the cosmetic kohl. The element is used in alloys, usually with lead, such as pewter, type metal, and Britannia metal. 自然锑 zì rán tī

Formula: (Symbol: Sb)

Origin: late Middle English (denoting stibnite, the most common ore of the metal): from medieval Latin antimonium, of unknown origin. The current sense dates from the early 19th century.

Example:

Any of numerous silver-gray alloys of tin with various amounts of antimony, copper, and sometimes lead, used widely for fine kitchen utensils and tableware.

白镴,一种银灰色的锡合金,带有多种数量的锑、铜,而且有时还有铅,广泛地用于优质的厨房器皿和餐具。

Extended Terms:

antimony trioxide 锑盐
sodium antimony 锑酸钠
pentachloride antimony 五氯化锑

---

apatite [ˈæpətait] *n.*

Definition: A widely occurring pale green to purple mineral, consisting of calcium phosphate with some fluorine, chlorine, and other elements. It is used in the manufacture of fertilizers.磷灰石 lín huī shí

Formula: $Ca_5[PO_4]_3(F,Cl,OH)$,含 P_2O_5 40%~42%.

Origin: early 19th century:coined in German from Greek apatē "deceit" (from the mineral's diverse forms and colours)

Example:

a sedimentary rock consisting predominantly of apatite and other phosphates
主要由磷灰石和其他磷酸盐组成的沉积岩

Extended Terms:

fluoridated apatite 氟化磷灰石
apatite crystal 磷灰石结晶
hydroxy apatite 羟(基)磷灰石
carbonate-apatite 碳酸磷灰石
uran-apatite 铀磷灰石

---

apophyllite [əˈpɔfilait] *n.*

Definition: A mineral occurring typically as white glassy prisms, usually as a secondary mineral in volcanic rocks. It is a hydrated silicate and fluoride of calcium and potassium.鱼眼石 yú yǎn shí

Formula: $KCa_4[Si_4O_{10}]_2(F,OH) \cdot 8H_2O$

Origin: early 19th century:from apo-+ Greek phullon "leaf" +-ite

Example:

Chemical analysis gave: SiO_2 51.22, CaO 27.35, K_2O 4.36, Na_2O 0.16, F 1.86, H_2O~+15.32, and H_2O~-0.17(on average). By comparing it with other apophyllites elsewhere in

the world, the apophyllite studied in this paper is relatively high in Ca, F but low in Si and K.
其化学成分为 SiO_2 51.22, CaO 27.35, K_2O 4.63, Na_2O 0.16, F 1.86, H_2O~+15.32, H_2O~-0.17,(平均),与国内外鱼眼石相比,它富 Ca、F,贫 Si、K。

aragonite [ə'rægənait, 'ærəgənait] *n.*

Definition: a mineral consisting of calcium carbonate and typically occurring as colourless prisms in deposits from hot springs 文石 wén shí

Formula: $CaCO_3$

Origin: early 19th century: from the place named Aragon +-ite

Example:

XRD analysis showed that $CaCO_3$ whiskers consisted of aragonite and calcite, with the mass content of aragonite being 88.75%.
XRD 分析表明,所制得的晶须 $CaCO_3$ 由文石相和方解石相组成,文石相质量分数为 88.75%。

Extended Terms:

aragonite stalagmite 文石石笋
biogenic aragonite 生物文石
aragonite inclusion 文石包裹体

arfvedsonite ['ɑːfvidsʌnait] *n.*

Definition: (*Mineralogy*) a rare sodium amphibole mineral 钠铁闪石 nà tiě shǎn shí

Formula: $Na_3Fe_2+4Fe_3+[Si_4O_{11}]_2(OH)_2$

Origin: named for Swedish chemist Johan August Arfwedson (1792-1841)

Example:

The Guangtoushan alkaline granile occurs in the Precambrian metamorphic rocks of the North China block and contains the assemblege of quartz + alkaline feldspar + arfvedsonite + aegirine augite+aenigmatite±astrophyllite; accessory minerals are zircon, ilmenite, and chevkinite.
光头山碱性花岗岩产出在华北北部的前寒武纪基底变质岩系之中,造岩矿物组合为石英+碱性长石+纳铁闪石+霓辉石+钠铁非石±星叶石,副矿物有锆石、钛铁矿、硅钛铈铁矿等。

Extended Term:

magnesium rich arfvedsonite 富镁亚铁钠闪石

argentite ['ɑːdʒəntait] *n.*

Definition: (*Mineralogy*) silver sulphide, (Ag_2S), a primary ore of silver, in cubic or hexoctahedral crystals 辉银矿 huī yín kuàng

Formula: Ag_2S

Example:

A soft white precious univalent metallic element having the highest electrical and thermal conductivity of any metal; occurs in argentite and in free form; used in coins and jewelry and tableware and photography.
白色柔软的珍贵单价金属元素，对电和热传导性最高的金属，在辉银矿中以自由状态存在，用于制造硬币、珠宝和餐具，亦用于摄影。

---

arsenolite [ɑːˈsenəulait] *n.*

Definition: (*Mineralogy*) A mineral crystallizing in the isometric system and usually occurring as a white bloom or crust. Also known as arsenic bloom.砷华 shēn huá

Formula: As_2O_3

Examples:

Effects of arsenolite on leukotriene B_4 and genic expression of 5-lipoxygenase in asthmatic mice.
砒石对哮喘小鼠白三烯 B_4 及 5-脂氧合酶基因表达的影响。
AIM: To investigate the role of leukotriene B_4(LTB_4) and the expression of 5-lipoxygenase (5-LO) in ovalbumin-induced asthmatic mice and study the effect of arsenolite on LTB_4 and the expression of 5-LO in those mice as well as the mechanism of arsenolite inhibiting asthma.
目的：探讨白三烯 B_4(LTB_4)水平、5 脂氧合酶(5 LO)基因表达在小鼠哮喘发病中的作用以及砒石(arsenolite，As)对哮喘的治疗作用与 LTB_4水平及 5 LO 基因表达的关系。

---

arsenopyrite [ˌɑːsinəuˈpairait] *n.*

Definition: a silvery-grey mineral consisting of an arsenide and sulphide of iron and cobalt 毒砂 dú shā

Formula: FeAsS

Example:

a very poisonous metallic element that has three allotropic forms; arsenic and arsenic compounds are used as herbicides and insecticides and various alloys; found in arsenopyrite and orpiment and realgar.
一种极具毒性的金属元素，有三种同素异性体，砷和砷的化合物用作除草剂、杀虫剂和各种合金，见于毒砂、雌黄、雄黄中。

Extended Terms:

cobltian arsenopyrite 钴毒砂

mispickel (mispickel arsenopyrite)砷黄铁矿毒砂

asbestos [æsˈbestəs] *n.*

Definition: a highly heat-resistant fibrous silicate mineral that can be woven into fabrics, and is used in brake linings and other fire-resistant and insulating materials 石棉 shí mián

The asbestos minerals include chrysotile (white asbestos) and several kinds of amphibole, notably amosite (brown asbestos) and crocidolite (blue asbestos). The danger to health caused by breathing in highly carcinogenic asbestos particles has led to more stringent control of its use.

Origin: early 17th century, via Latin from Greek asbestos "unquenchable" (applied by Dioscurides to quicklime)

Example:

Asbestos can be used to insulate a cooking stove.

石棉能用来使烹调用的炉灶绝热。

Extended Terms:

asbestos fibers 石棉纤维

asbestos substitutes 代石棉

asbolite [ˈæsbəlait] *n.*, asbolane

Definition: (*Mineralogy*) A black, earthy mineral aggregate containing hydrated oxides of manganese and cobalt. Also known as asbolane; black cobalt; earthy cobalt.钴土 gǔ tǔ

Example:

This kind of cobalt ores is generally known as "asbolite".

这种类型的钴矿是通常所谓的"钴土矿"。

Extended Term:

asbolite ore 钴土矿

atacamite [ˌætəˈkæmait, ˌeiˈtækəmait] *n.*

Definition: (*Mineralogy*) a green copper(Ⅱ) chloride hydroxide mineral, polymorphous with botallackite 氯铜矿 lǜ tóng kuàng

Formula: $Cu_2Cl(OH)_3$

Origin: Atacama Desert, Chile, and -ite

Examples:

In our study atacamite[$Cu_2(OH)_3Cl$] was discovered in Xinjiang. The mineral is similar to the standard atacamite with respect to its chemical composition, crystal powder data, differential thermal analysis (DTA), infrared absorption spectral fcatures and physical properties, etc. This is the first discovery of natural atacamite in China.
在新疆的研究工作中,发现了氯铜矿(Atacamite)$Cu_2(OH)_3Cl$,其化学成分、粉晶数据、红外吸收光谱、差热分析和物理特征等与标准氯铜矿(智利)基本一致。这是中国首次发现天然氯铜矿。
Results showed that in neutral aquatic solution atacamite released more chlorine ions than paratacamite did.
结果表明,在中性条件下,氯铜矿比副氯铜矿溶解电离出更多氯离子。

---

augite [ˈɔːdʒait] *n.*

Definition: a dark green or black aluminosilicate mineral of the pyroxene group. It occurs in many igneous rocks, including basalt, gabbro, and dolerite.(普通)辉石 huī shí

Formula: $(Ca,Na)(Mg,Fe,Al)(Si,Al)_2O_6$

Origin: early 19th century: from Latin augites, denoting a precious stone (probably turquoise), from Greek augitēs, from augē "lustre"

Example:

The high-pressure metamorphic rocks in northeastern Jiangxi Province include such types of rocks as jadeite-bearing aegirine-augite albite amphibole schist, jadeite-bearing aegirine-augite quartz albitite, aegirine-augite-bearing amphibole quartz albitite, aegirine-augite-bearing albite amphibole schist, winchite quartz albitite, and magnesio-riebeckite quartz albitite.
赣东北高压变质岩包括含硬玉霓辉石钠长角闪片岩、含硬玉霓辉石石英钠长石岩、含霓辉石角长闪石英钠长石岩、含霓辉石钠长角闪片岩、蓝透闪石石英钠长石岩、镁钠闪石石英钠长石岩等岩石类型。

Extended Terms:

augite nepheline 霓
ferriferous augite 斜铁辉石
augite megacryst 普通辉石巨晶

---

autunite [ˈɔːtənait] *n.*

Definition: A yellow mineral occurring as square crystals which fluoresce in ultraviolet light. It is a hydrated phosphate of calcium and uranium.钙铀云母 gài yóu yún mǔ

Formula: $Ca(UO_2)_2(PO_4)_2 \cdot 10\text{-}12H_2O$

Origin: mid 19th century: from Autun, the name of a town in eastern France, +-ite

Example:

But an unexpected fact was noted: certain minerals (pitchblende, chalcolite, autunite) had a greater activity than might be expected on the basis of their uranium or thorium content.
但是一个意外的事实引起了我的注意:某些矿物质(沥青铀矿物质、铜铀云母或钙铀云母)的活性比根据其铀或针含量所判断的要高。

Extended Term:

manganese autunite 锰铀云母

---

aventurine [ə'ventʃərin] *n.*

Definition: a translucent mineral containing small reflective particles, especially quartz containing mica or iron compounds, or feldspar containing haematite 砂金石 shā jīn shí

Origin: early 18th century: from French, from Italian avventurino, from avventura "chance" (because of its accidental discovery)

Example:

aventurine spangled densely with fine gold-colored particles
布满闪烁的金色小颗粒的砂金石

Extended Terms:

aventurine glaze 砂金釉,金星釉
aventurine feldspar 日长石
copper aventurine 铜砂金石

---

axinite ['æksinait] *n.*

Definition: a borosilicate of aluminum, iron, and lime, commonly found in glassy, brown crystals with acute edges 斧石 fǔ shí

Formula: $H_2(Ca,Fe,Mn)_4(BO)Al_2(SiO_4)_5$

Example:

This paper deals with the surface micromorphology and chemical composition, together with formatiom condition of axinite. Chemical analysis illustrated that it belongs to ferroaxinite and its formatiom is under the condition of medium-high temperature at $fo_2 = 10^{21} - 10^{31}$.
本文通过对(个旧)斧石晶体的微形貌、化学成分及生成条件的研究认为,(个旧)产出的斧石种类属于铁斧石,它是在 $fo_2 = 10^{-21} - 10^{-31}$ 中偏高温溶液中生成的。

Extended Term:

mn fe axinite 锰铁斧石

azurite ['æʒərait] *n.*

Definition: A blue mineral consisting of basic copper carbonate. It occurs as blue prisms or crystal masses, typically with malachite.蓝铜矿 lán tóng kuàng

Formula: $Cu_3(CO_3)_2(OH)_2$,含 Cu 55.2%

Origin: early 19th century:from azure +-ite

Examples:

The results indicate that malachite [$CuCO_3 \cdot Cu(OH)_2$] is the main substance of green corrosion product; azurite [$2CuCO_3 \cdot Cu(OH)_2$] is the main substance of blue corrosion product and parata-camite [$Cu_2(OH)_3Cl$] is the main substance of slightly green corrosion product.

结果表明,(九连墩楚墓出土的)青铜器上主要的锈蚀产物为孔雀石[$CuCO_3 \cdot Cu(OH)_2$],存在部分蓝铜矿[$2CuCO_3 \cdot Cu(OH)_2$]和少许副氯铜矿[$Cu_2(OH)_3Cl$]。

There are corrosion products of the bronze on the jade's surface, such as malachite, azurite and hydrocerusite etc.;it shows that the mosaic jade is not only eroded by the soil,groundwater and other environmental factors but also by corrosion products of the bronze ware.

镶嵌玉表面覆盖的锈蚀产物分析表明,在长期的埋藏过程中,镶嵌玉除受到土壤、地下水等外部环境因素侵蚀外,还受到镶嵌本体——青铜器锈蚀产物侵蚀。

Extended Terms:

azurite ore 曾青,扁青

azurite-malachite mixture 蓝铜矿—孔雀石混合材料

baddeleyite ['bædəliˌait] *n.*

Definition: a mineral consisting largely of zirconium dioxide, ranging from colourless to yellow, brown, or black 斜锆石 xié gào shí

Formula: ZrO_2

Origin: late 19th century:named after Joseph Baddeley, English traveller, +-ite

Examples:

To obtain AZS refractories which consist of baddeleyite, corundum and mullite, the firing temperature should be higher than 1520℃, and to ensure the dense structure, the sintering temperature must be higher than 1600℃.
要获得矿相组成为斜锆石、刚玉、莫来石的 AZS 耐火材料,烧成温度必须大于 1520℃,要使结构致密,烧结温度必须大于 1600℃。

Besides the single-grain zircon and baddeleyite U-Pb method, the ^{40}Ar-^{89}Ar and mineral Sm-Nd isochron methods are applicable to low-grade terrains.
在低级区除了单颗粒锆石和斜锆石 U-Pb 法外,^{40}Ar-^{89}Ar 法和矿物 Sm-Nd 等时线方法是适用的。

Extended Terms:

baddeleyite inclusion 斜锆石
baddeleyite glassy phase 锆硅质玻璃相

---

banalsite [ˈbænəlˌsait] *n.*

Definition: a rare barium, sodium aluminium silicate mineral 钠钡长石 nà bèi cháng shí

Formula: $BaNa_2Al_4Si_4O_{16}$

Origin: The name is derived from the chemical symbols of its composition.

---

baotite *n.*

Definition: A rare mineral recognized as having a unique four-fold silicate ring. Crystals are tetragonal, though commonly deformed to the extent of appearing monoclinic. 包头矿 bāo tóu kuàng

Formula: $Ba_4Ti_4(Ti,Nb,Fe)_4(Si_4O_{12})O_{16}Cl$

Origin: Named for the locality of first discovery, Baotou, China, baotite has been found in hydrothermal veins and alkalic rocks in various locations around the world.

Example:

Study on Raman Spectrum Characteristics of Hsianghualite, Baotite, Huanghoite and Jixianite.
关于香花石包头矿黄河矿及蓟县矿的拉曼光谱特征的研究。

---

barite [ˈbɛərait] *n.*

Definition: a mineral consisting of barium sulfate, typically occurring as colorless prismatic crystals or thin white flakes 重晶石 zhòng jīng shí

Formula: $BaSO_4$

Origin: From Greek, which means "weight".

Example:

The methods available for analyzing δ17O and δ18O of sulfates are extremely laborious and demand high-purity BrF_5. Here the authors present a new method which generates O_2 directly from barite ($BaSO_4$) for simultaneous analysis of δ17O and δ18O by isotope ratio mass spectrometry (IRMS).

现有的分析硫酸盐的 δ17O 和 δ18O 的方法极其复杂,而且要求高纯的 BrF_5。文中报道了从重晶石($BaSO_4$)中直接产生 O_2 供同位素比值质谱计(IRMS)同时分析 δ17O 和 δ18O 的新方法。

Extended Terms:

barite glaze 钡釉
insoluble barite 不溶性重晶石
barite rosettes 重晶石玫瑰花结
Artificial Barite 硫酸钡,沉淀硫酸钡[造影剂]

---

bassanite [bə'sɑːnait] *n.*

Definition: (*Mineralogy*) a saline evaporite, consisting of calcium sulphate, found at Vesuvius 烧石膏 shāo shí gāo

Formula: $Ca_2[SO_4]_2 \cdot H_2O$

Example:

Presented in this paper are the infrared spectra of gypsum and its thermal transformation products—bassanite and anhydrite.

本文给出了石膏及其热转变产物——烧石膏和硬石膏的红外光谱。

---

bastite ['bæstait] *n.*

Definition: (*Mineralogy*) a hydrated magnesium silicate, a variety of serpentine occurring from the alteration of orthorhombic pyroxenes such as enstatite 绢石 juàn shí

Origin: name for pseudomorphs of serpentine group minerals after enstatite

Example:

Electron microprobe analyses show that enstatite, En85.8 to En90.8, alters to "bastite" composed only of lizardite (5.0-12.0 weight percent FeO), whereas olivine, Fo90.8 to Fo91.6, forms lizardite+chrysotile+brucite with or without magnetite.

电子探针分析发现,顽辉石,En85.8 到 En90.8,改变成绢石,构成只有 lizardite(5.0-12.0 质量百分数 FeO)，而有没有磁铁矿，橄榄石、Fo90.8 到 Fo91.6，都可以形成利蛇纹石+ 温石棉+水镁石。

Extended Term:

schiller spar 绢石

---

bastnaesite [ˈbæstnəˌsait] *n.*

Definition: a yellow to brown mineral consisting of a fluoride and carbonate of cerium and other rare earth metals 氟碳铈矿 fú tàn shì kuàng

Formula: $Ce[CO_3]F$，含 $Ce_2O_3$74.6%

Origin: late 19th century：from Bastnäs， the name of a district in Västmanland， Sweden， +-ite

Extended Terms:

bastnaesite concentrate 氟碳铈矿精矿
bastnaesite-monazite concentrate 氟碳铈矿—独居石混合精矿

---

bauxite [ˈbɔːksait] *n.*

Definition: An amorphous clayey rock that is the chief commercial ore of aluminium. It consists largely of hydrated alumina with variable proportions of iron oxides. 铝土矿 lǚ tǔ kuàng

Formula: form of rock consists mostly of the minerals gibbsite $Al(OH)_3$， boehmite γ-AlO (OH)， and diaspore α-AlO (OH)， in a mixture with the two iron oxides goethite and hematite， the clay mineral kaolinite， and small amounts of anatase TiO_2

Origin: mid 19th century：from French， from Les Baux (the name of a village near Arles in SE France， near which it was first found) +-ite

Examples:

Aluminum is made from bauxite.
铝是从铝土矿中提炼出的。
Sintering the mixture of low-grade bauxite from Guizhou， Bayer red mud， soda and lime， more economical effect can be obtained on condition that A/S = 2.84±，[N]/[A]+[F] = 0.97±， [C]/[S] = 2.11±， t℃ = 1275±.
采用贵州低品位铝土矿和溶出拜耳赤泥、碱、石灰配制的混合料烧结，在 A/S = 2.84±、[N]/[A]+[F] = 0.97±、[C]/[S] = 2.11±、t℃ = 1275±时，可以获得较好的效果。

Extended Terms:

bauxite brick 铝土砖

bauxite chamotte 高铝矾土熟料
bauxite clay 铝质黏土

benitoite [bəˈniːtəuait] *n.*

Definition: (*Mineralogy*) a rare hard blue silicate mineral found in hydrothermally altered serpentinite 蓝锥矿 lán zhuī kuàng

Formula: $BaTi(SiO_3)_3$

Origin: named after San Benito County, California, where it can be found, and -ite

bertrandite [ˈbəːtrənˌdait] *n.*

Definition: (*Mineralogy*) a basic beryllium sorosilicate that is an ore of beryllium 羟硅铍石 qiǎng guī pí shí

Formula: $Be_4Si_2O_7(OH)_2$

Origin: named after French mineralogist Émile Bertrand (1844-1909) in 1883, +-ite

Example:

Beryllium element occurs in sericitization-kaolinization syenogranite as bertrandite single ore, and distributes in sericite as seperated type occurrence.
铍元素以羟硅铍石单矿物赋存于绢云母化高岭土化正长花岗岩中,其次呈分散状态分布于绢云母中。

beryl [ˈberil] *n.*

Definition: a transparent pale green, blue, or yellow mineral consisting of a silicate of beryllium and aluminium, sometimes used as a gemstone 绿柱石 lǜ zhù shí

Formula: $Be_3Al_2Si_6O_{18}$

Origin: Middle English: from Old French beril, via Latin from Greek bērullos

Examples:

There are three isomorphous series of beryl in nature, that is "octahedral" beryl (c/a= 0.991-0.996). "tetrahedral" beryl (c/a=0.999-1.003) and "normal" beryl (c/a=0.997-0.998).
自然界绿柱石存在三个类质同象系列,即"八面体"绿柱石(c/a=0.991~0.996),"正常"绿柱石(c/a=0.997~0.998)和"四面体"绿柱石(c/a=0.999~1.003)。

Channel-Water Molecular Pattern and ^{1}H, ^{23}Na NMR Spectra Representation in Synthetic Red Beryl.

合成红色绿柱石中通道水分子构型及^{1}H和^{23}Na核磁共振谱表征。

Extended Terms:

golden beryl 金色宝石
irradiated beryl 辐照绿柱石
synthetic beryl 合成绿柱石

B

---

biotite [ˈbaiətait] *n.*

Definition: a black, dark brown, or greenish black micaceous mineral, occurring as a constituent of many igneous and metamorphic rocks 黑云母 hēi yún mǔ

Formula: $K(Mg,Fe)_3[AlSi_3O_{10}](OH,F)_2$

Origin: mid 19th century: named after J.-B. Biot (1774-1862), French mineralogist

Examples:

Melting Experiment of Biotite Gneiss-H_2O System at 0.1-0.2 GPa Pressures
黑云母片麻岩—H_2O系统在0.1—0.2GPa压力下的熔融实验
The Dangreyongcuo trachyte yields a biotite isochron age of 13.2±0.3Ma, and three sanidine isochron ages of 13.0±0.3Ma, 13.7±0.3Ma and 13.0±0.3Ma, respectively.
当若雍错粗面岩的黑云母等时线年龄为13.2±0.3Ma，3个透长石的等时线年龄分别为13.0±0.3Ma、13.7±0.3Ma和13.0±0.3Ma。

Extended Terms:

biotite hornblende 黑云角闪
biotite granodiorite 黑云母花岗闪长岩
biotite gneiss 黑云母花岗岩

---

bismite [ˈbizmait], bismuth ocher

Definition: (*Mineralogy*) a monoclinic mineral composed of bismuth trioxide; native bismuth ore, occurring as a yellow earth. Also known as bismuth ocher. 铋华 bì huá

Formula: Bi_2O_3

Origin: for Bismuth in its composition

Example:

The late-stage gold mineralization contains bismite (Bi_2O_3), fluorine-bearing bismite, native bismuth, bismuthinite (Bi_2S_3), and joseite [$Bi_4(Te,S)_3$], and also chlorite, epidote, prehnite, chalcopyrite, and sphalerite.
金矿化的后期包含铋华、铋的氟化物、本土铋、辉铋矿、硫锑铋矿，也包含绿泥石、绿帘石、葡萄石、黄铜矿、闪锌矿。

bismuth [ˈbizməθ] *n.*

Definition: The chemical element of atomic number 83, a brittle reddish-tinged grey metal. Bismuth is usually obtained as a by-product from the smelting of tin, lead, or copper. For a metal bismuth has low thermal and electrical conductivity. Its main use is in specialized low melting point alloys; some bismuth compounds have been used medicinally. 自然铋 zì rán bì

Formula: (Symbol: Bi)

Origin: mid-17th century: from modern Latin bisemutum, Latinization of German Wismut, of unknown origin

Examples:

Theoretical Studies on Sodium Clusters, Bismuth Nanotubes and Si(15,3,23) Surface with First-Principles Methods
第一原理方法对钠团簇铋纳米管和 Si(15,3,23)表面的理论研究
Structures, Ferroelectric and Dielectric Properties of Bismuth Layer-structured Compounds
铋层状化合物的结构与铁电、介电性能

Extended Terms:

bismuth bronze 铋青铜合金
bismuth pectin 果胶铋
bismuth subcarbonate 次碳酸铋

bismuthinite [ˌbizməˈθainait] *n.*

Definition: (*Mineralogy*) A mineral consisting of bismuth trisulfide, which has an orthorhombic structure and is usually found in fibrous or leafy masses that are lead gray with a yellowish tarnish and a metallic luster. Also known as bismuth glance. 辉铋矿 huī bì kuàng

Formula: Bi_2S_3

Examples:

The formation of sulfur is through the redox reaction of Fe^{3+} and H_2S forming by non-oxidative acid dissolution of bismuthinite.
元素硫是由辉铋矿经酸络合分解生成 H_2S,H_2S 再被 Fe^{3+} 氧化而形成的。
Based on the mineralogical study, thermodynamics analysis, and anode electrochemistry study, the formation and oxidation mechanism of sulfur in the process of slurry electrolysis on bismuthinite had been studied.
基于辉铋矿矿浆电解渣的工艺矿物学研究,通过有关热力学的分析和阳极电化学的研究,探

讨了辉铋矿矿浆电解过程元素硫的形成及氧化机理。

bismutite [ˈbizməˌtait] *n.*

Definition: (*Mineralogy*) A dull-white, yellowish, or gray, earthy, amorphous mineral consisting of basic bismuth carbonate. Also known as bismuth spar. 泡铋矿 pào bì kuàng

Formula: $(BiO)_2CO_3$, 含 Bi 87%

Origin: in allusion to its bismuth-rich composition

Example:

The new mineral shows monoclinic holohedral symmetry and forms idiomorphic crystals (≤0.3 mm) and crystal aggregates (≤3 mm) which are frequently grown on crusts of eulytite; associated minerals are bismutite and bismutoferrite.
新的矿物显示出对称性,在闪铋矿的表面形成晶体(≤0.3 mm)和结晶集合体(≤3 mm),相关的矿物是泡铋矿和硅铋铁矿。
Bismuth vanadate (microprobe test) in varying shades of orange color and in well developed crystals (averaging 0.2 mm in size) occurs in bismutite in the Mutala granite pegmatite area, district of Zambezia, Mozambique.
不同深浅的橘色的铋钒酸盐(探针测试),在发达的晶体(场均 0.2 毫米大小)出现在 Mutala 伟晶岩的花岗岩地区,赞比亚区、莫桑比克的泡铋矿中。

blue asbestos [æsˈbɛstəs; æz-]

Definition: a fibrous, lavender-blue or greenish mineral, a sodium iron silicate that is used as a commercial form of asbestos 蓝石棉 lán shí mián

Example:

Geological Characteristics of Blue Asbestos Deposit in Southwest Part of Henan Province
豫西南蓝石棉成矿地质的特征

Extended Term:

asbestos 石棉

boehmite [ˈbeimait, ˈbəu-] *n.*

Definition: a crystalline mineral compound composed of aluminum oxide and hydroxide and found in bauxite 软水铝石 ruǎn shuǐ lǚ shí

Formula: $AlO(OH)$

Origin: named for Johann Böhm, German chemist

Example:

An investigation on the thermodynamic properties of α-Al_2O_3, γ-Al_2O_3, gibbsite, boehmite and diaspore: standard enthalpy of formation, standard entropy, standard free energy of formation, and heat capacity was made based on calorimetric determinations and the published literatures. The reliability and consistency of the recommended values of these thermodynamic properties were discussed in detail.
结合作者的量热测定结果和前人的工作,研究了氧化铝工业中最常见的几种铝化合物如 α-Al_2O_3,γ-Al_2O_3,三水铝石一水软铝石和一水硬铝石的基本热力学性质为:标准生成焓、标准熵、标准生成自由能和热容,得到了可靠、自洽的结果。

Extended Terms:

boehmite sols 铝溶胶
pseudo boehmite 拟薄水铝石
very light boehmite 低密度薄水铝石

---

boracite [ˈbɔːrəˌsait] *n.*

Definition: (*Mineralogy*) a mixed chloride and borate of magnesium that occurs as a white to green crystalline evaporite, with the chemical formula $Mg_3B_7O_{13}Cl$ 方硼石 fāng péng shí

Formula: $Mg_3B_7O_{13}Cl$, 含 MgO 25.71%、$MgCl_2$ 12.14%、B_2O_3 62.15%

Origin: The name is obviously derived from its boron content (19% to 20% boron).

Example:

Considering the speciality of the principal reciprocal susceptibility in trigonal boracite, the experiment scheme on the measurement of the principal reciprocal susceptibility components is presented.
鉴于铁电三方相倒极化率的特殊性,提出了其介电测量的实验方案。

Extended Term:

iron ammonium sulfate bath boracite 铁方硼石

---

borate minerals [ˈbɔːreit]

Definition: (*Mineralogy*) any of the large and complex group of naturally occurring crystalline solids in which boron occurs in chemical combination with oxygen 硼酸盐矿物 péng suān yán kuàng wù

Examples:

Partial melting model can be simplified as follows: primary enrichment of elementboron & formation of Mg-bearing ultrabasic rocks → boron-bearing acid hydrotherm bymigmatizitization

→ Mg-bearing borate minerals via replacing Mg-bearing ultrabasic rocksor boron-bearing silicate minerals via replacing Al-bearing ultrabasic rocks.
归纳硼矿床的成矿模式为部分熔融成矿模式:硼元素初始富集+镁质岩石形成→混合岩化形成硼酸热液→交代镁质岩石形成镁硼酸盐矿物或交代铝质岩石形成硼硅酸盐矿物电气石。
In this paper Raman spectroscopy was used to study the molecular vibration of borate minerals.
本文利用激光拉曼光谱研究了不同类型的硼酸盐矿物,讨论了它们的谱带特征及其归属。

Extended Terms:

borate 硼酸盐
a salt in which the anion contains both boron and oxygen, as in borax magnesium-borate minerals 镁硼矿

---

borax [ˈbɔːræks] *n.*

Definition: a white compound which occurs as a mineral in some alkaline salt deposits and is used in making glass and ceramics, as a metallurgical flux, and as an antiseptic 硼砂 péng shā

Formula: $Na_2B_4O_7(OH)_4 \cdot 8H_2O$

Origin: late Middle English: from medieval Latin, from Arabic būrak, from Pahlavi būrak

Example:

The best craft parameters we got are: xanthan gum 2% of starch weight, H_2O_2 6%-10%, pH value 9-10, temperature 40-60℃, alkali 6%-8%, borax 2%-3%.
最佳的工艺参数为:黄原胶为淀粉重量的 2%,H_2O_2 为 6%~10%,体系的 pH 值为 9~10,温度为 40℃~60℃,碱量为 6%~8%,硼砂为 2%~3%等。

Extended Terms:

calcined borax 煅硼砂
borax carmine 硼砂胭脂红显技
borax buffer 硼砂缓冲

---

bornite [ˈbɔːnait] *n.*

Definition: a brittle reddish-brown crystalline mineral with an iridescent purple tarnish, consisting of a sulphide of copper and iron 斑铜矿 bān tóng kuàng

Formula: Cu_5FeS

Origin: early 19th century: from the name of Ignatius von Born (1742-1791), Austrian mineralogist, +-ite

Examples:

This paper studies experimentally the kinetics of dissolution of chalcocite, chalcopyrite and

bornite. Activation energy of reaction is 36.46 kJ/mol for chalcocite, 44.05 kJ/mol for chalcopyrite and 51.96 kJ/mol for bornite, respectively.
本文对辉铜矿、黄铜矿和斑铜矿在 NaCl 溶液中的溶解动力学进行了实验研究,在 25℃~70℃,溶解反应活化能分别为辉铜矿 36.46kJ/mol,黄铜矿为 44.05kJ/mol,斑铜矿为 51.96kJ/mol。
The new process of lead and copper separetion is suitable for treatment of lead copper ore in wich bornite is dominant copper mineral.
这种新的铅铜分离法适于处理铜矿物以斑铜矿为主的铅铜矿石。

Extended Term:

bornite craftwork 斑铜工艺品

---

boulangerite [buː'lændʒərait] *n.*

Definition: (*Mineralogy*) a lead antimony sulfide mineral that forms metallic grey monoclinic crystals 硫锑铅矿 liú tī qiān kuàng

Formula: $Pb_5Sb_4S_{11}$

Origin: named after Charles Boulanger, French mining engineer, and -ite

Example:

The quartz from the gold-bearing quartz veins of boulangerite sample formed during the later stages of gold mineratization in the Mayum gold deposit, Tibet was dated by using the $^{40}Ar/^{39}Ar$ fast-neutron activation technique, which gave a plateau age of(22.46±)(1.20 Ma). This plateau age represents the age of quartz(crystallization).
西藏马攸木金矿床金矿化晚阶段含金硫锑铅矿脉石英样品 $^{40}Ar/^{39}Ar$ 快中子活化法坪年龄为 22.46±1.20Ma,坪年龄代表石英的结晶年龄。
Selenian boulangerite, a selenian variety of boulangerite, occurs in the mineral veins at Songchong Sb-Pb deposit, South Anhui.
含硒硫锑铅矿是硫锑铅矿的含硒变种,产于皖南宋冲锑—铅矿点的矿脉中。

---

bournonite ['bɔːnənait] *n.*

Definition: (*Mineralogy*) Steel-gray to black orthorhombic crystals; mined as an ore of copper, lead, and antimony. Also known as berthonite; cogwheel ore. 车轮矿 chē lún kuàng

Formula: $PbCuSbS_3$

Origin: The name given by Bournon himself (in 1813) was endellione, since used in the form endellionite, after the locality in Cornwall where the mineral was first found.

Example:

The paper reports the morphology of cinnabar, realgar, orpiment, stibnite, bismuthinite, arsenopyrite, bournonite and manganotantalite. All of them are measured strictly by the classical V. Goldschmidt's method for the first time in China.
论文论述了辰砂、雄黄、雌黄、辉锑矿、辉铋矿、毒砂、车轮矿、钽锰矿 8 种矿物的晶体形貌,文中所提供的全部形态测量结果是按 V.Goldschmidt 方法严格测量的新资料,这在中国尚属首次。

---

braunite [ˈbraunait] *n.*

Definition: (*Mineralogy*) Brittle mineral that forms tetragonal crystals; commonly found as steel-gray or brown-black masses in the United States, Europe and South America; it is an ore of manganese.褐锰矿 hè měng kuàng

Formula: $3Mn_2O_3 \cdot MnSiO_3$

Origin: It was named after the alderman Braun of Gotha, Thuringia, Germany.

Example:

According to correct crystal orientation, geometric constant is consistent with the unit cell parameter, but as to scapolite and stolzite in the "Atlas of Mineral Sands" and braunite in the upper volume of the "Systematic Mineralogy", the crystal face (100) is consistent with the unit cell (110).
正确的晶体定向,其晶体几何常数应与内部结构中的晶胞参数一致,而在《砂矿物图册》中的方柱石和钨铅矿、《系统矿物学》上册中的褐锰矿,却是晶形的(100)与单位晶胞的(110)相同。
The other dominated by braunite, then montmorillonite, mica, pyrolusite, psilomelane, is usually excellent Mn Ore with low phosphorous content, its utillization has important significance to metallurgical industry.
另一种以褐锰矿为主,次为蒙脱石、云母、软锰矿、黑锰矿等,这种锰矿多为低磷优质锰矿,其开发利用对我国冶金工业有重要意义。

---

brittle [ˈbritl] mica [ˈmaikə]

Definition: (*Mineralogy*) Hydrous sodium, calcium, magnesium, and aluminum silicates; a group of more or less related minerals that resemble true micas but cleave to brittle flakes and contain calcium as the essential constituent.脆云母 cuì yún mǔ

Example:

Based on the microscopic observations on tectonite, quartz and mica in the fracture belt, it is

suggested that it appears to have undergone a process of deformation from ductile, brittle, ductile-brittle to brittle stage.
通过构造岩及石英、云母的微观研究，认为本断裂带经历了韧性—脆性—脆韧性—脆性的交替变换变形过程。

Extended Terms:

brittle 脆的；易碎的，脆弱的
mica 云母

---

brochantite [ˌbrɔtʃænˈtait] *n.*

Definition: (*Mineralogy*) A monoclinic copper mineral, emerald to dark green, commonly found with copper sulfide deposits; a minor copper ore. Also known as brochanite; brochanthite; warringtonite.羟胆矾 qiǎng dǎn fán

Formula: $Cu_4(SO_4)(OH)_6$

Example:

It is present as green foliated aggregates. $D_m = 3.20$.The main composition of tyrolite is: CuO 43. 51%, CaO 6.96%, As_2O_5 30.16%, SO_3 0.73%.
该矿物呈绿色片状集合体产出，略具柔性，实测密度 3. 20，主要化学成分 CuO43.51%，CaO_6 96%，As_2O_5 30. 16%，$SO_3$0. 73%。

---

bronzite [ˈbrɔnzait] *n.*

Definition: (*Mineralogy*) an orthopyroxene mineral that forms metallic green orthorhombic crystals; a form of the enstatite-hypersthene series 古铜辉石 gǔ tóng huī shí

Formula: $(Mg, Fe)(SiO_3)$

Example:

Mineral assemblages of these rocks are made up of chrysolite, bronzite, endiopside, pargasite, labradorite-bytownite and a small amounts of phlogopite. The associated spinel is rich in iron and aluminium and deficient in magnesium and chromium.
岩石矿物组合为贵橄榄石、古铜辉石、顽透辉石、韭闪石、拉—倍长石和少量金云母，附生尖晶石矿物富铁铝、贫镁铬。
A meteorite fall occurred around Dongkou village, Juancheng county of Shandong province, China at 23 o'clock on Feburary 15th, 1997. The meteorite consists mainly of olivine (FO = 82), bronzite and opaque minerals which include camacite and meteorite.
1997 年 2 月 15 日 23 时，一场陨石雨降落在中国山东省鄄城县董口乡。陨石的主要组成矿物为橄榄石(FO = 82)、古铜辉石和金属矿物。

B

brookite [ˈbruːkait, ˈbru-] *n.*

Definition: (*Mineralogy*) a dark brown mineral form of titanium dioxide, with the chemical formula TiO_2板钛矿 bǎn tài kuàng

Formula: TiO_2

Origin: named after Henry James Brooke (1771-1857), British mineralogist

Examples:

The anatase and rutile could enhance the Raman scattering signal of α-Fe_2O_3. It was difficult for brookite to observe the Raman scattering effect on α-Fe_2O_3 because the positions of some of its vibrational phonon modes overlapped that of α-Fe_2O_3.

锐钛矿型及金红石型 TiO_2 对 α-Fe_2O_3 有一定的拉曼增强作用，当水解 pH 值较小时，得到板钛矿型 TiO_2，由于部分板钛矿型 TiO_2 振动声子模的峰位与 α-Fe_2O_3 的重叠，所以尚难观察它对 α-Fe_2O_3 的拉曼增强效应。

The results of TG/DTA and XRD show that Cu^{2+}/TiO_2 powder dried at 70℃ is a mixed phase of anatase and brookite.

TG/DTA 和 XRD 分析的结果表明未经高温处理的 Cu^{2+}/TiO_2 粉体是锐钛矿和板钛矿的混合晶型。

Extended Term:

anatase-brookite 锐钛矿—板钛矿

---

brucite [ˈbruːsait] *n.*

Definition: a white, grey, or greenish mineral typically occurring in the form of tabular crystals. It consists of hydrated magnesium hydroxide.氢氧镁石 qīng yǎng měi shí

Formula: $Mg(OH)_2$

Origin: early 19th century: named after Archibald Bruce (1777-1818), American mineralogist, +-ite

Example:

Valleriite is a flaky mineral. It has been determind to consist of alternate interlayering of layers of two kinds (a brucite layer and a sulfide layer). Its particular structure has been named "hybrid structure".

墨铜矿为层状矿物，是由两种层(氢氧镁石层和硫化物层)交叉堆积而成，形成特殊的结构名为"杂化结构"。

The results indicate that nano-fibriform silica is porous and that the nano-fibriform structure is attributed to the complete dissolution of the brucite octahedral sheets of chrysotile and the collapse of some Si-O tetrahedral sheets.

分析表明纤蛇纹石中氢氧镁石八面体层被酸溶蚀和硅氧四面体层的塌陷，是导致这种白炭黑具有多孔纳米纤维结构的直接原因。

Extended Terms:

brucite powders 水镁石粉体
fibrous brucite 纤维水镁石

---

bytownite [bai'taunait] *n.*

Definition: a mineral present in many basic igneous rocks, consisting of a calcic plagioclase feldspar 培长石 péi chánɡ shí

Formula: $Ca_xNa_xAlSi_3O_8$

Origin: mid 19th century: from Bytown, the former name of Ottawa, Canada, +-ite

Example:

One is zoning plagioclases, which has a more strongly sericitized core of bytownite and a rim of oligoclase. Another is andesine without zoning structure, but with small changes in compositions.
(桐庐杂岩体中的两种斜长石)一种为环带构造发育、中心绢云母化较强的斜长石，其中心成分有的已达培长石，而边缘主要为更长石，另一种为环带构造不发育的斜长石，其成分变化较小，主要为中长石。

calaverite [ˌkælə'vɛərait] *n.*

Definition: (*Mineralogy*) a mineral form of gold telluride $AuTe_2$碲金矿 dì jīn kuànɡ

Formula: $AuTe_2$

Origin: calaver +-ite

Example:

It is closely associated with pyrite, nature gold, calaverite, oxygen-bearing gold minera et al and appeared as irregular grained aggregates, the cubic crystals(<0.05 mm) are rare, brownish black, metals luster.
其共生和伴生矿物有黄铁矿、自然金、碲金矿、含氧金矿物等。常呈不规则粒状集合体，偶见立方体微晶(粒径小于 0.05mm)，棕黑色，金属光泽。

Extended Term:

calaverite alteration 碲金矿蚀变

calcite [ˈkælsait] *n.*

Definition: A white or colourless mineral consisting of calcium carbonate. It is a major constituent of sedimentary rocks such as limestone, marble, and chalk, can occur in crystalline form (as in Iceland spar), and is deposited in caves to form stalactites and stalagmites.方解石 fāng jiě shí

Formula: $CaCO_3$

Origin: mid 19th century:coined in German from Latin calx, calc- "lime" (See calx)

Example:

The calculation confirms that the density of Ca^{2+} on the surface of apatite is 5.26/nm^2, while that on the surface of calcite is 4.96/nm^2, so the floatability of apatlte is better than that of calcite.

经计算,磷灰石表面 Ca^{2+} 密度为 5.26 个/nm^2,方解石为 4.96 个/nm^2,故磷灰石的可浮性高于方解石。

Extended Terms:

calcite dolomite 灰质白云岩

calcite grain 方解石粒

cancrinite [ˈkæŋkrinait] *n.*

Definition: (*Mineralogy*) any of a family of feldspathoid minerals that are mixed carbonates and aluminosilicates of sodium and calcium 钙霞石 gài xiá shí

Formula: $Na_3CaAl_3Si_3O_{12}CO_3(OH)_2$

Origin: Found originally in 1839 in the Ural Mountains, it is named after Georg von Cancrin, a Russian minister of finance.

Examples:

Wadeite is paragenic with aegirine, nepheline, arfredsonite, microcline and apatite, and associated with natrolite, rinkite and cancrinite.

钾钙板锆石与霓石、霞石、钠铁闪石、微斜长石、磷灰石共生; 与钠沸石、层硅铈钛矿、钙霞石等伴生。

According to the study of the complex intrusions in morphological occurrence, petrological characteristics, chemical composition, accessory minerals, mineral associations of Nb and rare earth, the distribution and variance regularity of minor elements, it is thought that the genesis of

the complex intrusions results from the differentiation, evolution and multiple emplacements of deep-seated magma, i.e., from basaltic magma→(to) brown amphibole pyroxenite→(to) cancrinite syenite porphyry→(to) calcitic carbonatite.
根据对杂岩体的形态产状、岩石特征、化学成分、副矿物、铌、稀土矿物组合、微量元素分布及变化规律等方面的研究，认为杂岩体的成因是涉及岩浆分异演化和多次侵位的结果。即由玄武岩浆→棕闪辉石岩→钙霞石正长斑岩→方解石碳酸岩演化而成。

Extended Term:

jadeite-cancrinite syenite 硬玉钙霞正长岩

---

carbon nanotube [ˈkɑːbən ˈnænəutjuːb]

Definition: any nanostructure, a member of the fullerene family, having graphene layers wrapped into perfect cylinders 碳纳米管 tàn nà mǐ guǎn

Examples:

The screen printed carbon nano tube field emission display device fabrication process has been set up and a 2.2 inch CNT-FED device has been obtained.
建立了一整套较为完善的碳纳米管场发射冷阴极印刷制备及场发射显示器件制备的工艺流程,并成功获得2.2英寸碳纳米管动态显示原理型样机。
The paper review four main kinds of nanoparticle doping technologies in liquid crystal materials, such as carbon nano tube, metal nanoparticles, metal oxide nanoparticles and ferroelectric particles.
本文介绍了四种主要的掺杂纳米粒子类型:碳纳米管、金属纳米粒子、金属氧化物纳米粒子和铁电性纳米粒子。
carbon 碳(符号:C) the chemical element of atomic number 6, a non-metal which has two main forms (diamond and graphite) and which also occurs in impure form in charcoal, soot, and coal(Symbol:C)
late 18th century:from French carbone, from Latin carbo, carbon-"coal, charcoal"
nanotube 毫微管,纳米管

---

carbonado [kɑːbəˈneidəu] *n.*

Definition: a dark opaque diamond, used in abrasives and cutting tools 黑金刚石 hēi jīn gāng shí

Origin: mid 19th century:from Portuguese

Examples:

Study on Insert Process of Carbonado
卡邦金刚石拉丝模镶套工艺研究
Study and Application of Pressure Transmitting Medium in Synthesizing Carbonado Diamond

人造卡邦金刚石合成用传压介质的研制及应用

Extended Terms:

carbonado diamond 卡邦金刚石

C

carbonate [ˈkɑːbəneit] minerals

Definition: (*Mineralogy*) a mineral containing considerable amounts of carbonates 碳质矿物 tàn zhì kuàng wù

Examples:

Discovery of New Polytypes of the B_mS_n in the Calcium-Ceyium Fluoro CarbonaTe Minerals and their Microstructural Study.

钙-铈氟碳酸盐矿物中 B_mS_n 型新多型的发现及其微结构研究。

The Thermodynamics Equilibrium Condition of Carbonate Minerals in Strata Environment and Its Application in Kela 2 Gas field, Kuqua Depression.

地层水条件下碳酸盐矿物热力学平衡条件及其在克拉 2 气田的应用。

Extended Terms:

manganese carbonate minerals 碳酸锰矿物

genesis of carbonate minerals 碳酸盐矿物成因

carnallite [ˈkɑːnəlait] *n.*

Definition: a white or reddish mineral consisting of a hydrated chloride of potassium and magnesium 光卤石 guāng lǔ shí

Formula: $KMgCl_3 \cdot 6H_2O$

Origin: Carnallite was first described in 1856 from its type location of Stassfurt Deposit, Saxony-Anhalt, Germany. It was named for the Prussian mining engineer, Rudolf von Carnall (1804-1874).

Examples:

The constructional characteristic and the spectroscopic variation of dehydration processes of bischofite ($MgCl_2 \cdot 6H_2O$), carnallite ($KMgCl_3 \cdot 6H_2O$) and ammonium-carnallite ($NH_4MgCl_3 \cdot 6H_2O$) have been studied by using continuous heating in-situ infrared spectroscopy.

用连续加热原位红外光谱法研究了水氯镁石($MgCl_2 \cdot 6H_2O$)、光卤石($KMgCl_3 \cdot 6H_2O$)和铵光卤石($NH_4MgCl_3 \cdot 6H_2O$)脱水过程的光谱变化和结构特征。

After using the collector the content of KCl in carnallite concentrate was improved to more 25%, the content of NaCl being decreased to less 3%, the relative recovery of KCl being about 97%, the recovery of NaCl being over 91%.

使用该药剂后富集的精光卤石中 KCl 含量提高到 25%以上，NaCl 含量降到 3%以下,氯化钾相对收率 97%左右,氯化钠捕收率达 91%以上。

Extended Terms:

carnallite concentrate 精光卤石
carnallite distribution 光卤石分布
ammonium carnallite 铵光卤石

---

carnegieite [ˈkaːneigiˌeit, kəˈniːgieit] *n.*

Definition: (*Mineralogy*) An artificial mineral similar to feldspar; it is triclinic at low temperatures, isometric at elevated temperatures. 三斜霞石 sān xié xiá shí

Formula: $NaAlSiO_4$

Example:

The behavior of potassium, sodium and fluorine in the sintering process was studied by the cutting sintering tests. It is shown that in the sintering process, microcline, aegirine and fluorite in the primary ores have transformed into the kaliophitite, carnegieite and cuspldine in sinters, respectively.
通过烧结中断试验,研究了包钢烧结过程中钾、钠、氟的行为:原矿中的微斜长石、霓石和萤石分别形成了钾霞石、三斜霞石和枪晶石,并存在于烧结矿中。

---

carpholite [ˈkaːfəlait] *n.*

Definition: (*Mineralogy*) a straw-yellow fibrous mineral consisting of a hydrous aluminum manganese silicate occurring in tufts; specific gravity is 2.93. 纤锰柱石 xiān měng zhù shí

Formula: $MnAl_2Si_2O_6(OH)_4$

Examples:

The assemblage chloritoid + carpholite + phengite in low-grademetamorphic rocks in Kaishantun area, Yanbian, Jilin Province—the evidence of blueschist facies metamorphism.
吉林延边开山屯地区蓝片岩相变质作用——来自硬绿泥石+纤锰柱石+多硅白云母组合的证据。
Besides barroisite, actinolite, chalcedony, zoisite, phengite, chlorite and albite, the low-grade metamorphic minerals also include the chloritoid + carpholite + phengite assemblages discovered in grayish black metapelites.
低级变质作用矿物除冻蓝闪石、阳起石、玉髓、黝帘石、多硅白云母、绿泥石、钠长石外,还在灰黑色变泥质岩中发现硬绿泥石+纤锰柱石+多硅白云母组合。

Extended Term:

chloritoid carpholite 硬绿泥石+纤锰柱石

C

cassiterite [kəˈsitərait] *n.*

Definition: a reddish, brownish, or yellowish mineral consisting of tin dioxide. It is the main ore of tin. 锡石 xī shí

Formula: SnO_2

Origin: mid 19th century: from Greek kassiteros "tin" +-ite

Example:

It has been found that the content of REE in cassiterite varies considerably from 298.6 ppm high to 1.90 ppm low.

经研究锡石的 REE 含量相差悬殊，高者达 298.6 ppm，低者仅 1.90 ppm。

Extended Terms:

cassiterite flotation 锡石浮选

hunan cassiterite 钖

cassiterite slime 锡石细泥

---

celestine [ˈselistin, -tain; siˈlestin] *n.*

Definition: (*Mineralogy*) a mineral with orthorhombic crystals, $SrSO_4$, colourless or white with blue and sometimes red shades 天青石 tiān qīng shí

Formula: $Sr[SO_4]$

Origin: A blue fibrous vein material from Bellwood, Blair Co., Pennsylvania, described in 1791, was the original celestite, the first discovery of this mineral, which is named for its color.

Example:

Hechuan celestine deposit of Sichuan is situated in 2km west of Yanjing village in Hechuan county.

四川合川天青石矿床位于合川县盐井镇之西 2 千米处。

Extended Terms:

celestine ore 天青石矿

celestine deposit 天青石矿床

---

celsian [ˈselsiən] *n.*

Definition: (*Mineralogy*) colorless, monoclinic mineral consisting of barium feldspar 钡长石 bèi cháng shí

Formula: $BaAl_2Si_2O_8$

Examples:

An Experimental Research on Artificial Synthetic Celsian.
人工合成钡长石的试验研究。
Study on Carbon-Fiber-Reinforced Celsian-matrix Composites.
C 纤维增强钡长石基复合材料的研究。

---

cerussite [ˈsiərəsait] *n.*

Definition: (*Mineralogy*) a mineral form of lead carbonate, $PbCO_3$, that is an ore of lead 白铅矿 bái qiān kuàng

Formula: $PbCO_3$,含 Pb 77.6%

Origin: From Latin cērussa (“white lead”).

Example:

The results indicate that the collecting ability of diphenyl α-(3-phenylthioureido) alkanephosphonate for cerussite is stronger than that for calcite and quartz.
结果表明,α-(3-苯基硫脲基)烃基膦酸二苯酯对白铅矿的捕收能力强,而对方解石和石英的捕收能力弱。

---

cervantite [səːˈvæntait] *n.*

Definition: (*Mineralogy*) a white or yellow secondary mineral crystallizing in the orthorhombic system and formed by oxidation of antimony sulfide 黄锑矿 huáng tī kuàng

Formula: Sb_2O_4

Origin: after the supposed locality at Cervantes, Province Lugo, Spain

Example:

Cervantite has a Mohs hardness of 4-5.
黄锑矿的莫氏硬度为 4 至 5。

---

chabazite [ˈtʃæbəzait] *n.*

Definition: a colourless, pink, or yellow zeolite mineral, typically occurring as rhombohedral crystals 菱沸石 líng fèi shí

Formula: $Ca[Al_2Si_4O_{12}]\cdot 6H_2O$

Origin: early 19th century: from French chabazie, from Greek khabazie, a misreading of khalazie, vocative form of khalazios "hailstone" (from khalaza "hail", because of its form and colour), +-ite

Examples:

(The temperature and pressure is the same as that of one group.) Laumontite, Chabazite and NaPl zeolite can be synthesized from coal fly ash that producethe hydrothermal synthesizing reaction by radiation heating with microwave oven, each sample only produce a kind of zeolite in the synthesized sample, its transforming rate vary 20% to 40%.
利用微波炉辐射加热从粉煤灰进行水热合成反应可得到浊沸石、菱沸石、NaP1 沸石三种沸石,合成品种中每个样品只生成一种沸石,其转化率为 20%~40%。
NaOH water solution is chosen as an original reactant by changing the parameters such as activation temperature, thickness of NaOH water solution and synthesis time. In addition, the fly ash can be heated up and directly crystallized, on the condition of hydrothermal alkaline activation, by using microwave. As a result, three kinds of zeolites are synthesized: laumontite, chabazite and NaP1 zeolites.
着重选用 NaOH 水溶液为反应前驱物,通过改变反应温度、NaOH 浓度与合成时间等参数,在水热条件下利用微波加热直接对粉煤灰进行晶化,合成得到了浊沸石、菱沸石、NaP1 沸石三种沸石。

---

chaidamuite [ˌtʃeidəˈmuːt] *n.*

Definition: a sulfate mineral that was first found in Xitieshan mine south of Mt. Qilianshan in the Chaidamu basin, Qinghai (Chinghai) Province, China. It is a hydrated sulfate containing a hydroxyl and four dihydrogen monoxide molecules. It is a secondary mineral possibly formed due to mining process.柴达木石 chái dá mù shí

Formula: $ZnFe_3+[SO_4]_2(OH) \cdot 4H_2O$

Origin: It is named from locality.

---

chalcanthite [kælˈkænθait] *n.*

Definition: (*Mineralogy*) a mineral form of copper sulfate, $CuSO_4$胆矾 dǎn fán

Formula: $CuSO_4 \cdot 5H_2O$

Origin: From its old Latin name chalcanthum "flowers of copper".

Examples:

two sulphates—brochantite, chalcanthite
硫酸盐二种:水胆矾、胆矾
the beginning of the Production of Copper by Chalcanthite

关于胆铜生产的起始

chalcedony [kæl'sedəni] *n.*

Definition: a microcrystalline type of quartz occurring in several different forms including onyx and agate 玉髓 yù suí

Formula: Silica (silicon dioxide, SiO_2)

Origin: late Middle English: from Latin calcedonius, chalcedonius (often believed to mean "stone of Chalcedon", but this is doubtful), from Greek khalkēdōn

Example:

The quartzites as the most important host rocks exhibit massive and irregular banded structures, brownish yellow/light gray blastobedded structures, and consist of quartz (75%-80%), kaolinite (10%-15%), sericite (2%-3%), ilmenite (3%-4%), chalcedony (2%-3%), and a small amount of carbonaceous components.

其中最重要的赋矿岩石为石英岩,其主要特征为具块状、不规则条带状构造,呈褐黄色/浅灰色变余层状构造。由石英(75%~80%)、高岭石(10%~15%)、绢云母(2%~3%)、钛铁质(3%~4%)、玉髓(2%~3%)及少量碳质组成。

Extended Terms:

chrome chalcedony 铬玉髓

chrysocolla chalcedony 硅孔雀石玉髓

blue chalcedony 蓝玉髓

chalchuite ['kɑːltʃuəit] *n.*

Definition: A pyroxene mineral with composition $NaAlSi_2O_6$. It is monoclinic. It has a Mohs hardness of about 6.5 to 7.0 depending on the composition. 翡翠 fěi cuì

Formula: $NaAlSi_2O_6$

Origin: English name jadeite, from Spanish plcdode jade.

Example:

Analysis of Jadeite Market and Management.

浅析翡翠市场与经营。

The Synthesis and Spectral Properties of Jadeite doped with Ce^{3+}, Pr^{3+}, Tb^{3+}.

掺杂 Ce^{3+}、Pr^{3+} 和 Tb^{3+} 离子人工翡翠宝石的合成和光谱性质。

Extended Term:

jadeite 翡翠

chalcocite ['kælkəsait]; chalcosine ['kælkəsiːn] *n.*

Definition: cuprous sulfide, an ore of copper, usually occurring as black, fine-grained masses. 辉铜矿 huī tóng kuàng

Formula: Cu_2S

Origin: The term chalcocite comes from the alteration of the obsolete name chalcosine, from the Greek khalkos, meaning copper. It is also known as redruthite, vitreous copper and copper-glance.

Examples:

This paper studies experimentally the kinetics of dissolution of chalcocite, chalcopyrite and bornite. Activation energy of reaction is 36.46 kJ/mol for chalcocite, 44.05 kJ/mol for chalcopyrite and 51.96 kJ/mol for bornite, respectively.
本文对辉铜矿、黄铜矿和斑铜矿在 NaCl 溶液中的溶解动力学进行了实验研究,溶解反应活化能分别为辉铜矿 36.46 kJ/mol,黄铜矿 44.05 kJ/mol,斑铜矿 51.96 kJ/mol。
The typical mineral assemblage of the basalts-hostedcopper ore is native copper + bitumen + quartz and that of the carbonoliths-hosted ore is native copper + carbon matter + zeolite + quartz (+ chalcocite).
以玄武岩为主岩的铜矿石典型矿物组合为自然铜 +沥青 +石英及不含沥青等有机质的自然铜 +石英 +绿帘石,以含炭沉积岩为主岩的铜矿石典型矿物组合为自然铜 +炭质物 +沸石 +石英(+辉铜矿)。

Extended Terms:

beta chalcocite 辉铜矿
alpha chalcocite/ blue chalcocite 蓝辉铜矿

---

chalcopyrite [ˌkælkə'paiərait] *n.*

Definition: a yellow crystalline mineral consisting of a sulphide of copper and iron. It is the principal ore of copper. 黄铜矿 huáng tóng kuàng

Formula: $CuFeS_2$

Origin: mid 19th century: from modern Latin chalcopyrites, from Greek khalkos "copper" + puritēs (see pyrites)

Example:

The ranges of 206 Pb/ 204 Pb, 207 Pb/ 204 Pb and 208 Pb/ 204 Pb ratios for chalcopyrite are between 18.4426 ~ 18.5909, 15.5762 ~ 15.6145 and 38.5569 ~ 38.8568, respectively.
矿石矿物黄铜矿的 206 Pb/204 Pb、207 Pb/204 Pb、208 Pb/204 Pb 分别为 18.4426 ~ 18.5909、15.5762 ~ 15.6145、38.5569 ~ 38.8568。
Electronic and magnetic properties of 3d transition metal-doped Ⅱ-Ⅳ-V_2 chalcopyrite

semiconductor
3d 过渡金属掺杂Ⅱ-Ⅳ-V_2黄铜矿半导体的电磁性质

Extended Terms:

chalcopyrite disease 黄铜矿
chalcopyrite compound 黄铜矿类化合物
chalcopyrite concentrate 黄铜矿精矿

---

chamosite ['ʃæməˌzait] *n.*

Definition: (*Mineralogy*) a greenish-gray or black mineral consisting of silicate belonging to the chlorite group and having monoclinic crystals; found in many oolitic iron ores. 鲕绿泥石 ér lǜ ní shí

Formula: $(Fe,Mg)_3(Fe^{2+},Fe^{3+})_3[AlSi_3O_{10}](OH_8)$

Examples:

Chamosite, strawberry pyrite, collophanite and most of calcites indicate marine environment.
鲕绿泥石,草莓状黄铁矿、胶磷矿及大部分方解石等反映海相环境。
According to their mineral composition, two kinds of green grains can be distinguished: glauconitic smectite and berthierine-chamosite.
依据其矿物组成,它们为海绿石相蒙皂石和磁绿泥石—鲕绿泥石两种绿色颗粒。

---

chert [tʃəːt] *n.*

Definition: a hard, dark, opaque rock composed of silica (chalcedony) with an amorphous or microscopically fine-grained texture. It occurs as nodules (flint) or, less often, in massive beds. 燧石 suì shí

Origin: late 17th century. (originally dialect): of unknown origin

Examples:

The SiO_2 content ranges from 37.61% to 80.84%, Fe_2O_3 from 0.55% to 13.01%, CaO from 0.28% to 12.70% with S content reaching 7.74%.
硅质岩中 SiO_2 的质量分数为 37.61% ~ 80.84%, Fe_2O_3 为 0.55% ~ 13.1%, CaO 为 0.28% ~ 12.70%, S 可达 7.74%.
Generally, the chert formed in a midoceanic ridge area is rich in Fe_2O_3 and MnO, but very poor in Al_2O_3.
一般形成于大洋中脊的硅质岩在岩石化学成分上富于 Fe_2O_3 和 MnO,而 Al_2O_3 含量最低。

Extended Terms:

chert member 硅质岩段

thermal chert 热水硅质岩
chert layer 硅质岩

---

chiastolite [kaiˈæstəˌlait] *n.*

Definition: a form of the mineral andalusite containing carbonaceous inclusions which cause some sections of the mineral to show the figure of a cross 空晶石 kōng jīng shí

Formula: Al_2SiO_5

Origin: early 19th century: from Greek khiastos "arranged crosswise" +-ite

---

chlorapatite [klɔːˈræpətait] *n.*

Definition: (*Mineralogy*) a variety of apatite containing chloride instead of fluoride 氯磷灰石 lǜ lín huī shí

Formula: $Ca_5(PO_4)_3Cl$

Examples:

Effect of apatite on the direct electrochemistry of Cyt c in this paper, a novel favorite electron transfer promoter of apatite was developed and the direct electrochemistry of Cyt c on the hydroxyapatite, fluorapatite and chlorapatite modified glassy carbon electrodes were investigated.
磷灰石对 Cyt c 直接电化学的影响首次以氟磷灰石、羟基磷灰石、氯磷灰石纳米生物活性陶瓷作为电子传递促进剂,并在此基础上研究了 Cyt c 在磷灰石修饰玻碳电极上的直接电化学及反应机理。

A series of fluoridated chlorapatite solid solution were synthesized by solid-state reaction method and were investigated by XRD and FTIR. The results indicate that unit-cell parameter a value decreased, but c value increased as F~-content (x) increased gradually. Both had linear relationship with x: $a=0.182x+9.555$.
本文对合成的一系列氟氯磷灰石固溶体进行 XRD 分析,结果发现:随 F-含量(x)的增加晶胞参数 a 减小、c 增大,二者呈线性关系:$a=0.182x+9.555$。

Extended Term:

fluor chlorapatite 氟氯磷灰石

---

chlorite [ˈklɔːrait] *n.*

Definition: a dark green mineral consisting of a basic hydrated aluminosilicate of magnesium and iron. It occurs as a constituent of many rocks, typically forming flat crystals resembling mica. 绿泥石 lǜ ní shí

Formula: $(Mg,Fe,Al)_6(Si,Al)_4O_{10}(OH)_8$

Origin: late 18th century: via Latin from Greek khlōritis, a green precious stone

Examples:

The chemical analysis suggests that the chlorite contains w(MgO) 35.4%, w(Al_2O_3) 14.8%, w(SiO_2) 32.6%, w(FeO) 4.2% and w(H_2O) 13.0%.
能谱结果表明,该绿泥石晶体的化学组成 w(MgO)为 35.4%,w(Al_2O_3)为 14.8%,w(SiO_2)为 32.6%,w(FeO)为 4.2%,w(H_2O)为 13.0%。

Lawsonite blueschist is composed of glaucophane (35%-40%) + lawsonite (35%-40%) + chlorite(10%) + albite(10%) + garnet(1%-2%) + zoisite(<2%) + quarte(<1%).
硬柱石蓝片岩的矿物组合为蓝闪石(35%~40%)+硬柱石(35%~40%)+绿泥石(10%)+钠长石(10%)+石榴石(1%~2%)+黝帘石/斜黝帘石(<2%)+石英(<1%)。

Extended Terms:

potassium chlorite 亚氯酸钾
chlorite bleach 亚氯酸盐漂白
sodium chlorite 亚氯酸钠

---

chloritoid ['klɔːritɔid] *n.*

Definition: A greenish-grey or black mineral resembling mica, found in metamorphosed clay sediments. It consists of a basic aluminosilicate of iron, often with magnesium. 硬绿泥石 yìng lǜ ní shí

Formula: $(Fe^{2+},Mg,Mn)_2Al_4Si_2O_{10}(OH)_4$

Examples:

Besides barroisite, actinolite, chalcedony, zoisite, phengite, chlorite and albite, the low-grade metamorphic minerals also include the chloritoid + carpholite + phengite assemblages discovered in grayish-black metapelites.
低级变质作用矿物除冻蓝闪石、阳起石、玉髓、黝帘石、多硅白云母、绿泥石、钠长石外,还在灰黑色变泥质岩中发现硬绿泥石+纤锰柱石+多硅白云母组合。

Metamorphic minerals include kyanite, corundum, topaz, paragonite, chloritoid and muscovite.
变质矿物有蓝晶石、蓝刚玉、黄玉、钠云母、硬绿泥石和白云母等。

Extended Terms:

chloritoid schist 硬绿泥石片岩
chloritoid carpholite 硬绿泥石+纤锰柱石

---

chondrodite ['kɔndrədait] *n.*

Definition: (*Mineralogy*) a nesosilicate mineral with chemical formula $(Mg,Fe)_5(F,OH)_2$

$(SiO_4)_2$, the most frequently encountered member of the humite group 粒硅镁石 lì guī měi shí

Formula: $(Mg,Fe)_5(F,OH)_2(SiO_4)_2$

Examples:

In large scale melting process, superheating often occurs, but chondrodite forms only in a small quantity.

大吨位的熔制,过热现象是普遍存在的,但不会形成大量粒硅镁石。

During the collision-subduction process between Yangtze plate and North China plate, the phase transition from spinel lherzolite to garnet lherzolite(>2.0GPa), olivine + orthopyroxene to Ti-chondrodite, Ti-clinohumite and magnesite(>3.0GPa) and from Al-rich pyroxene to Al-poor pyroxene and garnet took place(~5.0GPa).

在扬子板块与华北板块碰撞俯冲过程中,高压下发生尖晶石相向石榴石相的转变(>2.0GPa)、橄榄石和斜方辉石向钛粒硅镁石、钛斜硅镁石和菱镁矿的转变(>3.0GPa)和超高压下发生富铝辉石向贫铝辉石和石榴石的转变(~5.0GPa)。

Extended Term:

fluor chondrodite 氟粒硅镁石

---

chrome-diopside [krəum-dai'ɔpsaid] *n.*

Definition: (*Mineralogy*) a bright green variety of diopside containing a small amount of Cr_2O_3铬透辉石 gè tòu huī shí

Examples:

Five samples of chrome spinel, forsterite, chrome-diopside, orthorhombic and whole rock were picked out under microscope from the broken pieces, were analyzed in the Isochronology Center, Tianjin Institute of Geology and Mineral Resouces, Geological Survey of China. A Sm-Nd whole rock-single mineral isochron age of 2702±19 Ma was got.

包体经人工碎样,镜下挑选了含铬尖晶石、镁橄榄石、含铬透辉石、斜方辉石、全岩5个样品,由国土资源部天津地质矿产研究所同位素室测试,得出Sm-Nd全岩—单矿物等时线同位素年龄2702±19 Ma。

Extended Terms:

chrome 铬

diopside 透辉石

---

chrome-spinel [krəum-spi'nel] *n.*

Definition: (*Mineralogy*) A dark-brown variety of hercynite that contains chromium and is commonly found in dunites. Also known as chrome spinel. 铬尖晶石 gè jiān jīng shí

Examples:

Cr/(Cr + Al) of chrome-spinel also increases in the rock system, which shows an evolutional feature of Cr-rich and Al-poor.

铬尖晶石的 Cr/(Cr+Al)亦同时升高,表现为富 Cr 贫 Al 的演化特征。

The characteristics of magmatic inclusions are systematically studied. The chemical compositions of such crystal-phase daughter minerals olivine, orthopyroxene, clinopyroxene and accessory chrome-spinel and gas-phase volatile components such as F_2, Cl_2, CH_4, H_2S, H_2O, SO_2 and CO_2 were measured.

本文系统研究了岩浆包裹体特征,阐述了包裹体中橄榄石、斜方辉石、单斜辉石、铬尖晶石等晶相子矿物的化学成分以及 F_2、Cl_2,CH_4、H_2S、H_2O、SO_2、CO_2 等气相挥发组分。

Extended Term:

spinel 尖晶石

chromite ['krəumait] *n.*

Definition: a brownish-black mineral which consists of a mixed oxide of chromium and iron and is the principal ore of chromium 铬铁矿 gè tiě kuàng

Formula: $(Fe,Mg)Cr_2O_4$

Origin: mid-19th century: from chrome or chromium +-ite

Examples:

This method fit for the determination of ω(Au)/10~(-9) ≥0.1, ω(Pt)/10~(-9) ≥1.0, ω(Pd)/10~(-9) ≥0.1 of Copper-Nickel minera, Chromite and other rocks and minerals.

本法适合于铜镍矿、铬铁矿及其他岩石和矿物中 ω(Au)/10-9≥0.1, ω(Pt)/10-9≥1.0, ω(Pd)/10-9≥0.1 的测定。

The podiform chromite depoisit of aluminum type is referred to an ore deposit occurring in mantle peridotite of ophiolite of PTG lineage and having an ore with Al_2O_3 high (>20%) and Cr_2O_3 low (<45%).

富铝型豆荚状铬铁矿床系指产于 PTG 系列蛇绿岩套地幔橄榄岩中的矿石,以富铝(Al_2O_3>20%)、低铬(Cr_2O_3<45%)为特征的铬铁矿床。

Extended Terms:

magnesium chromite 镁铬尖晶石

chromite residues 铬矿渣

chromium ['krəumiːəm] *n.*

Definition: the chemical element of atomic number 24, a hard white metal used in stainless steel and other alloys 自然铬 zì rán gè

Formula: (Symbol:Cr)

Origin: early 19th century:from chrome +-ium

Example:

The trimmings on the car are made of chromium.
车上装饰品是用铬合金做的。

Extended Terms:

chromium-copper 铬铜合金
chromium-plating 镀铬
hard chromium 硬铬镀层

---

chrysoberyl [ˈkrisəberil] *n.*

Definition: A greenish or yellowish-green mineral consisting of an oxide of beryllium and aluminium. It occurs as tabular crystals, sometimes of gem quality. 金绿宝石 jīn lǜ bǎo shí

Formula: $BeAl_2O_4$

Origin: mid 17th century:from Latin chrysoberyllus, from Greek khrusos "gold" + bērullos "beryl".

Examples:

From SRXRF microprobe analyses, it was concluded that the major trace elements in chrysoberyl are Fe, Ga, Ti, Ca and Fe, Cr, Ti, Ca in alexandrite.
SRXRF 微探针分析结果表明:金绿宝石和变石中主要的微量元素组成分别是 Fe、Ga、Ti、Ca 和 Fe、Cr、Ti、Ca。
Half quantitative analysis of SRXRF experiments were carried out on one chrysoberyl and one alexandrite; the results indicated that in chrysoberyl Fe concentration averaged to be 17800×10^{-6}, Cr 855×10^{-6}.
金绿宝石和变石各一块样品的同步辐射 X 荧光能谱半定量分析表明金绿宝石中 Fe 含量平均为 17800×10^{-6},Cr 为 855×10^{-6}。

Extended Terms:

synthetic chrysoberyl 合成金绿宝石
star chrysoberyl 星光金绿宝石

---

chrysocolla [ˌkrisəˈkɔlə] *n.*

Definition: a greenish-blue mineral consisting of hydrated copper silicate, typically occurring as opaline crusts and masses 硅孔雀石 guī kǒng què shí

Formula: $Cu_4H_4[Si_4O_{10}](OH)_8 \cdot nH_2O$

Origin: late 16th century (in the Greek sense): from Latin, from Greek khrusokolla, denoting a mineral used in ancient times for soldering gold

Examples:

Characterization of the catalyst precursor by X-ray diffraction (XRD), Fourier transform infrared spectroscopy(FT-IR), and Thermogravimetry(TG) indicated the formation of a highly dispersed copper hydrosilicate with structural properties similar to the mineral chrysocolla.
通过 XRD、FT-IR、热重(TG)等分析,证实了催化剂前体是一种具有矿物硅孔雀石结构的物质,它具有较高的热稳定性和低温催化活性。
The activation of non-polar organic chelators for the flotation of chrysocolla was investigated by the flotation tests and the mordern analytical techniques.
用浮选试验和现代测试技术,研究了有机螯合剂对硅孔雀石的活化作用。

Extended Terms:

chrysocolla chalcedony 硅孔雀石玉髓
blue green chrysocolla opal 蓝绿色硅孔雀石欧泊

---

chrysolite [ˈkrisəlait] *n.*

Definition: a yellowish-green or brownish variety of olivine, used as a gemstone 贵橄榄石 guì gǎn lǎn shí

Origin: late Middle English: from Old French crisolite, from medieval Latin crisolitus, from Latin chrysolithus, based on Greek khrusos "gold" + lithos "stone"

Examples:

Electron microprobe analyses showed the olivine (Fo = 71 ~ 90) being chrysolite, and clinopyroxene mostly was attributed to diopside.
电子探针结果显示:橄榄石为富镁质橄榄石(贵橄榄石)(Fo=71~90),单斜辉石为透辉石(次透辉石为主)。
In the cndogenic process, both Ferrum and Titanium are along with each other, they trend towards corresponding gathering or dispersing. With crystallization and separation of magnesium silicate-minerals (forsterite and chrysolite, etc.) from ultrabasic or basic magma at the early period of crystallization differentiation, the content of Ferrum is relatively going higher in magma, i.e. F $((Fe_2O_3+FeO)/(Fe_2O_3+FeO+MgO))$ is getting higher.
二者(Fe、Ti)有对应聚、散的趋势,在超基性—基性岩浆结晶分异早期,随着富镁硅酸盐矿物(镁橄榄石、贵橄榄石等)的结晶分离作用,岩浆中铁的含量增高,即 F 值($(Fe_2O_3+FeO)/(Fe_2O_3+FeO+MgO)$(重量%))增大,钛含量亦逐渐增加。

Extended Terms:

italian chrysolite 符山石
oriental chrysolite 东方贵橄榄石
brazilian chrysolite 黄蓝碧硒

---

chrysotile [ˈkrisəˌtail] *n.*

Definition: a fibrous form of the mineral serpentine 纤蛇纹石 xiān shé wén shí

Formula: $Mg_3(Si_2O_5)(OH)_4$

Origin: mid 19th century: from Greek khrusos "gold" + tilos "fibre".

Examples:

XRD shows that antigorite is characterized by the lines d202 = 0.2522nm (I/I0 = 19) and d203 = 0.2430nm (I/I0 = 18), while chrysotile only has the line 0.2446nm (I/I0 = 29) without the line near d020 = 0.250nm.

X 射线粉晶衍射结果表明:叶蛇蚊石具有 d202 = 0.2522nm(I/I0 = 19)和 d203 = 0.2430nm(I/I0 = 18)的特征谱线,而纤蛇纹石则具有 d202、006 = 0.2446nm(I/I0 = 29)的特征谱线,d020 > 0.245nm 近 0.249 nm 的特征谱线缺失。

It is found that the samples of chrysotile asbestos reach their perfection at 160℃ with a pH value of 13.00. The length of chrysotile nanotubes is 200 nm-400 nm with the inner and outer diameters of 6 nm-8 nm and 30 nm-35 nm respectively.

在 160℃、pH 值为 13 的条件下合成晶型最好、纤维最长的纤蛇纹石纳米管,其内径 6~8 nm、外径 30~35nm,长 200~400 nm。

Extended Terms:

chrysotile asbestos 温石棉
synthetical chrysotile 合成纤蛇纹石

---

cinnabar [ˈsinəbɑː] *n.*

Definition: a bright red mineral consisting of mercury sulphide. It is the only important ore of mercury and is sometimes used as a pigment. 辰砂 chén shā

Formula: mercury(II) sulfide, HgS

Origin: Middle English: from Latin cinnabaris, from Greek kinnabari, of oriental origin.

Examples:

LREE is enriched relative to HREE and the value of Eu depletion is <0.30. The rock shows higher Sn, W, F, As, Sb, Pb, Zn, Cu and U anomalies and the placer minerals are

cassiterite, wolframite, antimony minerals and cinnabar, as well as lead, zinc and copper minerals.
轻稀土相对重稀土富集,Eu 亏损值小于 0.30。岩石中 Sn、W、F、As、Sb、Pb、Zn、Cu、U 等有高的综合异常和重砂有锡石、黑钨矿、锑矿物、辰砂及铅、锌、铜矿物异常。
Most of mercury deposits beared on the slope facies of the platform edge are middle-high grade ores, the considerable parts of the deposits are the associated deposits of the mercury and the elements else (i.e. Se, Ti, Au, Mo, N, Ni, As), their alteration of the host rock is principal in silicification, and the cinnabar frequently keep paragenesis with the carbonate asphalts in the ores, it shows there occurs the oil-generating progress in the earth history.
与台地边缘斜坡相有关的汞矿,多为中—高品位矿石,相当部分为汞与其他元素(如 Se、Ti、Au、Mo、Ni、As 等)伴生的矿床,围岩蚀变以硅化为主,矿石中辰砂常与碳沥青共生,碳沥青的存在,表明地质历史上确实有过生油过程。

Extended Terms:

cinnabar school 丹鼎派
artificiel cinnabar 灵砂
hepatic cinnabar 肝辰砂

---

citrine [ˈsitriːn] *n.*

Definition: a glassy yellow variety of quartz 黄水晶 huáng shuǐ jīng

Origin: late Middle English: from Old French citrin "lemon-coloured", from medieval Latin citrinus, from Latin citrus "citron tree".

Examples:

On the Growth of Citrine
茶色水晶的生长研究
Research on Liquid Bright Citrine Gold
亮柠檬金水的研制

Extended Terms:

citrine ointment 柠檬色软膏
hemerocallis citrine baroni 黄花菜
amethyst-citrine 紫黄晶

---

claudetite [ˈklaudətait] *n.*

Definition: a mineral composed of arsenic trioxide, chemical formula AS_2O_3 白砷石 bái shēn shí

Formula: AS_2O_3

Origin: Named after F. Claudet, French chemist, who first described the natural material.

clay mineral

Definition: any of a group of minerals which occur as colloidal crystals in clay. They are all hydrated aluminosilicates having layered crystal structures 黏土矿物 nián tǔ kuàng wù

Examples:

Clay mineral assemblages include 1.4 nm transitional minerals(25%-45%)+Illite(10%-20%)+ Illite-Smectite layer-mixed minerals (20%-35%) + Kaolinite (15%-30%), with a mineral paragenetic sequence of:Feldspar,Biotite→Vermiculite→(1.4 nm transitional minerals)→Illite →Kaolinite.

黏土矿物组合为 1.4nm 过渡矿物(25%~45%)+伊利石(10%~20%)+伊蒙混层矿物(20%~35%)+高岭石(15%~30%),矿物演化系列是长石、黑云母→蛭石→(1.4nm 过渡矿物)→伊利石→高岭石。

The tight sandstone gas reservoir of the upper Triassic was in the center of Sichuan basin,the clay mineral content reached as high as 15.3%,the average porosity was 4.2%, and the average permeability was $0.05\times10^{-3}\mu m^2$,50% throat radius below 0.25 μm.

川中地区上三叠系致密砂岩气藏,黏土矿物含量高达 15.3%,平均孔隙度为 4.2%,平均渗透率为 $0.05\times10^{-3}\mu m^2$,其中 50%的喉道半径在 0.25μm 以下。

Extended Terms:

pillared clay mineral 柱撑黏土矿物
clay mineral characteristics 黏土矿物学特征
clay vanadium mineral 黏土钒矿

cleavelandite [ˈkliːvlənˌdait] *n.*

Definition: (*Mineralogy*) a white, lamellar variety of albite that is almost pure $NaAlSi_3O_8$ and has a tabular habit, with individuals often showing mosaic developments and tending to occur in fan-shaped aggregates 叶钠长石 yè nà cháng shí

Formula: $NaAlSi_3O_8$

Origin: Named after Parker Cleaveland, a professor of geology and mineralogy at Bowdoin College in Maine during the early 1800s.

Examples:

Bazzite, $Be_3Sc_2Si_6O_{18}$, the scandium analogue of beryl, is a rare accessory mineral together with ixiolite and pyrochlore of the cleavelandite-amazonite pegmatites at Heftetjern, Tordal, Telemark, Norway.

硅钪矿,$Be_3Sc_2Si_6O_{18}$,与绿宝石相似的一种钪,是一种罕见的锰钽矿副矿物和烧绿石一起的结晶花岗岩,叶钠长石—绿长石存在于 Heftetjern,Tordal,Telemark,挪威。

Solid masses often show textured, flat surfaces which develop when formed between plates of cleavelandite albite.

坚实的包块经常表现出纹理,当板与板之间的形成发展钠长石时,呈现平坦的表面。

C

---

clinochlore [ˈklainəˌklɔː, klaiˈnɔklɔː] *n.*

Definition: (*Mineralogy*) green mineral of the chlorite group, occurring in monoclinic crystals, in folia or scales, or massive 斜绿泥石 xié lǜ ní shí

Formula: $(Mg,Fe,Al)_3(Si,Al)_2O_5(OH)_4$

Example:

The jade is composed of almost clinochlore, with a chemical formula $Mg_6(Fe_{0.7}Mg_{3.8}Al_{1.5})(Si_{6.2}Al_{1.8})O_{20}(OH)_{16}$.

该绿泥石玉基本由斜绿泥石组成,斜绿泥石的晶体化学式为:$Mg_6(Fe_{0.7}Mg_{3.8}Al_{1.5})(Si_{6.2}Al_{1.8})O_{20}(OH)_{16}$。

---

clinoenstatite [ˌklainəˈenztətait] *n.*

Definition: (*Mineralogy*) a form of pyroxene, consisting of magnesium silicate, found in some meteorites 斜顽辉石 xié wán huī shí

Formula: $Mg_2(Si_2O_6)$

Example:

Caloivm Aluminum Oxide and Clinoenstatite Syn as main crystalline phase with the addition of slag and glass powder more than 70 percent in minor ingredients.

该材料以镁方柱石、铝酸三钙、斜顽辉石为主晶相,配料中矿渣和玻璃粉用量高于70%。

---

clinoptilolite [klaiˈnɔptiləˌlait] *n.*

Definition: (*Mineralogy*) a zeolite mineral that is considered to be a potassium-rich variety of heulandite 斜发沸石 xié fā fèi shí

Formula: $(Na,K,Ca)_{2\text{-}3}Al_3(Al,Si)_2Si_{13}O_{36}\cdot 12H_2O$

Origin: The name is derived from the Greek words klino (κλίνω;"oblique"), ptylon (φτερών;"feather"), and lithos (λίθος;"stone").

Example:

The selectivity sequence of Ca^{2+} for the adsorbed cations in the specially prepared clinoptilolite is: Na-Cp> NH_4-Cp>K-Cp>Mg-Cp>Ba-Cp.
不同阳离子型的斜发沸石对 Ca^{2+} 选择性大小的顺序是：Na-Cp>NH_4-Cp>K-Cp>Mg-Cp>Ba-Cp.

Extended Terms:

native clinoptilolite 天然沸石
modified clinoptilolite 改性斜发沸石

---

clinopyroxene [ˌklainəˈpirəziːn] *n.*

Definition: a mineral of the pyroxene group crystallizing in the monoclinic system 单斜辉石 dān xié huī shí

Origin: early 20th century: from clino-in the sense "monoclinic" + pyroxene

Example:

The magnetite-amphibolite consists of amphibole (40%-45%), magnetite (35%-45%), plagioclase (10%-15%), and minor clinopyroxene and accessory minerals such as zircon and apatite.
磁铁角闪岩主要由角闪石(40%~45%)、磁铁矿(35%~45%)、斜长石(10%~15%)和次要的单斜辉石及副矿物锆石、磷灰石组成。

Extended Terms:

clinopyroxene megacryst 单斜辉石巨晶
clinopyroxene phenocryst 单斜辉石斑晶

---

clinozoisite [klinɔˈzɔizait] *n.*

Definition: (*Mineralogy*) a mineral found in crystalline schists, a metamorphic product of calcium feldspar 斜黝帘石 xié yǒu lián shí

Formula: $Ca_2Al_3(SiO_4)_3(OH)$

Example:

Mild pervasive hydrothermal alteration which affects much of the pluton comprises 3 distinct varieties: early propylitic alteration (chlorite-sericite ± clinozoisite, sphene) was concentrated above the NW-SE trending, residually warm pluton core, while later intermediate argillic alteration (kaolinite, illite-tourmaline) developed in the vicinity of roof contacts due to trapping and condensation of acid volatiles.
影响很大一部分岩体的弱弥漫型热液蚀变由三种有明显区别的类型组成：早期绿磐岩化

(绿泥石-绢云母±斜黝帘石、榍石)集中于北西—南东走向的、当时尚有余温的岩体核部上方,而其后的中性泥质蚀变(高岭石、伊利石—电气石)发育于顶板接触带附近,这是由于捕获和凝聚酸性挥发造成的。

---

cobaltite [ˈkəubɔːltait] *n.*

Definition: (*Mineralogy*) a rare, gray mineral, a mixed sulfide and arsenide of cobalt and iron with the chemical formula CoAsS; it is an ore of cobalt. 辉砷钴矿 huī shēn gǔ kuàng

Formula: CoAsS

Origin: The name is from the German, Kobold, "underground spirit" in allusion to the "refusal" of cobaltiferous ores to smelt as they are expected to.

Examples:

The ore-forming material sources of the Baiyangping copper-cobalt-silver polymetallic deposit have been studied by the characteristics of isotopic compositions of S, Pb, C, O, and H, and the Co/Ni ratio of cobaltite.
利用白秧坪矿床的 S,Pb,C,O,H 同位素组成特征和辉砷钴矿的 Co/Ni 比值特征,对成矿物质来源进行了分析。
The composition of and ratio of Co/Ni in cobaltite suggest cobalt ore-formation is related to basic-ultrabasic magma.
辉砷钴矿的成分和 Co/Ni 比值揭示钴的成矿作用与基性超基性岩浆有关。

Extended Terms:

strontium cobaltite 钴酸锶
cobaltite ceramics 辉钴矿型陶瓷
cobaltite spinel 钴尖晶石

---

coesite [ˈkəusait] *n.*

Definition: (*Mineralogy*) a high-pressure polymorph of silica found in extreme conditions such as the impact craters of meteorites, with the chemical composition of silicon dioxide, SiO_2 柯石英 kē shí yīng

Formula: SiO_2

Examples:

coesite + kyanite + rutile + apatite and coesite +kyanite + phengite+rutile in zircons from kyanite quartzites.
蓝晶石英岩为柯石英+蓝晶石+金红石+磷灰石、柯石英+蓝晶石+多硅白云母+金红石。
Free SiO_2 is represented by quartz, coesite and stishovite polymorphs.

自由 SiO_2 系指石英及其同质多型物柯石英、斯石英等。

Extended Terms:

coesite eclogite 柯石英榴辉岩
coesite pseudomorph 柯石英假象
coesite-bearing zircon 含柯石英锆石

C

colemanite [ˈkəulmənait] *n.*

Definition: a white crystalline mineral, typically occurring as glassy prisms, consisting of hydrated calcium borate 硬硼钙石 yìng péng gài shí

Formula: $Ca[B_3O_4(OH)_3] \cdot H_2O$

Origin: named after William T. Coleman (1824-1893) +-ite

Extended Term:

Turkey's colemanite 土耳其硬硼钙石

copper [ˈkɔpə] *n.*

Definition: a red-brown metal, the chemical element of atomic number 29 自然铜 zì rán tóng

Copper was the earliest metal to be used by humans, first by itself and then later alloyed with tin to form bronze. A ductile easily worked metal, it is a very good conductor of heat and electricity and is used especially for electrical wiring.

Formula: (Symbol: Cu)

Origin: Old English copor, (related to Dutch koper and German Kupfer), based on late Latin cuprum, from Latin cyprium as "Cyprus metal" (so named because Cyprus was the chief source).

Example:

Copper conducts electricity well.
铜是电的良导体。

Extended Terms:

aluminum copper 铝铜合金
chrome copper 铬铜合金

coquimbite [kəuˈkiːmˌbait] *n.*

Definition: (*Mineralogy*) a white mineral that crystallizes in the hexagonal system; it is dimorphous with paracoquimbite 针绿矾 zhēn lǜ fán

Formula: $Fe_2(SO_4)_3 \cdot 9H_2O$

Origin: Named after the province of Coquimbo, Chile.

---

cordierite [ˈkɔːdiərait] *n.*

Definition: a dark blue mineral occurring chiefly in metamorphic rocks. It consists of an aluminosilicate of magnesium and iron, and also occurs as a dichroic gem variety. 堇青石 jǐn qīng shí

Formula: $Al_3(Mg,Fe)_2[Si_5AlO_{18}]$

Origin: early 19th century: named after Pierre L. A. Cordier (1777-1861), French geologist, +-ite

Example:

A series of cobalt-based catalysts had been prepared with supports γ-Al_2O_3, OSM (oxygen storage materials, CeO_2-Y_2O_3-ZrO_2), YSZ-Al_2O_3 (γ-Al_2O_3 modified by Y_2O_3 and ZrO_2) and mixed YSZ-Al_2O_3+OSM using impregnation method and were coated on the surface of cordierite.

以 γ-Al_2O_3、储氧材料(OSM)、Y_2O_3,ZrO_2 来稳定的 γ-Al_2O_3(YSZ-Al_2O_3)及 YSZ-Al_2O_3+OSM 等为载体,以 Co_3O_4 为主要活性组分,Fe_2O_3 和 MnO_2 为助剂,堇青石蜂窝陶瓷为基体制备了整体式甲烷燃烧催化剂。

Extended Terms:

cordierite ceramics 堇青石陶瓷
cordierite honeycomb 堇青石蜂窝
synthesizing cordierite 合成堇青石

---

cordylite [ˈkɔːdiːlait] *n.*

Definition: 氟碳钡铈矿 fú tàn bèi shì kuàng

Formula: $(Ce,La)_2Ba(CO_3)_3F_2$,含 Ce_2O_3,La_2O_3 各约 25%

Origin: From the Greek κορδύ, club, in allusion to the shape of the crystals.

Example:

Syntactic intergrowth or syntaxy between cebaite and cordylite, as well as cordylite and huanghoite were observed.

观察了氟碳铈钡矿和氟碳钡铈矿以及氟碳钡铈矿和黄河矿的体衍交生关系。

---

corundum [kəˈrʌndəm] *n.*

Definition: extremely hard crystallized alumina, used as an abrasive. Ruby and sapphire are varieties of corundum. 刚玉 gāng yù

Formula: Al_2O_3

Origin: early 18th century: from Tamil kuruntam and Telugu kuruvindam

Example:

The results clearly showed that the basic oxide (Na_2O) was the main component of the corrosion, Na_2O in glass could decompose the mullite in brick to form α-Al_2O_3 and SiO_2, and Na_2O transformed the corundum into β-Al_2O_3. At high temperature, because of the volatiliy of alkali, β-Al_2O_3 progressively dissolved into the glass.

结果表明，碱性氧化物 Na_2O 仍然是侵蚀的主要成分，玻璃中 Na_2O 分解砖中莫来石，形成 α-Al_2O_3 和 SiO_2，Na_2O 把刚玉(α-Al_2O_3)转变成 β-Al_2O_3，β-Al_2O_3 在高温下由于碱的挥发，逐步溶解在玻璃中。

Extended Terms:

black corundum 黑刚玉

common corundum 普通刚玉

synthetic corundum 人造刚玉，合成刚玉

---

covelline [kəu'velain] *n.*

Definition: a native sulphide of copper, occuring in masses of a dark blue color-hence called indigo copper, another term for covellite 铜蓝 tóng lán

Origin: named after Covelli, the discoverer

Examples:

Research on Changes Principles of Iron Metabolism Indexes of Athletic Low Haemoglobin Serum of Rats, Covelline Albumen in Tissues, and Blood Serum Iron.

关于运动性低血色素大鼠血清、组织铜蓝蛋白及血清铁等铁代谢指标变化规律的研究。

The studies which draw on the impact of sports training on the level of covelline albumen in tissues, in particular, the influence of the sports training on the athletic low haemoglobin serum of rats and the changes of covelline albumen in various tissues are rarely seen in China.

运动训练对铜蓝蛋白的影响，尤其是运动性低血色素发生时血清及不同组织铜蓝蛋白的变化目前在中国的研究中尚未见报道。

Extended Term:

covelline albumen 铜蓝蛋白

---

cristobalite [kris'təubəlait] *n.*

Definition: a form of silica which is the main component of opal and also occurs as small octahedral crystals 方英石 fāng yīng shí

Formula: SiO_2

Origin: late 19th century: named after Cerro San Cristóbal in Mexico, where it was discovered, +-ite

Example:

By using $Na_2O \cdot mSiO_2$ as the raw material, and applying the deposition method and heat processing for 6h at 850℃ or 1350℃, quartz phase or cristobalite phase SiO_2 powders were obtained respectively.

以 $Na_2O \cdot mSiO_2$ 为原料,用沉淀的方法并经 850℃和 1350℃热处理 6 小时后,可以分别制得石英相和方石英相的 SiO_2 粉末。

Extended Terms:

cristobalite phase 方石英相
cristobalite crystallite 方石英晶体
cristobalite content 方石英含量

---

crocoite [ˈkrəukəuait] *n.*

Definition: a rare bright orange mineral consisting of lead chromate 铬赤铅 gè chì qiān

Formula: $Pb[CrO_4]$

Origin: mid 19th century: originally as French crocoise, from Greek krokoeis "saffron-coloured", from krokos "crocus". The spelling was altered to crocoisite, then crocoite.

---

cryolite [ˈkraiəulait] *n.*

Definition: a white or colourless mineral consisting of a fluoride of sodium and aluminium. It is added to bauxite as a flux in aluminium smelting. 冰晶石 bīng jīng shí

Formula: Na_3AlF_6

Origin: early 19th century: from cryo-"cold, frost" (because the main deposits are found in Greenland) +-ite

Examples:

When the molecular ratio of cryolite was 2.1, the content of ScF_3 9%, the solubility of Sc_2O_3 was more than 5%.

在冰晶石分子比为 2.1 、ScF_3含量为 39% 时,Sc_2O_3的溶解度可达 5% 以上。

Interaction between Y_2O_3 and cryolite has been suggested by considering various factors influencing the solubility.

通过考察影响溶解度的各种因素推断了 Y_2O_3 与冰晶石熔体相互作用的反应式。

Extended Terms:

cryolite compound 冰晶石类氟化物
molten cryolite 冰晶石熔体
potassium cryolite 钾冰晶石

---

cummingtonite [ˈkʌmiŋtəˌnait] *n.*

Definition: a mineral occurring typically as brownish fibrous crystals in some metamorphic rocks. It is a magnesium-rich iron silicate of the amphibole group. 镁铁闪石 měi tiě shǎn shí

Formula: $(Mg,Fe)_7Si_8O_{22}(OH)_2$

Origin: early 19th century: named after Cummington, a town in Massachusetts, US, +-ite

Extended Term:

cummingtonite-amphibolite 镁铁闪煌岩

---

cuprite [ˈkjuːprait] *n.*

Definition: a dark red or brownish black mineral consisting of cuprous oxide 赤铜矿 chì tóng kuàng

Formula: Cu_2O

Origin: Cuprite was first described in 1845 and the name derives from the Latin cuprum for its copper content.

Examples:

Study and analysis for the frescoes in caves of ancient China have shown that red pigments are mainly cinnabar, red iron oxide, iron-containing clay, red lead, red orpiment, zinc iron gahnite ($ZuFe_2O_4$), cuprite.
我国古代洞窟壁画的研究和分析结果表明,古代使用的红色颜料是朱砂、铁红和含铁黏土矿物颜料、红丹、雄黄、锌铁尖晶石($ZnFe_2O_4$)、赤铜矿.
The secondary oxide enrichment zone consisted mainly of cuprite and native copper was formed under weak oxidizing to reducing($Eh<0.16$) conditions, which explain why cuprite and native copper always appear in the middle to lower part of oxidized zone.
以赤铜矿、自然铜组合为主的次生氧化物富集带形成环境为弱氧化还原条件($Eh< 0.16$),它们常靠近氧化带的中、下部分布。

---

cyanite [ˈsaiənait], kyanite [ˈkaiənait] *n.*

Definition: a blue or green crystalline mineral consisting of aluminium silicate, used in heat-resistant ceramics 蓝晶石 lán jīng shí

Formula: Al_2SiO_5

Origin: late 18th century: from Greek kuanos, kuaneos "dark blue" +-ite

Examples:

The sintering performance of the samples composed of two different sizes cyanite and different amount of bauxite has been investigated at 1600℃ and 1650℃.

采用两种不同粒度的蓝晶石为主要原料,加入不同数量的轻烧高铝矾土,并分别于1600℃和1650℃下烧结,研究讨论所得试样的烧结性能。

Study on Processing Flowsheet and Multipurpose Utilization of Yinshan Cyanite Mine

隐山蓝晶石矿选矿工艺流程及综合利用研究

danburite [ˈdænberait] *n.*

Definition: (*Mineralogy*) a crystalline mineral similar to topaz 赛黄晶 sài huáng jīng

Formula: $Ca(B_2Si_2O_8)$

Origin: named after Danbury, Connecticut, USA, where it was first discovered, and -ite

Extended Term:

treatment of danburite 赛黄晶的优化处理

---

datolite [ˈdeitəlait] *n.*

Definition: (*Mineralogy*) a borosilicate of lime commonly occurring in glassy, greenish crystals, with the chemical formula $CaBSiO_4(OH)$ 硅硼钙石 guī péng gài shí

Formula: $CaBSiO_4(OH)$

Origin: from Greek to divide +-lite; in allusion to the granular structure of a massive variety

Example:

This paper studies overall the mineralogical features and formative condition of the prehnite and

datolite by the new methods of the chemical analysis, X-ray powder diffraction and infrared spectrum analysis, differential thermal analysis.
本文运用化学分析、X 射线粉末衍射、红外吸收光谱和差热分析等现代测试方法,对葡萄石与硅硼钙石矿物学特征及形成条件进行了全面研究。

Extended Terms:

acicular datolite 针状硅灰石
datolite short-fiber 硅灰石短纤维
superfine acicular datolite short-fiber 超细针状硅灰石短纤维

---

descloizite [deiˈklɔizait] *n.*

Definition: (*Mineralogy*) a rare orthorhombic mineral consisting of basic lead and zinc vanadate, isomorphous with olivenite 钒铅锌矿 fán qiān xīn kuàng

Formula: $(Pb,Zn)_2(OH)VO_4$

Origin: named after Alfred Des Cloizeaux (1817-1897), French mineralogist, and -ite

---

diallage [ˈdaiəlidʒ] *n.*

Definition: (*Mineralogy*) a green form of pyroxene 异剥石 yì bō shí

Example:

Hashatubei ultramafic complex, which located in the northwestern margin of North China craton, is characterized by a small exposed range and various petrographical faces, and dominated by serpentinizated peridotite, olivine websterite, diallage pyroxenolite, hornblende pyroxenolite, and less grabbro and quartz diorite porphyrite.
哈沙图北超镁铁杂岩体位于华北克拉通西北缘,具小规模多岩相的特征,主要由蛇纹石化橄榄岩、橄榄二辉岩、角闪辉石岩、异剥辉石岩及石英闪长玢岩等组成铁质超镁铁杂岩。

Extended Term:

green chloroplast diallage 绿异剥石

---

diamond [ˈdaiəmənd] *n.*

Definition: a precious stone consisting of a clear and colourless crystalline form of pure carbon, the hardest naturally occurring substance 钻石 zuàn shí
Diamonds occur in some igneous rock formations (kimberlite) and alluvial deposits. They are typically octahedral in shape but can be cut in many ways to enhance the internal reflection and refraction of light, producing jewels of sparkling brilliance. Diamonds are also used in cutting

tools and abrasives.

Formula: C

Origin: Middle English: from Old French diamant, from medieval Latin diamas, diamant-, variant of Latin adamans (See adamant)

Example:

Methods: The ODS-C_{18}(Diamond)(250 mm×4.6 mm, 5 μm) column was selected, the mobile phase was composed of acetonitile -1% acetic acid(20:80), the flow rate was 1.0 mL/min, detection wavelength was 322 nm, the injection volume was 10 μL.

方法采用 ODS-C_{18}(钻石)色谱柱(250mm×4.6mm,5μm),流动相为乙腈-1%冰醋酸(20:80),流速为 1.0mL/min,检测波长为 322nm,进样量为 10μL。

Extended Terms:

diamondite 赛金刚石合金
diamondoid 钻石形的
Irish diamond 水晶

---

diaspore ['daiəspɔː] *n.*

Definition: a natural hydrate of aluminium, sometimes forming stalactites 硬水铝石 yìng shuǐ lǚ shí

Formula: AlO(OH)

Origin: from the Greek διασπειρειν (to scatter), describing its decrepitaton on heating

Example:

The isoelectric points of diaspore, kaolinite, pyrophyllite and illite are at pH 6.0, 3.4, 2.3 and 3.2 respectively and the electrokinetic potential on their surface will negatively increase with the pH increase.

一水硬铝石、高岭石、叶蜡石及伊利石的等电点分别为 pH 6.0,3.4,2.3,3.2,随着矿浆 pH 值提高,这些矿物的表面动电位均呈负增加。

Extended Terms:

roasted diaspore 水硬铝石型铝土矿
out diaspore 水硬铝石

---

dickite ['dikait] *n.*

Definition: (*Mineralogy*) a mineral of the Kaolin group found crystallized in clay in hydrothermal veins; it is polymorphous with kaolinite and nacrite. 地开石 dì kāi shí

Formula: $Al_2Si_2O_5(OH)_4$

Example:

The data of hydrogen and oxygen isotopic analyses are reported in 7 dickite clay deposits of Zhejiang Province.
报道了浙江7个地开石黏土矿床的氢和氧同位素分析数据。

Extended Terms:

hard rock dickite 硬质地开石
kaolinite dickite mixed layer 高岭石—地开石混层矿物

---

diopside [dai'ɔpsaid] *n.*

Definition: a mineral occurring as white to pale green crystals in metamorphic and basic igneous rocks. It consists of a calcium and magnesium silicate of the pyroxene group, often also containing iron and chromium. 透辉石 tòu huī shí

Formula: $CaMg(Si_2O_6)$

Origin: early 19th century: from French, formed irregularly from di-"through" + Greek opsis "aspect", later interpreted as derived from Greek diopsis "a view through"

Example:

Pyroxenes are diopside and baicalite, enriched in Ca ((CaO) = 19.70% ~ 24.64%) and depleted in Na ((Na_2O) = 0.28 ~ (2.72)).
辉石为透辉石和次透辉石，富 Ca(CaO = 19.70% ~ 24.64%)，贫 Na(Na_2O = 0.28% ~ 2.72%)。

Extended Terms:

diopside glass 透辉石玻璃
diopside porcelain 透辉石质瓷
diopside-iadeite 玉质透辉石

---

dioptase [dai'ɔpteis] *n.*

Definition: a rare mineral occurring as emerald green or blue-green crystals. It consists of a hydrated silicate of copper. 透视石 tòu shì shí，绿铜矿 lǜ tóng kuàng

Formula: $CuSiO_2(OH)_2$

Origin: early 19th century: from French, formed irregularly from Greek dioptos "transparent"

Examples:

In addition, a band at 27,000 cm^{-1} have been also observed, it may be caused by the exchange

interaraction of adjacent Cu^{2+} ions in the dioptase structure.
此外,我们还观察到 27,000cm^{-1}处的一条吸收带,它可能是由透视石结构中邻近 Cu^{2+}离子的交换相互作用引起的。
The Room-temperature optical absorption spectrum of dioptase from Zaire and the theoretical analysis of its crystal-field are reported in this paper.
本文报告了扎伊尔透视石的室温光吸收谱及其晶体场理论分析结果。

---

dolomite [ˈdɔləmait] *n.*

Definition: a translucent mineral consisting of a carbonate of calcium and magnesium, usually also containing iron 白云石 bái yún shí

Formula: $CaMg(CO_3)_2$

Origin: late 18th century: from French, from the name of Dolomieu (1750-1801), the French geologist who discovered it, +-ite

Examples:

Forgrain dolostone of shoal facies the order degree of dolomite δ is 0.857, $MgCO_3$ = 50.57%, Sr/Ba = 0.206, less than 1. This type of dolomite is suggested to be product of mixed water dolomitization.
浅滩相的颗粒白云岩,其白云石有序度为 0.857,$MgCO_3$ = 50.574%,Sr/Ba = 0.206,小于 1,为混合水白云石化模式产物。
The results show that the δ 13C values of the calcite and dolomite vary in the same range of −3.5‰ to −7.3‰, being within the normal mantle δ 13C values of −5‰± 2‰.
结果表明,方解石和白云石的 δ 13C 值变化范围一致,均为−3.5‰~−7.3‰,落在正常地幔 δ 13C 值范围 −5‰±2‰内。

Extended Terms:

dolomite marble 粗粒白云石
friable dolomite 易碎白云岩
calcite dolomite 灰质白云岩

---

dravite [ˈdrɑːvait] *n.*

Definition: a complex crystalline silicate containing aluminum, boron, and other elements, used in electronic instrumentation and, especially in its green, clear, and blue varieties, as a gemstone 镁电气石 měi diàn qì shí

Origin: French, from Sinhalese toramalli, carnelian

Examples:

It is characterized by bright dark green, short prism crystal (0.5-0.8 cm), high transparency, high birefringence (0.021), strong pleochroism and 3.15 of specific gravity. Hence, the mineral is considered as dravite.
经鉴定为含镁电气石宝石。它具有鲜艳浓绿色,0.5～0.8cm 短柱状晶体,透明度高,双折射率大(0.021),二色性明显,比重为 3.15 等特征。
According to electronic microprobe and X-ray diffraction analysis results, it is iron-magnesium tourmalinite which belongs to transitional type between dravite and iron tourmalinite.
电子探针和 X 衍射结果表明,该电气石矿主要矿物成分为铁镁电气石,属镁电气石和铁电气石的过渡类型。

Extended Term:

dravite schorl 黑色电气石

---

dumortierite [djuːˈmɔːtiərait] *n.*

Definition: a rare blue or violet mineral occurring typically as needles and fibrous masses in gneiss and schist. It consists of an aluminium and iron borosilicate. 蓝线石 lán xiàn shí

Formula: $Al_8BSi_3O_{19}OH$

Origin: late 19th century: from the name of V.-E. Dumortier (1802-1876), French geologist, +-ite

Example:

Lanhuaxing and Ziluolan belong to the dumortierite andalusite pyrophyllite type.
蓝花星和紫罗兰为蓝线石、红柱石、叶蜡石型。

Extended Term:

dumortierite quartz 含蓝线石包裹体的石英

eastonite [ˈiːstəˌnait] *n.*

Definition: (*Mineralogy*) a mineral consisting of basic silicate of potassium, magnesium, and aluminum; it is an end member of the biotite system. 铁叶云母 tiě yè yún mǔ

Formula: $K_2Mg_5AlSi_5Al_3O_{20}(OH_4)$

Example:

In the peraluminous system, phlogopite is a solid solution (ss) of phlogopite, muscovite, talc and eastonite components.
在过铝质的系统里,金云母是一种固体的白云母、滑石和富镁黑云母的混合物。

---

edenite [ˈiːdəˌnait] *n.*

Definition: a variety of amphibole. See Amphibole. 浅闪石 qiǎn shǎn shí

Origin: from Edenville, New York

Example:

Cr-bearing edenite, just like emerald, is emerald-green, translucent to transparent, with high hardness, vitreous luster, well-formed crystal, and its single crystal size is large.
含 Cr 浅闪石呈翠绿色,半透明—透明,硬度较高,玻璃光泽,晶体完好,单个晶体较大,可用作宝石材料,外观极像祖母绿。

Extended Term:

Cr-bearing edenite 含 Cr 浅闪石

---

elbaite [elˈbeit] *n.*

Definition: (*Mineralogy*) a sodium-lithium-aluminium borosilicate mineral of the tourmaline group, prized as a gemstone 锂电气石 lǐ diàn qì shí

Formula: $Na(LiAl)_3Al_6Si_6O_{18}(BO_3)_3(OH)_4$

Origin: Elba +-ite, named after the island where it was first discovered.

Examples:

The Elbaite powder has higher Zeta potential than other species of tourmaline
锂电气石粉体的 Zeta 电位高于其他类型的电气石
The Action of Metallic Ions on Quartz, Elbaite and Calcite
金属阳离子对石英、锂电气石、方解石作用的研究

---

electrum [iˈlektrəm] *n.*

Definition: a natural or artificial alloy of gold with at least 20 percent of silver, used for jewellery, especially in ancient times(尤指古人用的天然或人造的)金银合金 jīn yín hé jīn

Origin: late Middle English: via Latin from Greek ēlektron "amber, electrum"

Examples:

$H = 161.9kg/mm^2$, $D = 7.1$. It usually intergrows with wittichenite, tetradymite, nessite, native

gold and electrum, and occurs in explosion breccia-type and quartz vein-type gold deposits formed at relatively high temperatures.
硬度为 161.9kg/mm^2，比重为 7.1。该针硫铋铅矿常和硫铋铜矿、辉碲铋矿、碲银矿、自然金、银金矿等密切共生，产在形成温度相对较高的爆破角砾岩型和石英脉型金矿中。
The major metal minerals of ore are arsenopyrite, pyrrhotine, electrum and native gold. Arsenopyrite and pyrrhotine are the main carrier minerals. The useful elements mainly are Ag, Zn, Pb, Cu, Sb.
矿石的主要金属矿物有毒砂、磁黄铁矿、银金矿和自然金。毒砂、磁黄铁矿是 Au 主要载体矿物，矿石有益元素主要有 Ag、Zn、Pb、Cu、Sb。

Extended Term:

electrum deposit 银金矿床

---

enargite [iˈnɑːdʒait] *n.*

Definition: a dark grey mineral consisting of a sulphide of copper and arsenic 硫砷铜矿 liú shēn tóng kuàng

Formula: Cu_3AsS_4

Origin: mid 19th century: from Greek energēs "clear" (referring to evident cleavage) +-ite

Example:

There are vertical zoning of mineralization, that is chalcocite in the upper, enargite at the elevation between 580 to 400m, and chalcopyrite below 400m.
矿化具垂直分带，上部以蓝辉铜矿为主，标高 580~400 米间以硫砷铜矿为主，400 米以下以黄铜矿为主。

---

enstatite [ˈenstətait] *n.*

Definition: a translucent crystalline mineral of varying colours that occurs in some igneous rocks and stony meteorites. It consists of magnesium silicate and is a member of the pyroxene group. 顽辉石 wán huī shí

Formula: $MgSiO_3$

Origin: mid 19th century: from Greek enstatēs "adversary" (because of its refractory nature) +-ite

Example:

The results show that with the increasing temperature magnesium aluminotitanate (MAT), β-quartz solid solution (β-QSS), sapphirine, spinel, α-cordierite, α-quartz, cristobalite, enstatite are precipitated in glass in sequence.

结果表明:随温度升高,玻璃中依次析出镁铝钛酸盐、β-石英固溶体、假蓝宝石、尖晶石、α-堇青石、α-石英、方石英、顽辉石等晶体。

Extended Terms:

chrome enstatite 铬顽火辉石
enstatite chondrite 顽火辉石球粒陨石

epidote ['epidəut] *n.*

Definition: a lustrous yellow-green crystalline mineral, common in metamorphic rocks. It consists of a basic, hydrated silicate of calcium, aluminium, and iron. 绿帘石 lǜ lián shí

Formula: $Ca_2(Al, Fe)_3(SiO_4)_3OH$

Origin: early 19th century: from French épidote, from Greek epididonai "give additionally" (because of the length of the crystals)

Example:

The overall mineral assemblages for these rocks are amphibole + plagioclase (An = : 0 ~ 91) + chlorite + epidote, accessory phases: ilmenite + titanite ± rutile ± zircon ± tourmaline.
主要矿物组合为角闪石+斜长石(An=0~91)+绿泥石+绿帘石,副矿物为:钛铁矿+榍石±金红石±锆石±电气石。

Extended Terms:

epidotization 绿帘石化作用
epidote quartz 绿帘石石英
manganese epidote 锰绿帘石

epsomite ['epsəmait] *n.*

Definition: (*Mineralogy*) a saline evaporite, consisting of magnesium sulphate, also found in fumaroles 泻利盐 xiè lì yán

Formula: $MgSO_4 \cdot 7H_2O$

Origin: named after the place Epsom in England in 1824

Example:

The evaporating experimental results show that, hydromagnesite is first separated with the lake-water being concentrated, and then halite, thenardite, bloedite, picromerite, epsomite, sylvite and carnallite precipitated successively.
蒸发实验的结果表明:首先析出水菱镁矿,随着湖水继续浓缩,进一步析出石盐、无水芒硝、白钠镁矾、软钾镁矾、泻利盐、钾石盐和光卤石。

erionite [ˈeriənait] *n.*

Definition: (*Mineralogy*) a zeolite mineral with a molecular structure similar to chabazite, usually found in weathered volcanic ash 毛沸石 máo fèi shí

Formula: $KNaCa[Al_2Si_6O_{16}]_2 \cdot 12H_2O$

Example:

Asbestos, erionite, wollastonite, sepiolite, palygorskite, sepiolite and brucite are classified as carcinogenic or potentially carcinogenic and bioactive fibrous minerals by IARC. This paper summarizes new advances in research on bioactivity of these fibrous minerals.

石棉、毛沸石、硅灰石、坡缕石、海泡石、水镁石是被国际癌症研究所(IARC)列为具有致癌、潜在致癌作用或具有较强生物活性的六种纤维矿物。本文综述了目前国内外对这几种纤维矿物生物活性的研究状况及最新进展。

Extended Term:

sythetic erionite 合成毛沸石

---

erythrine, erythrite [iˈriθrait] *n.*

Definition: a reddish secondary cobalt mineral, found in veins bearing cobalt and arsenic and used in coloring glass 钴华 gǔ huá

Formula: $CO_3(AsO_4)_2 \cdot 8H_2O$

Origin: Erythrite was first desctibed in 1832 for an occurrence in Grube Daniel, Schneeberg, Saxony and takes its name from the Greek ἔρυθρος (erythros), meaning red.

---

eschynite [ˈeskəˌnait] *n.*

Definition: (*Mineralogy*) a black mineral, occurring in prismatic crystals; a rare oxide of cesium, titanium, and other metals, which is isomorphous with priorite 易解石 yì jiě shí

Formula: $(Ce,Ca,Fe,Th)(Ti,Cb)_2O_6$

Extended Term:

gd-dy-eschynite 富钆镝易解石

---

euclase [ˈjuːkleis] *n.*

Definition: (*Mineralogy*) a monoclinic beryllium aluminium hydroxide silicate mineral, a product of the decomposition of beryl in pegmatites 蓝柱石 lán zhù shí

Formula: $BeAlSiO_4(OH)$

Origin: from the Greek for easy and fracture, due to good cleavage

Example:

Natural blue and colorless rare-gem mineral specimens of euclase from Brazil are investigated by electron paramagnetic resonance (EPR).
通过电子顺磁核磁共振(EPR)对来自巴西的自然的蓝色和无色稀有宝石矿物蓝柱石标本进行调查。

---

eudialyte [juːˈdaiəlait] *n.*

Definition: a mineral of a brownish red color and vitreous luster, consisting chiefly of the silicates of iron, zirconia, and lime 异性石 yì xìng shí

Formula: $(Na,Ca)_6ZrSi_6O_{17}(OH,Cl)_2$

Examples:

In addition to gillespite, eudialyte is another kind of mineral so far knowm, in which Fe^{2+} is square planar coordinated.
除硅铁钡矿之处,异性石是迄今仅知的另一种含有平面四配位 Fe^{2+} 的矿物。
The eudialyte sample, glossy and transparent in nossy, was found in a specimen of alkali granite.
异性石晶体样品为玫瑰色,透明,有油脂光泽,取自一花岗岩标本。

---

euxenite [ˈjuːksinait] *n.*

Definition: (*Mineralogy*) a dark brown lustrous mineral that is a mixed oxide of cerium, erbium, titanium, uranium, yttrium and other more common metals, with the chemical formula $(Y,Ca,Ce,U,Th)(Nb,Ta,Ti)_2O_6$黑稀金矿 hēi xī jīn kuàng

Formula: $(Y,Ca,Ce,U,Th)(Nb,Ta,Ti)_2O_6$

Origin: from Greek euksenos "friendly to strangers" or "hospitable"

Example:

The complex relations are expected to be involved in morphotropy and phase transition between aeschynite and euxenite, depending on the average ionic radii of Group-A atoms in $(Nb,Ti)_2O_8$ crystals.
结果表明,易解石相和黑稀金矿相为晶变和相转变的复杂关系,这种复杂关系受 A 组阳离子平均离子半径控制。

fayalite [faiˈɑːlait, fei-] *n.*

Definition: a black or brown mineral which is an iron-rich form of olivine and occurs in many igneous rocks 铁橄榄石 tiě gǎn lǎn shí

Formula: Fe_2SiO_4

Origin: mid 19th century: from Fayal (the name of an island in the Azores) +-ite

Example:

The Mossbauer spectrum of fayalite at 298K consists of one doublet, which is assigned to Fe^{2+} ions at M_1 and M_2 sites.

298K 温度下铁橄榄石的穆斯堡尔谱由一组双峰组成，它被指派给 Fe^{2+}(M_1, M_2)离子。

Extended Term:

mangan fayalite 锰铁橄榄石

---

feldspar [ˈfeldspɑː] *n.*

Definition: an abundant rock-forming mineral typically occurring as colourless or pale-coloured crystals and consisting of aluminosilicates of potassium, sodium, and calcium 长石 chánɡ shí

Formula: $KAlSi_3O_8$-$NaAlSi_3O_8$-$CaAl_2Si_2O_8$

Origin: mid-18th century: alteration of German Feldspat, Feldspath, from Feld "field" + Spat, Spath "spar" (See spar). The form felspar is by mistaken association with German Fels "rock".

Example:

From the heating experiments it is concluded that Fe^{3+}, Al^{3+} and Fe^{3+}, Si^{4+} exchange among the distinct T sites in the lattice of alkali feldspar is distinctly more sluggish than Al^{3+} and Si^{4+}.

实验表明，在碱性长石中不同四面体 T 点间 Fe^{3+}、Al^{3+} 和 Fe^{3+}、Si^{4+} 与 Al^{3+}、Si^{4+} 交换比较，前者速度要缓慢得多。

Extended Terms:

aventurine feldspar 日长石

barium feldspar 钡长石
calcium feldspar 钙长石
orthoclase feldspar 正长石
plagioclase feldspar 斜长石

feldspathoid(s) [feld'spæθɔid] *n.*

Definition: any of a group of minerals chemically similar to feldspar but containing less silica, such as nepheline and leucite 似长石 sì cháng shí

Origin: feld, (from Middle High German veld, from Old High German feld) + Spath, Spar + Oid, like or resembling

Example:

The author points out that only the igneous rocks fitting into one of the following conditions can be defined as the alkaline rocks:(1) containing feldspathoid or alkaline melaminerals…
笔者认为符合下列条件之一的火成岩才能定为碱性岩:(1)含似长石或碱性暗色矿物……

fergusonite ['fəːgəsəˌnait] *n.*

Definition: a mineral composed of oxides of various rare earth elements 褐钇铌矿 hè yǐ ní kuàng

Formula: $Y_2O_3 \cdot (Nb,Ta)_2O_5$

Examples:

The luminescence of fergusonites resulted from Er^{3+}, whereas that of brocenites (Fergusonite Ce) is due to the Er^{3+} and Eu^{3+}.
分析表明,该族矿物的发光中心是 Er^{3+} 和 Eu^{3+},其中褐钇铌矿的发光主要由 Er^{3+} 产生,而 Eu^{3+} 和 Er^{3+} 的共同发光构成褐铈铌矿的发光谱。
Fluorescence spectra in the Raman spectra of the sesquioxides and the annealing recrystallization fergusonite and brocenite have been studied when 488.0nm and 514.5nm laser lines are used as excitations.
研究了三价稀土氧化物及退火结晶褐钇铌矿和褐铈铌矿在488.0nm和514.5nm激光激发下所得Raman光谱中的荧光带。

Extended Terms:

fergusonite group 褐钇铌矿族
fergusonite group minerals 褐钇铌矿族矿物

ferrosilite [ˌferɔ'silait] *n.*

Definition: (*Mineralogy*) a mineral in the orthopyroxene group; the iron analog of enstatite; occurs in hypersthene, but is not found separately in nature. 铁辉石 tiě huī shí

Formula: $FeSiO_3$

Origin: originally a hypothetical mineral named by Henry Stevens Washington(1867-1934) in 1903. First used to name natural material in 1935 by Norman L. Bowen.

Example:

The incorporation of hydrogen into ferrosilite, Fe-bearing enstatite and orthopyroxene containing different trivalent cations (Cr^{3+} and Al^{3+}, Cr^{3+} and Fe^{3+}) was investigated experimentally.
在实际工作中,对氢到铁灰矿,含铁的顽辉石和斜方辉石包含不同的三价阳离子(三价铬和Al^{3+}、三价铬和Fe^{3+})进行了试验研究。

---

fersmite [ˈfəːzmait] *n.*

Definition: (*Mineralogy*) a black mineral composed of an oxide and fluoride of calcium and columbium with cerium and titanium 铌钙矿 ní gài kuàng

Formula: $(Ca,Ce)(Cb,Ti)_2(O,F)_6$

Examples:

The experimental result shows that the diphosphonic acid is a good collector of fersmite; its recovery can reach 83.27% when the pH value of the pulp is 2.5 ~ 5.0 and the dosage of diphosphonic acid is 20 mg/L.
试验结果表明双膦酸是铌钙矿的良好捕收剂,而且铌钙矿的回收率在双膦酸用量为20mg/L且矿浆pH值为2.5~5.0时,达到了83.27%。
And the XPS indicates that the binding energy of P2P peak of fersmite treated by diphosphonic acid has a change of 3.85eV, showing that the adsorption is mainly the chemical adsorption.
同时XPS检测结果表明经双膦酸处理后的铌钙矿的P2P峰位键合能变化了3.85eV,证明了该吸附主要为化学吸附。

---

fibrolite [ˈfaibrəlait] *n.*

Definition: another term for sillimanite. A usually white, hard mineral, occurring in highly metamorphosed rock as long, slender, fibrous crystals. Also called *fibrolite*. 硅线石 guī xiàn shí,发夕线石 fā xī xiàn shí

Formula: Al_2SiO_5

Origin: from its fibrous crystals

Example:

Contemporaneously with the coarsening of fibrolite to prismatic sillimanite, the myrmekite texture and albitic plagioclase rim occurred.

发夕线石粗粒化即进一步转化形成柱状夕线石的同时形成蠕英结构和斜长石生长边。

fluorapatite [fluə'ræpətait] *n.*

Definition: (*Mineralogy*) a calcium halophosphate mineral, in which fluoride replaces the hydroxide of apatite, that is mined as a phosphate ore, and occurs in the enamel of teeth. 氟磷灰石 fú lín huī shí

Formula: $Ca_5(PO_4)_3F$

Example:

The electron transfer rate constants ko of Cyt c at fluorapatite, hydroxyapatite and chlorapatite modified glassy electrodes were 3.0×10^{-6}, 3.9×10^{-6} and 4.2×10^{-6} cm · s^{-1} according to the Nicholson's equation, respectively.

由 Nicholson 方程求得 Cyt c 在氟磷灰石、羟基磷灰石、氯磷灰修饰玻碳电极表面电子传递速率常数 ko 分别为 3.0×10^{-6}、3.9×10^{-6}和 4.2×10^{-6} cm · s^{-1},

Extended Terms:

carbonate fluorapatite 碳氟磷灰石

hydroxyl-fluorapatite 天然氟-羟磷灰石

reduction of fluorapatite 氟磷灰石还原

fluorite ['flu(:)ərait] *n.*

Definition: a mineral consisting of calcium fluoride which typically occurs as cubic crystals, colourless when pure but often coloured by impurities 萤石 yíng shí

Formula: CaF_2

Origin: mid 19th century: from fluor (See fluorspar) +-ite

Example:

The XRD result shows that the CeO_2 powder doped with Sm_2O_3 and Gd_2O_3, calcined at 750℃, is cubic fluorite structure, implies that Sm_2O_3 and Gd_2O_3 dissolve into CeO_2 and forms CeO_2 based solid solution.

由粉末 XRD 分析可知,经 750℃焙烧的二元稀土掺杂 CeO_2 粉末为立方萤石结构,说明 Sm_2O_3 与 Gd_2O_3 已完全固溶到 CeO_2 中形成了 CeO_2 基固溶体。

Extended Terms:

fluorite granite 萤石花岗岩

fluor 氟石, 萤石

forsterite ['fɔːstərait] *n.*

Definition: a magnesium-rich variety of olivine, occurring as white, yellow, or green crystals 镁橄榄石 měi gǎn lǎn shí

Formula: Mg_2SiO_4

Origin: early 19th century: from the name of J. R. Forster (1729-1798), German naturalist, +-ite

Example:

Up to 700℃, the characteristic Raman bands 824 and 854 cm^{-1} are clearly observed, indicating the appearance of forsterite, along with the appearance of the overlapped bands 686 and 340 cm^{-1} of noncrystalline enstatite.

700℃时,镁橄榄石的拉曼特征峰 824cm^{-1} 和 854cm^{-1} 出现,同时出现非晶质顽火辉石 686cm^{-1} 和 341cm^{-1} 附近的包络峰。

Extended Terms:

forsterite ceramic 镁橄榄石瓷
forsterite ophicalcite 镁橄榄石大理岩
forsterite refractory 镁橄榄石耐火材料

---

franklinite ['fræŋklinait] *n.*

Definition: (*Mineralogy*) Black, slightly magnetic mineral member of the spinel group; usually possesses extensive substitution of divalent manganese and iron for the divalent zinc, and limited trivalent manganese for the trivalent iron. 锌铁尖晶石 xīn tiě jiān jīng shí

Formula: $ZnFe_2O_4$

Example:

Nanocrystalline $ZnFe_2O_4$ particles are prepared by thermal decomposed method. The morphology and structure of the particles are characterised by XRD and TEM. The $ZnFe_2O_4$ powder roasted at 300℃ for 1 h is proved to be spherical even nanocrystal particles with franklinite structure and an average diameter of 10nm.

研究了草酸盐热分解法制备 $ZnFe_2O_4$ 纳米粒子,并用 TRD 和 TEM 技术进行了初步表征。在 300℃焙烧 1 小时后,所制备的球形粒子粒径约 10nm,大小均匀。

---

fullerite ['fuləriːn], fullerence ['fulərəns] *n.*

Definition: any of various cagelike, hollow molecules composed of hexagonal and pentagonal groups of atoms, and especially those formed from carbon, that constitute the third form of carbon after diamond and graphite 富勒烯(C60 晶体)fù lè xī

Formula: C60

Origin: named after Richard Buckminster FULLER (from the resemblance of their configurations to his geodesic domes) +ite

Example:

It is found by FTIR, FABMS, ^{1}H-NMR and ^{13}C-NMR spectra that the water-soluble fullerite is multi-hydroxylate fullerite.
通过研究分析其红外吸收谱、快原子轰击电离质谱和^{1}H 及^{13}C 核磁共振谱,发现该衍生物为富勒醇。

gadolinite ['gædəlinait] *n.*

Definition: a rare dark brown or black mineral, consisting of a silicate of iron, beryllium, and rare earths 硅铍钇矿 guī pí yǐ kuàng

Formula: $(Ce,La,Nd,Y)_2FeBe_2Si_2O_{10}$

Origin: early 19th century: named after Johan Gadolin (1760-1852), the Finnish mineralogist who first identified it

Example:

The Mossbauer spectrum of gadolinite consists of one Fe^{3+} doublet and two Fe^{2+} doublets, and the multiple Fe^{2+} doublets were attributed to the next nearest neighbour effect.
结果表明,硅铍钇矿的穆斯堡尔谱是由一对 Fe^{3+} 双线和两对 Fe^{2+} 双线构成,其中的多重 Fe^{2+} 双线来源于次近邻效应。

---

galena [gə'liːnə] *n.*

Definition: a bluish, grey, or black mineral of metallic appearance, consisting of lead sulphide. It is the chief ore of lead. 方铅矿 fāng qiān kuàng

Formula: PbS

Origin: late 17th century: from Latin, lead ore (in a partly purified state)

Example:

The experiments on galena dissolution kinetics are performed in 1mol /L NaCl solutions from

pH 0.43 to 2.45 at 25~75℃.
在25~75℃、pH 0.43~2.45的1mol/L NaCl溶液中进行了方铅矿的溶解动力学实验。

Extended Terms:

argentiferous galena 银方铅矿
false galena 闪锌矿

---

garnet ['gaːnit] *n.*

Definition: a precious stone consisting of a deep red vitreous silicate mineral 石榴石 shí liú shí

Formula: $A_3B_2(SiO_4)_3$

Origin: Middle English: probably via Middle Dutch from Old French grenat, from medieval Latin granatus, perhaps from granatum (See pomegranate), because the garnet is similar in colour to the pulp of the fruit.

Example:

a deep red garnet consisting of iron aluminum silicate
一种深红色的石榴石,由铁铝硅酸盐矿物组成

Extended Terms:

crushed garnet 碎石榴石
iron garnet 铁榴石
precious garnet 贵榴石
syrian garnet 沙廉榴石

---

garnierite ['gaːniərait] *n.*

Definition: a bright green amorphous mineral consisting of a hydrated silicate of nickel and magnesium 硅镁镍矿 guī měi niè kuàng,镍纤蛇纹石 niè xiān shé wén shí

Formula: $Ni_4(Si_4O_{10})(OH)_4 \cdot 4H_2O$

Origin: named after Jules Garnier (1839-1904), French geologist

Example:

a hard malleable ductile silvery metallic element that is resistant to corrosion; used in alloys; occurs in pentlandite and smaltite and garnierite and millerite.
一种硬而软的银色金属元素,可锻造,抗腐蚀,用于合金中,见于硫镍铁矿、砷钴矿、硅镁镍矿、针镍矿中。

---

gehlenite [ˈgeihliˌnait] *n.*

Definition: (*Mineralogy*) a particular kind of sorosilicate; a mineral of the melilite group. Minerals of the group are solid solutions of several endmembers, the most important of which are gehlenite and akermanite. 钙铝黄长石 gài lǚ huáng cháng shí

Formula: $(CaNa)_2(AlMgFe^{2+})[(AlSi)SiO_7]$

Origin: The name derives from the Greek words meli (μέλι) "honey" and lithos (λίθους) "stone".

Example:

The hydrates of plain CAS glass are mainly gehlenite hydrate and CSH gel; and the preponderant hydrates or the mixture of CAS glass and gypsum are ettringite and CSH gel.
纯 CAS 玻璃相的水化产物主要为水化钙铝黄长石晶体和 CSH 凝胶;而掺有适量石膏的 CAS 玻璃相的水化产物则为钙矾石晶体和 CSH 凝胶。

Extended Terms:

ferri-gehlenite 铁黄长石
iron gehlenite 铁黄长石
manganese-gehlenite 锰钙黄长石

---

gersdorffite [ˈgəːzˌdɔːfait] *n.*

Definition: a nickel arsenic sulfide mineral with formula NiAsS. It crystallizes in the isometric system showing diploidal symmetry. It occurs as euhedral to massive opaque, metallic grey-black to silver white forms. 辉砷镍矿 huī shēn niè kuàng

Formula: NiAsS

Origin: named after Herr von Gersdorff, owner of Schladming Mine, Austria

---

gibbsite [ˈgibzait] *n.*

Definition: a colourless mineral consisting of hydrated aluminium hydroxide, occurring chiefly as a constituent of bauxite or in encrustations 三水铝石 sān shuǐ lǚ shí

Formula: $Al(OH)_3$

Origin: early 19th century: named after George Gibbs (1776-1833), American mineralogist, +-ite

Examples:

Studies on the Kinetics of Digestion Process of Synthetic Gibbsite by DSC
用 DSC 研究合成三水铝石溶出动力学

Prospects for Comprehensive Utilization of Gibbsite from Accumulated Bauxite in Western Guangxi Province
桂西堆积型铝土矿中三水铝石矿综合利用前景

Extended Term:

gibbsite layer 氢氧化铝八面体层

glauconite [ˈglɔːkənait] *n.*

Definition: a greenish clay mineral of the illite group, found chiefly in marine sands 海绿石砂 hǎi lǜ shí shā

Formula: $(K,Na)(Al,Fe,Mg)_2(Al,Si)_4O_{10}(OH)_2$

Origin: mid 19th century: from German Glaukonit, from Greek glaukon (neuter of glaukos "bluish-green") +-ite

Example:

A sand or sediment having a dark greenish color caused by the presence of glauconite.
海绿石砂是因含有海绿石而呈深绿色的砂子或沉积物。

Extended Term:

glauconite sandstone 海绿砂岩

glaucophane [ˈglɔːkəˌfein] *n.*

Definition: a bluish sodium—containing mineral of the amphibole group, found chiefly in schists and other metamorphic rocks 蓝闪石 lán shǎn shí

Formula: $Na_2Mg_3Al_2Si_8$

Origin: mid 19th century: from German Glaukophan, from Greek glaukos "bluish-green" +-phanēs "shining"

Example:

The Upper Archean Enclavement of Magnesioriebeckites-Bearing Gneiss in northern Liaoning Province: Old Glaucophane-Schist Relict
辽北晚太古宙含镁钠闪石的片麻岩包体:古老蓝闪片岩残留

Extended Terms:

glaucophane schist 蓝闪石片岩
glaucophane schist facies 蓝闪石片岩相
glaucophane schist series 蓝闪石片岩系列

goethite [ˈɡəuθait] *n.*

Definition: a dark or yellowish-brown mineral consisting of hydrated iron oxide, occurring typically as masses of fibrous crystals 针铁矿 zhēn tiě kuàng

Formula: α-FeO(OH)

Origin: early 19th century: from the name of J.W. von Goethe +-ite

Example:

At lower pH (pH<5.4 for soil, pH<5.4 for kaolinite, pH<5.5 for goethite, pH<6.4 for montmorillonite), BSM promoted Cd's adsorption on soil and soil claymiaerals and reduced the desorption ratio of corresponding adsorbed Cd.

在低 pH 值时(土壤 pH<5.4,高岭石 pH<5.4,针铁矿 pH<5.5,蒙脱石 pH<6.4),苄嘧磺隆对相应的吸附态镉的解吸起阻碍作用。

Extended Terms:

gothite 针铁矿
goethite growth 铁黄生长
acicular goethite 针状铁黄

---

gold [ɡəuld] *n.*

Definition: a yellow precious metal, the chemical element of atomic number 79, valued especially for use in jewellery and decoration, and to guarantee the value of currencies 自然金 zì rán jīn

Gold is quite widely distributed in nature but economical extraction is only possible from deposits of the native metal or sulphide ores, or as a by-product of copper and lead mining. The use of the metal in coins is now limited, but it is also used in electrical contacts and (in some countries) as a filling for teeth.

Formula: Au

Origin: Old English, of Germanic origin; related to Dutch goud and German gold, from an Indo-European root shared by yellow

Example:

an area containing abundant deposits of gold or gold ore

蕴藏大量黄金或金矿的地区

Extended Terms:

gold-bearing 含金的,产金的
goldbeating 制金箔

golddust 砂金，金泥，金粉
goldfield 金矿区，黄金产地

goldmanite [ˈgəudmənait] *n.*

Definition: vanadium rich clay found in a metamorphosed uranium-vanadium deposit.钙钒石榴子石 gài fán shí liú zǐ shí

Origin: The name honors Macus I. Goldman, a sedimentary petrologist with the U. S. Geological Survey.

graphite [ˈgræfait] *n.*

Definition: a grey crystalline allotropic form of carbon which occurs as a mineral in some rocks and can be made from coke. It is used as a solid lubricant, in pencils, and as a moderator in nuclear reactors. 石墨 shí mò

Formula: C

Origin: late 18th century: coined in German (Graphit), from Greek graphein "write" (because of its use as pencil "lead")

Examples:

Research on the Polypropylene/Nylon 66/Grafted Polypropylene/Graphite Binary and Multicomponent Electrically Conductive Nanocomposites
聚丙烯/尼龙 66/接枝聚丙烯/石墨二元及多元导电纳米复合材料的研究
Studies on Solide State Shear Compounding Technology & Preparation and Properties of the Electric and Thermal Conductive Polypropylene/Graphite Nanocomposites
磨盘碾磨固相剪切复合技术及导电导热聚丙烯/石墨纳米复合材料的制备与性能研究

Extended Terms:

blocky graphite 块状石墨
globular graphite 球状石墨
native graphite 天然石墨
eutectic graphite 共晶石墨
white graphite 白石墨，氮化硼
black lead 石墨,笔铅

greenockite [ˈgriːnəˌkait] *n.*

Definition: a mineral consisting of cadmium sulphide which typically occurs as a yellow crust on zinc ores 硫镉矿 liú gé kuàng

Formula: CdS

Origin: mid 19th century: from the name of Lord Greenock, who later became Earl Cathcart (1783-1859), +-ite

Example:

Mineral identification reveals 33 minerals of which greenockite is first identified.
在氧化带内鉴别出33种矿物,首次发现了硫镉矿。

---

grossular ['grəusjulə] *n.*

Definition: a mineral of the garnet group, consisting essentially of calcium aluminium silicate 钙铝榴石 gài lǚ liú shí

Formula: The general formula is $X_3Y_2(SiO_4)_3$/ $(Mg, Mn, Fe)_3Al_2Si_3O_{12}$ and $Ca_3(Cr, Al, Fe)_2Si_3O_{12}$

Origin: early 19th century: from modern Latin grossularia "gooseberry". The yellow-green variety is sometimes known as gooseberry garnet.

Examples:

They are rich in Cr_2O_3 and MgO and high in $Cr/(Cr^+Al)$, and knorringite and uvarovite molecules, but low in Fe_2O_3+FeO, $Fe/(Fe^+Mg)$, and grossular and almandine molecules as compared with those in the Cenozoic basalts of eastern China.
它们与我国东部地区新生代玄武岩中的镁铝榴石相比,前者Cr_2O_3及MgO含量、$Cr/(Cr^+Al)$比值,镁铬榴石分子和钙铬榴石分子高;而Fe_2O_3+FeO含量、$Fe/(Fe^+Mg)$比值,钙铝榴石分子及铁铝榴石分子低。

The garnet is composed of 51%~59% almandine, 26%~31% pyrope and 13%~19% grossular respectively.
石榴石中铁铝、镁铝和钙铝榴石分子含量分别为51%~59%、26%~31%和13%~19%。

Extended Term:

massive grossular 块状钙铝榴石

---

grunerite ['gru:nərait] *n.*

Definition: (*Mineralogy*) an iron-rich amphibole related to cummingtonite 铁闪石 tiě shǎn shí

Formula: $Fe_7Si_8O_{22}(OH)_2$

Origin: It was discovered in 1853 and named for Louis Gruner, a Swiss-French chemist who first analysed it.

Example:

The theoretical results suggest that there are given oxygen isotope fractionations among chemically distinct amphiboles. The obtained order of 18O-enrichment is as follows: riebeckite> glaucophane> grunerite>actinolite= cummingto nite > anthophyllite > tremolite > hornblende> gedrite> pargasite.

理论结果表明,不同化学成分的角闪石之间存在一定的氧同位素分馏,其 18O 富集顺序为:钠闪石>蓝闪石>铁闪石>阳起石=镁铁门石>直闪石>透闪石>普通角闪石>铝直闪石>韭闪石。

---

gypsum ['dʒipsəm] *n.*

Definition: a soft white or grey mineral consisting of hydrated calcium sulphate. It occurs chiefly in sedimentary deposits and is used to make plaster of Paris and fertilizers, and in the building industry. 石膏 shí gāo

Formula: $CaSO_4 \cdot 2H_2O$

Origin: late Middle English: from Latin, from Greek gupsos

Example:

a mixture of lime or gypsum with sand and water

石灰或石膏、沙子和水的混合物

Extended Terms:

anhydrous gypsum 无水石膏

calcined gypsum 煅烧石膏

crystal gypsum 结晶石膏

hemihydrate gypsum 熟石膏，半水石膏

halite ['hælait], rock salt

Definition: sodium chloride as a mineral, typically occurring as colourless cubic crystals; rock salt 石盐 shí yán

Formula: NaCl

Origin: mid 19th century: from Greek hals "salt" +-ite

Examples:

the preliminary study of fluid inclusions in salt from five halite deposits of four provinces in Southeast China

我国东南四省五个岩盐矿床石盐中流体包裹体的初步研究

It was found that halite (NaCl) was closely cemented with soil particles, thenardite (Na_2SO_4) was transformed from mirabilite $Na_2SO_4 \cdot 10H_2O$ by dehydration, and the glanberite[$Na_2Ca(SO_4)_2$] was formed by calcium sulfate and sodium sulfate.

扫描电子显微镜的照片，展示出某些盐类结晶的自然特征，它们是：与土粒胶结紧密的石盐(NaCl)、由芒硝($Na_2SO_4 \cdot 10H_2O$)脱水转变成的无水芒硝(Na_2SO_4)，或由硫酸钠、硫酸钙形成的复盐——钙芒硝 $Na_2Ca(SO_4)_2$。

Extended Terms:

halite water 高咸水

salt and halite deposit 盐和石盐矿床

---

halloysite [həˈlɔisait] *n.*

Definition: porcelainlike clay mineral whose composition is like that of kaolinite but contains more water and is structurally distinct; varieties are known as metahalloysites(又称多水高岭石、叙永石)埃洛石 āi luò shí

Formula: $Al_2Si_2O_5(OH)_4 \cdot 2H_2O$

Origin: It was first described in 1826 and named for the Belgian geologist Omalius d'Halloy.

Example:

Hotochromic sodalite has been prepared separately with two different methods, solid-state reaction and hydrothermal reaction, and have been compared with each other. It is easy to obtain the pure photochromic sodalites doped respectively with NaCl, $NaCl \cdot Na_2SO_4$, Na_2S, and $NaCl \cdot NH_4Cl$, NaF, NaBr, crystallized hydrothermally with halloysite as original material. Being activated under innert atomosphere, these sodalites show clear photochromic effect, irradiated with X-ray radiation.

用固相反应及水热合成两种方法制成了光致变色材料方钠石，并作比较。以多水高岭土为原料，经水热反应容易制得高纯度的、掺有 NaCl, $NaCl \cdot Na_2SO_4$, Na_2S, $NaCl \cdot NH_4Cl$, NaF 或 NaBr 的方钠石。它们在惰性气氛中活化后，对 X 光照射有明显的光致变色效应。

Extended Terms:

white halloysite 白石脂

red halloysite 赤石脂

halloysite clay 埃洛石黏土

harmotome [ˈhɑːmətəum] *n.*

Definition: (*Mineralogy*) a rare zeolite, a hydrated barium silicate that forms vitreous white monoclinic crystals 交沸石 jiāo fèi shí

Formula: $Ba[Al_2Si_6O_{16}] \cdot 6H_2O$

Origin: from the Greek for "joint" and "cutting", as the twin crystals

Example:

Separate species are recognized in topologically distinctive compositional series in which different extra framework cations are the most abundant in atomic proportions. To name these, the appropriate chemical symbol is attached by a hyphen to the series name as a suffix, except for the names harmotome, pollucite and wairakite in the phillipsite and analcime series.
在具特定结构的组成系列中，格架外的不同阳离子中的原子比值最大者可确定为独立的矿物种，命名时采用相应的元素符号作为后缀，用连字符加在系列名称之后[钙十字沸石类(philipsite)和方沸石(analcime)系列中的交沸石(harmotome)、铯沸石(pollucite)和斜钙沸石(wairakite)除外]。

Extended Term:

baryta harmotome 重十字石

---

hastingsite [ˈheistiŋsait] *n.*

Definition: (*Mineralogy*) a mineral of the amphibole group crystallizing in the monoclinic system and composed chiefly of sodium, calcium, and iron, but usually with some potassium and magnesium 绿钙闪石 lǜ gài shǎn shí

Formula: $NaCa_2(Fe,Mg)_5Al_2Si_6O_{22}(OH)_2$

Origin: named after its locality

Example:

Different skarn types have different definite amphibole variaties and associated mineralizations: amphiboles in calcic skarns mostly belong to calcic amphiboles such as hastingsite, chlorian hastingsite, potassian hastingsite, potassic hastingsite, magnesio hastingsite, ferrohornblende, chlorian ferrohornblende, fluorian ferrotschermakite, ferroedenite, actinolite, and potassian ferropargasite.
根据夕卡岩类型及其伴生金属矿化的不同，把角闪石分为四大类：钙夕卡岩中的角闪石多属钙角闪石，包括绿钙闪石、铁角闪石、镁绿钙闪石、铁浅闪石、阳起石、铁阳起石、铁镁钙闪石和铁韭闪石等。

Extended Term:

hastingsite respectively 绿钙闪石

---

hausmannite [həusmə'nit] *n.*

Definition: (*Mineralogy*) brownish-black, opaque mineral composed of manganese tetroxide 黑锰矿 hēi měng kuàng

Formula: Mn_3O_4

Origin: named after the German mineralogist, J. F. L. Hausmann (1782-1859)

Example:

ECDs were calculated at a well depth of 4150m for these muds, and the results showed that when mud density is greater than 1.46 g/cm^3, lower ECD can be obtained using hausmannite.
模拟计算出了井深为4150m时的当量循环密度,表明黑锰矿粉在密度为1.46 g/cm^{3}以上可以获得低于其他加重材料的当量循环密度。

Extended Term:

zinc hausmannite 锌黑锰矿

---

hauyne ['hɔːain] *n.*

Definition: a tectosilicate mineral with sulfate and chloride. It is a feldspathoid and a member of the sodalite group. Hauyne crystallizes in the isometric system forming translucent, vitreous typically twinned crystals with highly variable color (blue, white, gray, yellow, green, pink).蓝方石 lán fāng shí

Formula: $(Na,Ca)_{4\text{-}8}Al_6Si_6(O,S)_{24}(SO_4,Cl)_{1-2}$

Origin: It was named for the French crystallographer, René Just Haüy (1743-1822).

---

hedenbergite ['hedənbəːgait] *n.*

Definition: (*Mineralogy*) a black mineral consisting of calcium-iron pyroxene and occurring at the contacts of limestone with granitic masses 钙铁辉石 gài tiě huī shí

Formula: $CaFeSi_2O_6$

Origin: It was named in 1819 after M.A. Ludwig Hedenberg, who was the first to define hedenbergite as a mineral.

Example:

Jadeite is made up of pyroxene, amphibole and feldspar group minerals, in which the existence, quantity and change of chemical composition of pyroxene group minerals, such as jadeite, aegirine, vryite and hedenbergite, largely influence the quality of jadeite.
翡翠主要由辉石族矿物、闪石族矿物、长石族矿物等组成。其中辉石族的矿物如硬玉、霓石、钠铬辉石、钙铁辉石,它们的存在和数量的多少及化学成分的变化对翡翠的质量影响很大。

hematite [ˈhemətait] *n.*

Definition: a reddish-black mineral consisting of ferric oxide. It is an important ore of iron. 赤铁矿 chì tiě kuàng

Formula: Fe_2O_3

Origin: late Middle English: via Latin from Greek haimatitēs (lithos) "bloodlike (stone)", from haima, haimat-"blood".

Example:

Hematite mainly occurs in cleavage cracks and microfissures of such minerals as feldspar, and the higher the hematite content, the deeper the red color.
赤铁矿主要赋存在长石等矿物的解理缝及其显微缝隙中,赤铁矿的含量越多,颜色越红。

Extended Terms:

haematite 赤铁矿,赤铁石
micaceous hematite 云母状赤铁矿
specular hematite 镜铁矿

hemimorphite [ˌhemiˈmɔːfait], calamine [ˈkæləmain] *n.*

Definition: a usually white or colorless mineral, an important ore of zinc, also called *calamine* 异极矿 yì jí kuàng

Formula: $Zn_4Si_2O_7(OH)_2 \cdot H_2O$

Origin: Hemimorphite was originally named calamine but this name had been used for another mineral and hemimorphite was proposed and is now in widespread use. The hemi means half while the morph means shape and thus hemimorphite is aptly named.

Examples:

An oxidized zinc ore from Inner Mongolia was found to be very difficult to treat due to its features of high oxidation rate, heavy weathering and the intimate association of hemimorphite and limonite.
内蒙某氧化锌矿石氧化率高,风化严重,异极矿和褐铁矿共生关系紧密且复杂,是一个难选的氧化锌矿。
It has specific gravity of 3.34~3.46 g/cm^3, refractive index of 1.60~1.62 and hardness of 4~4.5. The main component is hemimorphite, and some carbonate such as smithsonite in small amount.
玉石密度为3.34~3.46g/cm^3,折射率为1.60~1.62,硬度为4~4.5。它的主要组成矿物为异极矿,另有少量的菱锌矿等碳酸盐矿物。

Extended Terms:

electric calamine 电异极矿
calamine lotion 炉甘石液
calamine cerate 炉甘石蜡膏

hercynite ['hə:sənait] *n.*

Definition: A spinel mineral. It occurs in high-grade metamorphosed iron rich argillaceous sediments as well as in mafic and ultramafic igneous rocks. 铁尖晶石 tiě jiān jīng shí

Formula: $FeAl_2O_4$

Origin: from the old Latin, Hercynia Silva, "Forested Mountains"

Examples:

formations of ferrous oxide and hercynite
氧化亚铁与铁铝尖晶石的形成
sintering synthesis of hercynite clinker
铁铝尖晶石料的烧结合成

hessite ['hesait] *n.*

Definition: A mineral form of disilver telluride. It is a soft, dark grey telluride mineral which forms monoclinic crystals. 碲银矿 dì yín kuàng

Formula: Ag_2Te

Origin: named after the Swiss chemist, G. H. Hesse (1802-1850)

Examples:

The contents of gold in hessite and petzite vary largely from trace to 26.1wt%, which is believed to occur as an isomorphic form of Ag in hessite.
碲银矿和碲金银矿中 Au 含量变化较大,从微量到 26.1 wt%,可能是以类质同象形式替代 Ag 进入到碲银矿中。
Afterwards, regional metumorphism and sturnization caused two elements to combine onto hessite(Ag_2Te) which later liquated out of the galena as its unmiaingor secondary inclusion.
区域变质-矽卡岩化过程中,Ag,Te 在方铅矿中也随之结合成碲银矿(Ag_2Te),熔离出来成为方铅矿的不混熔包体或次生包体。

heulandite ['hju:ləndait] *n.*

Definition: (*Mineralogy*) a common mineral of the zeolite group with monoclinic crystals,

片沸石 piàn fèi shí

Formula: $Ca(Ca,Na)_{2\text{-}3}Al_3(Al,Si)_2Si_{13}O_{36} \cdot 12H_2O$

Origin: named after British mineral collector John Henry Heuland (1778-1856)

Examples:

The determined closure temperatures of argon for heulandites are lower than 70℃. The LG-2 heulandite without chlorine disturbing yield a stable 40Ar/39Ar plateau age of (134.0±1.7) Ma, and in agreement with the isochron ages of U-Th-Pb isotopic system, which indicates that there existed the second hydrothermal activation and native copper mineralization during Early Cretaceous.

片沸石的 Ar 封闭温度低于 70℃,无 Cl 干扰的片沸石 LG-2 给出了(134.0±1.7)Ma 的 40Ar/39Ar 稳定坪年龄,并与 U-Th-Pb 等时线定年结果相一致,表明该区在白垩纪早期存在第二次低温热液作用和自然铜矿化。

The revised 40Ar/39Ar ages for the LD-14 actinolite and heulandite are 235.7~238.6 Ma and 149.1 Ma, respectively, which are comparable with those of the zeolites free from chlorine.

修正后的阳起石年龄为 235.7~238.6Ma,片沸石年龄为 149.1Ma,可与无氯沸石的年龄相比较。

---

hexahydrite [ˌheksəˈhaidrait] *n.*

Definition: (*Mineralogy*) a saline evaporite, consisting of magnesium sulphate hexahydrate, a white or greenish-white monoclinic mineral composed of hydrous magnesium sulfate 六水泻盐 liù shuǐ xiè yán

Formula: $MgSO_4 \cdot 6H_2O$

Origin: named after the number of water molecules hexa, "six" and hydor, "water"

Hingganite [ˈhingænit, ˈhingənait] *n.*

Definition: 又称羟硅铍钇铈矿(yttroceberysite)兴安石 xīng ān shí

Formula: $(Y,Ce)[BeSiO_4](OH)$

Origin: named for it's locality. Originally called yttroceberysite, later redefinition while applying Levinson rule.

---

hornblende [ˈhɔːnblend] *n.*

Definition: a dark brown, black, or green mineral of the amphibole group consisting of a hydroxyl alumino-silicate of calcium, magnesium, and iron, occurring in many igneous and metamorphic rocks 普通角闪石 pǔ tōng jiǎo shǎn shí

Formula: $NaCa_2(Mg,Fe,Al)_5[(Si,Al)_4O_{11}]_2(OH)_2$

Origin: late 18th century: from German, from "horn" + blende (See blende)

Example:

As the age of the later metamorphism, 40Ar/ 39Ar plateauand isochron age of hornblende and plagooclase is between 23 and 24Ma, the same as the K-Ar apparent age of hornblende.

晚期变质作用矿物角闪石的 K-Ar 法得出 23Ma,闪石和斜长石的 40 Ar/ 39Ar 年龄谱图和等时线年龄均显示 23~24Ma 的变质年龄。

Extended Terms:

hornblende gabbro 角闪辉长岩
hornblende diorite 角闪闪长岩
hornblende syenite 角闪正长岩

---

humite [ˈhjuːmait] *n.*

Definition: an orthorhombic mineral of a transparent vitreous brown to orange color 硅镁石 guī měi shí

Formula: $(Mg,Fe)_7(SiO_4)_3(F,OH)_2$

Origin: It was first described in 1813 and named for Abraham Hume (1749-1838).

Examples:

Characteristics of Humite Group Minerals Formed in Metasomatic Experimentation and Physicochemical Conditions for Their Formation

交代实验形成的硅镁石族矿物特征和物理化学条件

It is emphasized that the amount of K_2O in the batch composition not only has an effect on the growth of mica crystals but also is a determining factor to the segregation of humite in the melt.

本文强调:配方中 K_2O 含量,不仅影响着云母长大,而且对熔体是否会析出硅镁石起着关键作用。

Extended Terms:

humite group 硅镁石族矿物
garnet-humite-hornblende schist 石榴硅镁石角闪片岩
characteristics of humite group minerals 硅镁石族矿物特征

---

hydrochlorborite [ˌhaidrəklɔːˈbɔrait] *n.*

Definition: white botryoidal crust of hydrochlorborite on massive white anhydrite. 多水氯硼钙石 duō shuǐ lǜ péng gài shí

Formula: $Ca_4B_8O_{15}Cl_2 \cdot 22H_2O$

Origin: named for the essential chemical components, water, chlorine and borate

Example:

Hydrochlorborite is a new hydrous chlor-borate mineral in China.
多水氯硼钙石是一种在我国发现的硼酸盐矿物。

hydromica [ˌhaidrəˈmaikə] *n.*

Definition: (*Mineralogy*) an alternate name for illite 水云母 shuǐ yún mǔ

Example:

Fluvoaquic soil with an abundant hydromica content decreased slightly, and the decline value ranging from 35 to 274 mg · kg^{-1} was 6% to 8%。
水云母含量较丰富的潮土固钾量和固钾率略有降低,分别降低了 35~274mg · kg^{-1}和 6%~8%.
It can make 425 or 325 fluidized cinder cement by combining cinder of fluidized bed, which burns from coal spoil consisting of mainly the hydromica of clay and 40% or 12% to 18% ordinary portland cement.
采用黏土矿物以水云母为主的劣质煤燃烧的沸腾炉渣和 40%或 12%~18%的普通硅酸盐水泥,可配制 425 号或 325 号沸渣水泥。

Extended Terms:

hydromica-schist 水云母片岩
illite hydromica 伊里水云母

hydroxylapatite [haidrɔksiˈlæpətait] *n.*

Definition: (*Mineralogy*) a rare form of the apatite group that crystallizes in the hexagonal system 羟磷灰石 qiǎng lín huī shí

Formula: $Ca_5(PO_4)_3OH$

Example:

It is shown that during the former process, $Ca_2P_2O_7$ and $Ca_{10}(PO_4)_6(OH)_2$ (hydroxylapatite, abbreviated to OHAp) are synthesized from the reactions between $Ca(NO_3)_2 \cdot 4H_2O$ and H_3PO_4, which deposits on the surface of silica gel.
结果表明:在多组分溶胶的形成过程中,由 $Ca(NO_3)_2 \cdot 4H_2O$ 和 H_3PO_4 反应生成 $Ca_2P_2O_7$ 和 $Ca_{10}(PO_4)_6(OH)_2$(羟基磷灰石,简写为 OHAp),两者均在硅溶胶颗粒的表面上析晶长大。

Extended Terms:

hydroxylapatite crystal 羟基磷灰石结晶

hydroxylapatite deposition disease 羟磷酸灰石沉积病

hydrozincite [ˌhaidrəuˈziŋkait] *n.*

Definition: (*Mineralogy*) a white carbonate mineral with the chemical formula $Zn_5(CO_3)_2(OH)_6$又称羟碳锌石.水锌矿 shuǐ xīn kuàng

Formula: $Zn_5(CO_3)_2(OH)_6$

Origin: early 19th century: coined in French, from hyper-“exceeding” + Greek sthenos “strength” (because it is harder than hornblende.)

Example:

The analysis of crystallogram of precursors revealed that the crystals are mainly composed of Malachite, Aurichalcite and Hydrozincite.
前驱体的晶相分析表明:前驱体主要由孔雀石、绿铜锌矿和水锌矿三种晶相组成。

hypersthene [ˈhaipəːsθiːn] *n.*

Definition: a greenish rock-forming mineral of the orthopyroxene class, consisting of a magnesium iron silicate 紫苏辉石 zǐ sū huī shí

Formula: $(Mg,Fe)SiO_3$

Origin: Hypersthene is the name given to the mineral when a significant amount of both elements are present.

Examples:

The cathodo luminescene (CL) images and mineral inclusions from the eclogite reveal that there are two types of inherited zircons, one is magmatic origin with clear oscillatory growth zonation and the other is metamorphic origin with relic garnet, hypersthene and plagioclase minerals of granulite-facies.
根据阴极发光图像及矿物包体组合,(罗田榴辉岩中)继承锆石可分为两种,即具有岩浆结晶环带的岩浆锆石和含有石榴子石+紫苏辉石+斜长石等麻粒岩相变质矿物包体的变质锆石。

Single Grain Zircon U-Pb Ages of the “Early Hercynian” Miantian Granites and Zhongping Hypersthene Diorite in the Yanbian Area
延边“早海西期”棉田花岗岩和仲坪紫苏辉石闪长岩的单颗粒锆石 U-Pb 定年

Extended Terms:

hypersthene granite 紫苏辉石花岗岩
hypersthene-gabbro 紫苏辉卡岩

iceland spar ['aislənd spɑː], icespar

Definition: a transparent variety of calcite, showing strong double refraction, formerly known as iceland crystal, is a transparent variety of calcite, or crystallized calcium carbonate, originally brought from Iceland, and used in demonstrating the polarization of light. 冰洲石 bīng zhōu shí

Extended Terms:

iceland spar deposit 冰洲石矿床
natural iceland spar 天然冰洲石
iceland spar crystal 冰洲石晶体

---

idocrase ['aidəukreis] *n.*

Definition: a mineral consisting of a silicate of calcium, magnesium, and aluminium, occurring typically as dark-green to brown prisms in metamorphosed limestone 符山石 fú shān shí

Formula: $Ca_{10}(Mg, Fe)_2Al_4(SiO_4)_5(Si_2O_7)_2(OH,F)_4$

Origin: early 19th century: from Greek eidos "form" + krasis "mixture"

Example:

Apart from magnetite and almandine garnet, the skarn also contains these minerals: idocrase, actinolite, epidote, fluorite and andradite.
主要构成矿物除磁铁矿和铁铝榴石外,矽卡岩还包含其他矿物,如符山石、阳起石、绿帘石、萤石和钙铁榴石。

Extended Terms:

chrome idocrase 铬符山石
idocrase jade 符山石玉

---

illite ['ilait] *n.*

Definition: a clay mineral of a group resembling micas, with a lattice structure which does not expand on absorption of water 又称水白云母(hydromuscovite),伊利石 yī lì shí

Formula: $K<1Al_2[(Al,Si)Si_3O_{10}](OH)_2 \cdot nH_2O$

Origin: 1930s:named after Illinois +-ite

Example:

K-richillite. Mg-, Ca-, Na-, Zn-rich smectite, Mn- and Fe-rich chlorite produce a clear coupling relationship between clay minerals and chemical compositions of clay fraction.
富 K 的伊利石,富 Mg,Ca,Na,Zn 的蒙皂石,以及富 Mn,Fe 的绿泥石等使得黏土矿物与黏土级组分的化学元素之间的耦合关系十分明显。

Extended Terms:

sodium illite 钠伊利石
degraded illite 退化伊利石
illite structure 伊利石结构
illite hydromica 伊里水云母
illite clay 白云, 母石黏土

---

ilmenite [ˈilmənait] *n.*

Definition: a black mineral consisting of oxides of iron and titanium, of which it is the main ore 钛铁矿 tài tiě kuàng

Formula: $FeTiO_3$

Origin: early 19th century:named after the Ilmen mountains in the Urals +-ite

Example:

The Microstructure Control and Process Optimization of Fe/Ti (C, N) Composites from Carbothermic Reduction of Ilmenite
用钛铁矿制备基 Ti(C,N)复合材料的组织控制和工艺优化

Extended Terms:

ilmenite type 钛铁矿型
ilmenite black 钛铁黑(将钛铁矿直接粉碎筛选制得的黑色颜料)
ilmenite loaded concrete 钛铁混凝土

---

ilvaite [ilˈveit] *n.*

Definition: a mineral consisting of a basic silicate of calcium and iron, typically occurring as black prisms 黑柱石 hēi zhù shí

Formula: $CaFe^{2+}_2Fe^{3+}Si_2O_7O(OH)$

Origin: early 19th century: from Latin Ilva "Elba" +-ite

Examples:

Hydrogen isotope fractionation between ilvaite and water has been studied experimentally at the temperature range 250℃-750℃.

在250℃~750℃温度范围内，对黑柱石、水之间的氢同位素平衡分馏进行了实验研究。

Ilvaite is a rare mineral of calcium-iron-bearing silicate.

黑柱石是一种少见的含钙铁硅酸盐矿物。

---

inderite [ˈindəːrait] *n.*

Definition: pale green translucent crystalline pinnoite on chalky white inderite 多水硼镁石 duō shuǐ péng měi shí

Formula: $Mg[B_3O_3(OH)_5]\cdot 5H_2O$，含 B_2O_3 37%

Origin: named after its source, the type locality, the Inder Lake, Kazakhstan

Example:

There are four fields in this diagram, corresponding to H_3BO_3, $MgO\cdot 2B_2O_3\cdot 9H_2O$ (Hungchaoite), $MgO\cdot 3B_2O_3\cdot 7.5H_2O$ (Mcallisterite) and $2MgO\cdot 3B_2O_3\cdot 15H_2O$ (Inderite), respectively.

该相图存在四个相区，分别与 H_3BO_3，$MgO\cdot 3B_2O_3\cdot 7.5H_2O$，$MgO\cdot 2B_2O_3\cdot 9H_2O$ 和 $2MgO\cdot 3B_2O_3\cdot 15H_2O$（多水硼镁石）相对应。

---

inyoite [ˈinjəuait] *n.*

Definition: (*Mineralogy*) a colourless monoclinic mineral 板硼钙石 bǎn péng gài shí

Formula: $Ca(H_4B_3O_7)(OH)\cdot 4H_2O$，含 B_2O_3 37%

Origin: named after Inyo County, California, where it was discovered, +-ite

---

iridium [iˈridiəm] *n.*

Definition: the chemical element of atomic number 77, a hard, dense silvery-white metal 自然铱 zì rán yī

Iridium is a member of the transition series and is one of the densest metals. Iridium-platinum alloys are hard and corrosion-resistant and are used in jewellery and for electrical contacts; an alloy with osmium is used in fountain-pen nibs.

Formula: Ir

Origin: early 19th century: modern Latin, from Latin iris, irid-"rainbow" (so named because it forms compounds of various colours)

Extended Terms:

iridium lamp 铱灯
iridium tetrachloride 四氯化铱
iridium source 铱(放射)源

iron ['aiən] *n.*

Definition: a strong, hard magnetic silvery-grey metal, the chemical element of atomic number 26, much used as a material for construction and manufacturing, especially in the form of steel 自然铁 zì rán tiě

Iron is widely distributed as ores such as haematite, magnetite, and siderite, and the earth's core is believed to consist largely of metallic iron and nickel. Besides steel, other important forms of the metal are cast iron and wrought iron. Chemically a transition element, iron is a constituent of some biological molecules, notably haemoglobin.

Formula: Fe

Origin: Old English īren, īsen, īsern, of Germanic origin; related to Dutch ijzer and German Eisen, and probably ultimately from Celtic

Example:

Iron that has not combined with carbon occurs commonly in steel, cast iron, and pig iron below 910=C.

未与碳混合过的铁通常存在于钢、铸型铁及910=C以下的铣铁中。

Extended Terms:

iron-oilite 多孔铁
iron-oxidizer 铁氧化剂
ironsand 铁矿砂

jadeite ['dʒeidait] *n.*

Definition: A green, blue, or white mineral which is one of the forms of jade. It is a silicate of sodium, aluminium, and iron and belongs to the pyroxene group.硬玉 yìng yù

Formula: $Na(Al,Fe^{3+})Si_2O_6$

Origin: derived (via French:l'ejade and Latin:ilia) from the Spanish phrase piedra de ijada which means "stone of the side"

Example:

Daintiness, texture, light or translucence and shades of color became then of the first importance, as also in the case of stone, jade and jadeite snuff bottles, which came later.
所以雅致、构造、半透明和色泽变成最重要的质素;关于后来盛行的石鼻烟壶、玉鼻烟壶和硬玉鼻烟壶,情形也是如此。

Extended Terms:

jadeite-diopside 硬玉透辉石
imagined absorption of jadeite 翠霞

---

jamesonite [ˈdʒeimsəˌnait] *n.*

Definition: (*Mineralogy*) a dark grey monoclinic sulfosalt mineral 脆硫锑铅矿 cuì liú tī qiān kuàng

Formula: $Pb_4FeSb_6S_{14}$

Origin: named after Robert Jameson (1774-1854), Scottish mineralogist, +-ite

Examples:

The Producing Practice of Smelting Jamesonite Concentrate with Volatilization Process
挥发法熔炼脆硫锑铅矿精矿生产实践
Selective flotation separation of jamesonite from pyrrhotite by potassuim cyanide
氰化脆硫锑铅矿和磁黄铁矿选择性分离

Extended Terms:

silver jamesonite 银毛矿
jamesonite concentrate 脆硫锑铅矿精矿
jamesonite group 脆硫锑铅矿族

---

jarosite [ˈdʒærəˌsait] *n.*

Definition: (*Mineralogy*) a mineral with rhombohedral crystals. 黄钾铁矾 huáng jiǎ tiě fán

Formula: $KFe_3^{3+}(SO_4)_2(OH)_6$

Origin: named after the place Jaroso ravine in Spain in 1852

Example:

The results indicate that both the precipitates produced by Leptospirillum ferriphium strain DY

and Acidithiobacillus ferrooxidans strain GF are composed of jarosite and ammonium jarosite.
结果表明:无论是嗜铁钩端螺旋菌 DY 菌株还是嗜酸氧化亚铁硫杆菌 GF 菌株合成的沉淀都是黄钾铁矾和黄铵铁矾的混合物。
However, the content of jarosite in the precipitates synthesized by strain GF is 5.53% higher than that synthesized by strain DY, but the content of ammonium jarosite in the precipitates synthesized by strain DY is 15.24% higher than that synthesized by strain GF.
但是,GF 合成的沉淀混合物中黄钾铁矾的含量比 DY 合成的黄钾铁矾约高出 5.53%; 而 DY 合成的沉淀中,黄铵铁矾的含量比 GF 合成的沉淀中的黄铵铁矾的含量约高出 15.24%。

Extended Terms:

jarosite process 黄钾铁矾法
sodium jarosite 黄钠铁矾

---

jedrite (gedrite) ['dʒedrait]) *n.*

Definition: 铝直闪石 lǚ zhí shǎn shí

Formula: $[Mg_2][Mg_3Al_2][(OH)_2|Al_2Si_6O_{22}]$

Origin: named for the locality at Héas, Gedres, France

Example:

Through garnet biotite thermometry and internally consistent thermodynamic calculation, we get that garnet, stourolite and rutile assemblage formed at 680℃ and 0.8 GPa, corundum and gedrite present mineral assemblage at a little P-T space of 730℃-750℃ and about 0.7 GPa; The temperature of sillimanite and hercynite present mineral assemblage are a little higher than that of second association, yet pressures are less than one of second assemblage.
矿物温度压力计和内部一致性热力学计算结果表明早期矿物组合形成于 680℃和 0.8GPa 左右,刚玉、铝直闪石和毛发状夕线石组合形成于 730℃~750℃和 0.7GPa 左右的很小的 P-T 范围内。夕线石和尖晶石矿物组合的形成温度与前一组合相似或略高,压力偏低。

Extended Term:

kyanite gedrite schist 蓝晶石铝直闪石片岩

---

jinshajiangite [ˌdʒinʃədʒiənˈdʒait] *n.*

Definition: reddish elongated junshajiangite grain to 0.3mm in matrix 金沙江石 jīn shā jiāng shí

Formula: $Na_2K_5BaCa(Fe,Mn)_8(Ti,Fe,Nb,Zr)_4Si_8O_{32}(O,F,H_2O)_6$

Origin: named for the locality, Jinshajiang River, Sichuan Province, China

---

jixianite [dʒiksiən'nait] *n.*

Definition: reddish brown crystal druse of jixianite with minor tan cuprotungstite 蓟县矿 jì xiàn kuàng

Formula: $Pb(W,Fe^{3+})_2(O,OH)_7$

Origin: named for the locality. Pan-shan stock, Jixian, Hebei Province, China

Example:

Study on Raman Spectrum Characteristics of Hsianghualite, Baotite, Huanghoite and Jixianite
香花石包头矿黄河矿及蓟县矿的拉曼光谱特征

kainite ['kainait] *n.*

Definition: a white mineral consisting of a double salt of hydrated magnesium sulphate and potassium chloride 钾盐镁矾 jiǎ yán měi fán

Formula: $KMg[SO_4]Cl \cdot 3H_2O$,含 K_2O 15.70%,MgO 16.19%

Origin: mid 19th century:from German Kainit, from Greek kainos "new, recent", because of the mineral's recent formation

Example:

A light soft silver-white metallic element of the alkali metal group; oxidizes rapidly in air and reacts violently with water; is abundant in nature in combined forms occurring in sea water and in carnallite and kainite and sylvite.
一种轻而软的银白色碱金属元素,空气中能迅速氧化,能与水发生剧烈反应,以天然化合物状态大量存在于海水、光卤石、钾盐镁矾和钾盐中。

---

kalsilite ['kælsəˌlait] *n.*

Definition: a rare mineral, a form of $KAlSiO_4$, found in volcanic rocks in parts of Uganda 六方钾霞石 liù fāng jiǎ xiá shí

Formula: $KAlSiO_4$

Origin: from the letters in its chemical formula, +-ite

Examples:

The Cenozoic kamafugites in Lixian County of West Qinling are poor in SiO_2 and Al_2O_3 but rich in MgO, CaO, TiO_2 and K_2O+Na_2O. The modal mineral assemblage is composed mainly of olivine, clinopyroxene, nepheline/ kalsilite, melilite/leucite and Ti-phlogopite.
西秦岭礼县地区新生代钾霞橄黄长岩系具有贫 SiO_2、Al_2O_3，富 MgO、CaO、TiO_2及 K_2O + Na_2O 的特征，矿物组合中除橄榄石、透辉石外，普遍含有霞石 /钾霞石、黄长石 /白榴石和钛金云母等矿物。
Synthesis and characterization of kalsilite powder using a fast sol-gel method
快速溶胶凝胶法制备钾霞石及其反应机理

---

kaolinite [ˈkeiəlinait] *n.*

Definition: a white or grey clay mineral which is the chief constituent of kaolin. 高岭石 gāo lǐng shí

Formula: $Al_2Si_2O_5(OH)_4$

Origin: The name is derived from Chinese：高陵/高嶺; pinyin：Gaoling or Kao-ling (“High Hill”) in Jingdezhen, Jiangxi Province, China. The name entered English in 1727 from the French version of the word：“kaolin”, following Francois Xavier d’Entrecolles’s reports from Jingdezhen.

Example:

Clay fraction is dominated by illite and kaolinite with small proportions of smectite, chlorite and vermiculite.
黏粒组成主要为伊利石和高岭石，也有少量蒙皂石、绿泥石和蛭石。

Extended Terms:

kaolinite clay 高岭土
kaolinite structure 高岭石结构

---

kieserite [ˈkiːzərait] *n.*

Definition: a fine-grained white mineral consisting of hydrated magnesium sulphate, occurring often in salt mines 水镁矾 shuǐ měi fán

Formula: $Mg[SO_4]\cdot H_2O$

Origin: mid 19th century：from the name of Dietrich G. Kieser (1779-1862), German physician, +-ite

Example:

fractional crystallization of boric acid and kieserite from the liquor of ascharite leached by

sulfuric acid
从硼镁矿硫酸分解液中分级结晶硼酸和水镁矾

kimzeyite [ˈkimziːˌait] *n.*

Definition: black isometric crystal of kimzeyite garnet 锆榴石 gào liú shí

Formula: $Ca_3(Zr,Ti)_2(Si,Al,Fe^{3+})_3O_{12}$

Origin: named for the Kimzey family, long associated with the mineralogy at Magnet Cove, Arkansas

kurnakovite [kuˈnɑːkəˌvait] *n.*

Definition: (*Mineralogy*) a hydrated borate mineral 富水镁硼石 fù shuǐ měi péng shí

Formula: $Mg[B_3O_3(OH)_5]\cdot 5H_2O$，含 B_2O_3 37%

Origin: named after Nikolai S Kurnakov (1860-1941), Russian mineralogist, +-ite

Example:

It was asserted that the ore is mainly composed of minerals of kurnakovite, inderite and pinnoite.
据推断该硼镁矿的主要矿物组成为库水硼镁石、多水硼镁石及柱硼镁石。

labradorite [ˈlæbrəˌdɔːrait] *n.*

Definition: a mineral of the plagioclase feldspar group, found in many igneous rocks 拉长石 lā cháng shí

Formula: $Ca_xNa_xAlSi_3O_8$

Origin: early 19th century: from Labrador Peninsula, where it was found, +-ite

Example:

Dentrites of sample 1-6, 1-8, 1-7 are diopside. Dentrites of sample 4-6, 4-8, 4-7 are plagioclase, in which exist less diopside. Comparing the samples "X-ray diffraction pattern and that of all plagioclase end members", we consider the dentrites of samples 4-6, 4-8, 4-7 are labradorite.
1 号成分点(位于透辉石首晶区)样品 1~6、1~8、1~7 的枝晶为透辉石枝晶，4 号成分点(位于斜长石首晶区)样品 4~6、4~8、4~7 的枝晶主要为斜长石枝晶,对比不同斜长石端元组分的 X 射线粉晶衍射图谱,最终确定为拉长石。

Extended Terms:

labradorite sunstone 拉长日光石
sunstone labradorite 拉长日光石

---

laihunite [lei'hʌnait] *n.*

Definition: (*Mineralogy*) an iron neosilicate of the olivine group 莱河矿 lái hé kuàng

Formula: $Fe^{2+}Fe^{3+}2(SiO_4)_2$

Origin: named after the locality, Little Lai-He Village, Liaoning Province, China

Example:

Single laihunite layers and laihunite-forsteritic olivine intergrowths increase the resistance of crystals to weathering.
单高铁橄榄石层和高铁橄榄石交互生长,增加水晶风化阻力。

---

larnite ['lɑːnait] *n.*

Definition: a polymorph of calcium silicate, found in limestone or chalk in contact with semimolten basalts 斜硅钙石 xié guī gài shí

Formula: Ca_2SiO_4

Origin: named after its locality Scawt Hill, near Larne, Co., Antrim, Ireland

Example:

Larnite has a Mohs hardness of 6..
斜硅钙石的莫氏硬度为 6。

---

laumontite [ləu'mɔntait] solution pore

Definition: (*Mineralogy*) a mineral, of a white color and vitreous luster. It is a hydrous silicate of alumina and lime. Exposed to the air, it loses water, becomes opaque, and crumbles. 浊沸石 zhuó fèi shí

Formula: $CaAl_2Si_4O_{12} \cdot 4H_2O$

Origin: named after the discoverer François Pierre Nicholas Gillet de Laumont (1747-1834) as lomonite, renamed as laumonite in 1809 and as laumontite in 1821

Example:

the heulandite+laumontite zone during Stage A of late diagenesis, and the laumontite+albite

zone during Stages B to C of late diagenesis.
片沸石+浊沸石带主要为晚成岩期 A 阶段形成；浊沸石+钠长石带为晚成岩期 B-C 阶段形成。

Extended Term:

laumontite solution pore 浊沸石溶孔

L

lawsonite [ˈlɔːsənait] *n.*

Definition: a metamorphic silicate mineral related to the epidotes; it is a sorosilicate, based on the dimeric anion $Si_2O_7^{6-}$. 硬柱石 yìng zhù shí

Formula: $CaAl_2Si_2O_7(OH)_2 \cdot H_2O$

Origin: It was first described in 1895 for occurrences in the Tiburon peninsula, Marin County, California. It was named for geologist Andrew Lawson (1861-1952) of the University of California.

Example:

The characteristic mineral assemblage of pumpellyite blueschist is glaucophane (40%) + pumpellyite (30%) + chlorite (10%) + albite (8%) + quartz (5%) + lawsonite (3%) calcite (1%).
绿纤石蓝片岩的特征变质矿物组合为蓝闪石(40%)+绿纤石(30%)+绿泥石(10%)+钠长石(8%)+石英(5%)+硬柱石(3%)+方解石/文石(1%)。

Extended Terms:

while lawsonite 硬柱石
glaucophane-lawsonite 蓝闪石+硬柱石

lazulite [ˈlæzjuˌlait] *n.*

Definition: an azure-blue mineral with a glasslike luster 天蓝石 tiān lán shí

Formula: $(FeMg)A_{12}P_2O_8(OH)_2$

Origin: It was first described in 1795 for deposits in Austria. Its name comes from the Arabic for heaven.

Extended Term:

iron-lazulite 铁天蓝石

lazurite [ˈlæzjuˌrait] *n.*

Definition: a bright blue mineral which is the chief constituent of lapis lazuli and consists chiefly of a silicate and sulphate of sodium and aluminium 青金石 qīng jīn shí

Formula: $(Na,Ca)_8[AlSiO_4]_6(SO_4,S,Cl)_2$

Origin: named for the distinctive blue coloration

Example:

Transmission electron microscopic studies on the azurite from the contact metamorphic zone of pegmatites and dolomitic marbles in South Baikal, Russia, revealed a new type of domain structure intergrown with ordered and disordered domains on a very small scale(≤100nm).
通过透射电镜研究,作者在产于俄罗斯南贝加尔的青金石中,观察到了一种由尺度都很小(长径约在 100m 以内)的有序结构畴与无序结构畴相互交生的畴结构新类型。

---

leadamalgam [ledə'mælgəm] *n.*

Definition: Prepared by rubbing lead filings with mercury in a mortar or by pouring molten lead into mercury. Has no definite composition. Possesses a brilliant white color and remains liquid with as much as 33% of lead. A 50:50 lead-mercury amalgam can be crystallized, and a piece of clean lead plunged into this will be found to be covered with crystals of this amalgam when withdrawn 汞铅矿 gǒng qiān kuàng

Formula: $HgPb_2$

---

lepidocrocite [ˌlepidəu'krəusait] *n.*

Definition: a red to reddish-brown mineral consisting of ferric hydroxide, typically occurring as scaly or fibrous crystals 纤铁矿 xiān tiě kuàng

Formula: γ-FeO(OH)

Origin: early 19th century: from Greek lepis, lepid "scale" + krokis "fibre"

Examples:

a widely occurring iron oxide ore; a mixture of goethite and hematite and epidocrocite
一种常见的氧化铁矿石,是针铁矿、赤铁矿和纤铁矿的混合物
By ferrihydrite and lepidocrocite addition, the contribution (%) of electron-transfer belong to iron reduction were respectively increased to 63.32% and 46.90% from 18.30% (control), however, the contribution (%) by methane production were respectively decreased to 35.85% and 52.32% from 80.92% (control). Because of election competing consummation of iron reduction, the methanogenesis was greatly inhibited.
添加水铁矿和纤铁矿后,Fe(Ⅲ)还原占总电子传递的贡献率分别由对照的 18.30%增至 63.32%和 46.90%,而形成甲烷的电子传递贡献率由对照的 80.92%降至 35.85%和 52.32%,Fe(Ⅲ)还原对电子的竞争消耗,使土壤产甲烷过程被强烈抑制。

Extended Term:

raft-like lepidocrocite 筏状

---

lepidolite [leˈpidəlait] *n.*

Definition: a mineral of the mica group containing lithium, typically grey or lilac in colour 锂云母 lǐ yún mǔ (又称鳞云母, 红云母)

Formula: $KLi_2Al(Al,Si)_3O_{10}(F,OH)_2$

Origin: late 18th century: from Greek lepis, lepid-"scale" +-ite

Example:

The Yashan topaz lepidolite granite was emplaced at the latest stage of the composite batholith, and has a high phosphorus content (0.15~0.55 wt%).

雅山黄玉锂云母花岗岩是雅山复式岩体的最晚阶段岩体,具有较高的磷含量(0.15~0.55wt%)。

Extended Terms:

lepidolite granite 锂云母花岗岩
topaz lepidolite granite 黄玉锂云母花岗岩
lepidolite glass-ceramics 锂云母微晶玻璃
CS lepidolite 铯锂云母
lepidolite petalite bearing pegmatite 锂云母—透锂长石伟晶岩

---

leucite [ˈljuːsait] *n.*

Definition: a potassium aluminosilicate mineral, crystallizing in the tetrahedral system and typically found as grey or white glassy trapezohedra in volcanic rocks 白榴石 bái liú shí

Formula: $K[AlSi_2O_6]$

Origin: late 18th century: from Greek leukos "white" +-ite

Example:

As one of the main raw materials, the content of K_2O or orthoclase ($K_2O \cdot Al_2O_3 \cdot 6SiO_2$) is the key factor in the formation of leucite.

作为长石瓷原材料的主要组分之一,K_2O 或钾长石($K_2O \cdot Al_2O_3 \cdot 6SiO_2$)的含量对白榴石的生成有决定性作用。

Extended Terms:

leucite trachyte 白榴粗面岩
leucite basalt 白榴石玄武岩

leucite phonolite 白榴响岩
leucite syenite 白榴正长岩
noselite leucite phonolite 黝方白榴响岩

liberite ['libərait] *n.*

Definition: a new lithium-beryllium silicate mineral 锂铍石 lǐ pí shí

Formula: Li_2BeSiO_4

Origin: named in 1964 for the component elements:lithium and beryllium

Example:

Liberite has a Mohs hardness of 7.
锂铍石的莫氏硬度是7。

limonite ['laimənait] *n.*

Definition: an amorphous brownish secondary mineral consisting of a mixture of hydrous ferric oxides, important as an iron ore 褐铁矿 hè tiě kuàng

Formula: $FeO(OH) \cdot nH_2O$

Origin: early 19th century:from German Limonit, probably from Greek leimōn "meadow" (suggested by the earlier German name Wiesenerz, literally "meadow ore")

Example:

Their colors are also generated after the formation of jadeite crystals, often found in the red layer, caused by disseminated limonite.
它们的颜色也是硬玉晶体生成后才形成,常常分布于红色层之上,是由褐铁矿浸染所致。

Extended Terms:

pitchy limonite 沥青褐铁矿
relief limonite 多孔褐铁矿
massive limonite 块状褐铁矿
fibrous limonite 纤褐铁矿
limonite rock 褐铁岩

linarite ['lainərait] *n.*

Definition: a somewhat rare, crystalline mineral that is known among mineral collectors for its unusually intense, pure blue color. It is formed by the oxidation of galena and chalcopyrite and other copper sulfides. 青铅矿 qīng qiān kuàng

Formula: $PbCuSO_4(OH)_2$

Origin: named after its locality: Linares, Spain

Example:

Linarite has a Mohs hardness of 2.5 and a specific gravity of 5.3-5.5.
青铅矿的莫氏硬度为 2.5,相对密度为 5.3~5.5。

---

linnaeite [liˈniːait] *n.*

Definition: a mineral of pale steel-gray color and metallic luster, occurring in isometric crystals, and also massive. It is a sulphide of cobalt containing some nickel or copper. 硫钴矿 liú gǔ kuàng

Formula: $Co^{+2}Co_2^{+3}S_4$

Origin: named after Carl von Linné (Carolus Linnaeus; Original Swedish name: Carl Nilsson Linnæus) (1707-1778), Swedish taxonomist, botanist, physician, geologist and zoologist

Example:

Linnaeite has a Mohs hardness of 4.5-5.5 and a specific gravity of 4.8.
硫钴矿的莫氏硬度为 4.5~5.5,相对密度为 4.8。

Lishizhenite [liʃiˈtʃenait]

Definition: a new zinc sulphate mineral, found in the oxidation zone of a Pb-Zn deposit at Xitieshan, Qinghai Province, China 李时珍石 lǐ shí zhēn shí

Formula: $ZnFe_2^{3+}(SO_4)_4 \cdot 14\ H_2O$

Origin: named for Li Shizhen (1518-1593), famous Chinese pharmacologist

Example:

Lishizhenite was found in a crack or cavity of anhydrite associated with roemerite, copiapite, sulphur, gypsum, pyrite and quartz.
李时珍石产于硬石膏裂隙或孔洞中,与粒铁矾、叶绿矾、自然硫、石膏、黄铁矿、石英等共生。

---

litharge [ˈliθɑːdʒ] *n.*

Definition: lead monoxide, especially a red form used as a pigment and in glass and ceramics 密陀僧 mì tuó sēng (又称一氧化铅,铅黄,黄丹)

Formula: PbO

Origin: Middle English: from Old French litarge, via Latin from Greek litharguros, from lithos "stone" + arguros "silver"

Example:

Litharge itself has been used as a stabilizer.

一氧化铅本身就可作稳定剂使用。

Extended Terms:

litharge stock 一氧化铅混合剂

yellowish litharge 黄相黄丹

red litharge 红相黄丹

litharge-glycerincement 铅黄甘油胶合剂

lizardite ['lizədait] *n.*

Definition: (also called orthoantigorite) a very fine-grained, platy variety of the serpentine group 利蛇纹石 lì shé wén shí

Formula: $Mg_6[Si_4O_{10}](OH)_8$

Origin: named after its type locality on the Lizard Peninsula, Cornwall, UK

Example:

Anla-jade is a serpentine jade of the lizardite type.

安绿玉属于利蛇纹石型的蛇纹石玉。

loparite ['lɔpərait] *n.*

Definition: a granular, brittle oxide mineral of the perovskite class 铈铌钙钛矿 shì ní gài tài kuàng

Formula: $(Ce,Na,Ca)(Ti,Nb)O_3$

Origin: named from the Russian name for the Lapps, "Lopar", inhabitants of the Kola Peninsula, Russia

Example:

Loparite has a Mohs hardness of 5.5-6.0 and a specific gravity of 4.60-4.89.

铈铌钙钛矿的莫氏硬度为5.5~6.0,相对密度为4.60~4.89。

lopezite ['ləupəzait] *n.*

Definition: a rare red triclinic chromate mineral 铬钾矿 gè jiǎ kuàng

Formula: $K_2Cr_2O_7$

Origin: named from the name of Emiliano López Saa (1871-1959), Chilean mining engineer

Example:

Lopezite was first described in 1937 for an occurrence in Iquique Province, Chile.
铬钾矿最早记载于 1937 年,出现在智利的伊基克省。

---

ludlamite [ˈluːdləmait] *n.*

Definition: A mineral occurring in small, green, transparent, monoclinic crystals. It is a hydrous phosphate of iron. 板磷铁矿 bǎn lín tiě kuàng

Formula: $(Fe,Mg,Mn)_3(PO_4)_2 \cdot 4(H_2O)$

Origin: named for Henry Ludlam (1824-1880), English mineralogist and collector

Example:

Ludlamite has a Mohs hardness of 3.5.
板磷铁矿的莫氏硬度是 3.5。

---

ludwigite [ˈlʌdwiˌgait] *n.*

Definition: a magnesium-iron borate mineral 硼镁铁矿 péng měi tiě kuàng

Formula: Mg_2FeBO_5

Origin: late 19th century (1874): named for Ernst Ludwig (1842-1915), an Austrian chemist at the University of Vienna, +-ite

Example:

Tinan ludwigite, a new Sn-rich subspecies of ludwigite, occurs in a magnesian skarn borate deposit at Qiliping, Changning County, Hunan Province.
富锡硼镁铁矿是硼镁铁矿的含锡亚种,产于湖南常宁县七里坪镁质矽卡岩硼矿床中。

Extended Terms:

mangan ludwigite 硼镁锰矿
ludwigite-magnetite 硼镁铁矿—磁铁矿
ludwigite ore 硼铁矿
tin-ludwigite 富锡硼镁铁矿
tin-rich ludwigite 富锡硼镁铁矿

---

lunijianlaite [ˌjuːniˈdʒaiənlait] *n.*

Definition: a regularly interlayered mineral discovered in 1990 in China, colorless and transparent, occurring in the openings of corundum balls of a pyrophyllite deposit, being of volcanic hydrothermal origin 绿泥间蜡石 lǜ ní jiān là shí

Formula: $LiAl_6(Si_7Al)O_{20}(OH)_{10}$

Origin: named for "chlorite alternating with pyrophyllite"

Example:

Lunijianlaite has a Mohs hardness of 2.
绿泥间蜡石的莫氏硬度是2。

maghemite ['mægəmait] *n.*

Definition: a magnetic mineral with a grey blue, white, or brown shade 磁赤铁矿 cí chì tiě kuàng

Formula: Fe_2O_3, γ-Fe_2O_3

Origin: early 20th century(1927): after the beginnings of magnetite and hematite, because of magnetism and similar arrangement of elements

Example:

The rock magnetic results demonstrate that magnetite, hematite and thermal-unstable maghemite are the most important magnetic minerals in the Dachai sediments.
岩石磁学研究结果显示,磁铁矿、赤铁矿和热不稳定性磁赤铁矿是大柴剖面沉积物中主要的磁性矿物。

---

magnesite ['mægnəsait] *n.*

Definition: a whitish mineral consisting of magnesium carbonate, used as a refractory lining in some furnaces 菱镁矿 líng měi kuàng

Formula: $MgCO_3$

Origin: named after its chemical composition

Example:

Chromite and/or magnesite mineralization is usually related to the ophiolites.
铬铁矿及(或)菱镁矿化通常与蛇绿岩有关。

Extended Terms:

caustic magnesite 轻烧菱镁砂
fused magnesite 熔成菱镁石
chrome magnesite 铬镁砖
magnesite mortar 镁火泥
refractories magnesite 菱镁耐火材料

---

magnetite [ˈmæɡnitait] *n.*

Definition: a grey-black magnetic mineral which consists of an oxide of iron and is an important form of iron ore 磁铁矿 cí tiě kuàng

Formula: $Fe^{2+}Fe^{3+}2O_4$

Origin: mid 19th century: from magnet +-ite

Example:

Magnetite and hematite both are the high thermal-stable magnetic carriers.
磁铁矿和赤铁矿同为高温稳定性磁性载体。

Extended Terms:

magnetite sand 磁矿砂
titaniferous magnetite 钛铁磁铁矿
magnetite bronzitite 磁铁古铜岩
magnetite diallagite 磁铁异剥岩
magnetite pyroxenite 磁铁辉岩

---

malachite [ˈmæləˌkait] *n.*

Definition: A bright green mineral consisting of hydrated basic copper carbonate. It typically occurs in masses and fibrous aggregates and is capable of taking a high polish. 孔雀石 kǒng què shí（又称石绿）

Formula: $Cu_2CO_3(OH)_2$

Origin: late Middle English: from Old French melochite, via Latin from Greek molokhitis, from molokhē, variant of malakhē "mallow"

Example:

The mineral malachite contains the elements copper, hydrogen, carbon, and oxygen.
孔雀石矿石包含有铜、氢、碳和氧元素。

Extended Terms:

tar malachite 星光孔雀石

shimmer malachite 闪光孔雀石
prase malachite 玉髓孔雀石
malachite green 孔雀石绿
siliceous malachite 硅孔雀石

manganotantalite [ˌmæŋgənəuˈtæntəlait] *n.*

Definition: the manganese-rich form of tantalite 钽锰矿 tǎn měng kuàng（又称锰钽铁矿）

Formula: $MnTa_2O_6$

Origin: named after its chemical composition containing manganese and tantalum

Example:

Manganotantalite has a Mohs hardness of 6.25.
钽锰矿的莫氏硬度是6.25。

manganite [ˈmæŋgənait] *n.*

Definition: a mineral consisting of basic manganese oxide, typically occurring as steel-grey or black prisms 水锰矿 shuǐ měng kuàng

Formula: $MnO(OH)$

Origin: named after its chemical composition

Example:

Pyrolusite, cryptomelane, coronadite, psilomelane, nsutite, manganese spar, manganite and manganocalcite were found in the E'rentaolegai silver deposit.
额仁陶勒盖银矿床出现的锰矿物主要为软锰矿、隐钾锰矿、铅硬锰矿、硬锰矿、六方锰矿、菱锰矿、水锰矿及含钙菱锰矿。

Extended Terms:

perovskite manganite 钙钛矿锰氧化物
bilayer manganite 双层锰氧化物
manganite oxides 锰氧化合物
silver manganite 银锰
manganite film 锰氧化物薄膜

manganosite [ˌmæŋgəˈnəusait] *n.*

Definition: a rare mineral composed of manganese(II) oxide MnO 方锰矿 fāng měng kuàng

Formula: MnO

Origin: named as an oxide of manganese

Example:

Manganosite has a Mohs hardness of 5-6 and a specific gravity of 5.18.
方锰矿的莫氏硬度为 5~6,相对密度为 5.18。

---

marcasite [ˈmɑːkəsait] *n.*

Definition: a semi-precious stone consisting of iron pyrites 白铁矿石(一种次贵重宝石)bái tiě kuàng shí

Formula: FeS_2

Origin: late Middle English: from medieval Latin marcasita, from Arabic mar kaŝta, from Persian

Example:

Element As occurs in arsenopyrite, partly in pyrite, pyrrhotite, marcasite and tetrahedrite in the form of isomorphous admixture.
As 以毒砂形式存在,部分以类质同像形式存在于黄铁矿、雌黄铁矿、白铁矿和黝铜矿中。

Extended Terms:

melnikovite-marcasite 胶铁矿,胶白铁矿
marcasite particles 白铁矿颗粒

---

margarite [ˈmɑːgərait] *n.*

Definition: a calcium-rich member of the mica group of phyllosilicates, forming white to pinkish or yellowish-gray masses or thin laminae 珍珠云母 zhēn zhū yún mǔ

Formula: $CaAl_2(Al_2Si_2)O_{10}(OH)_2$

Origin: from the Greek margaritos-"pearl"

Example:

The primary margarite in blastomylonite belt was first discovered in Xinyang area, Henan Province, China.
产于变余糜棱岩带中的原生珍珠云母首见于中国河南省信阳地区。

Extended Terms:

margarite pearl mica 珍珠云母

soda-margarite 钠珠云母
margarite hydrolysate 珍珠层粉水解液

marialite [məˈriəlait, ˈmæriəlait] *n.*

Definition: A member of the scapolite group and a solid solution exists between marialite and meionite, the calcium endmember. It is a rare mineral usually used as a collector's stone. It has a very rare but attractive gemstones and cat's eye. 钠柱石 nà zhù shí

Formula: $Na_4Al_3Si_9O_{24}Cl$

Origin: named by von Rath in honor of his wife, Maria Rosa vom Rath (1830-1888).

Example:

Marialite has a Mohs hardness of 5.5-6.
钠柱石的莫氏硬度为 5.5~6。

Extended Terms:

marialite orebody 钠柱石矿体
gem marialite deposit 宝石级钠柱石矿床

massicot [ˈmæsikɔt] *n.*

Definition: a yellow form of lead monoxide, used as a pigment 铅黄 qiān huáng（又称黄丹,天然一氧化铅）

Formula: PbO

Origin: late 15th century: from French (influenced by Italian marzacotto "unguent"), ultimately from Arabic martak

Example:

Massicot has a Mohs hardness of 2.
铅黄的莫氏硬度为 2。

mayingite [ˌmɑːjiˈnait, ˌmeiiŋait] *n.*

Definition: a member of the Pyrite Group of minerals, Iridium Bismuth Telluride 马营矿 mǎ yíng kuàng

Formula: IrBiTe

Origin: named after its locality: near the village of Maying, about 230 km north-north-east of Beijing, China

Example:

Mayingite has a Mohs hardness of 4.
马营矿的莫氏硬度是 4。

---

meerschaum [ˈmiəʃəm] *n.*

Definition: a soft white clay-like material consisting of hydrated magnesium silicate, found chiefly in Turkey 海泡石 hǎi pào shí

Formula: $Mg8[Si_{12}O_{30}](OH)_4 \cdot 12H_2O$

Origin: late 18th century: from German, literally sea-foam, from Meer "sea" + Schaum "foam", translation of Persian kef-i-daryā (alluding to the frothy appearance of the silicate)

Example:

The modified meerschaum was first used to recover gallium from acid lixivium of zinc residues and the experimental results were satisfactory.
采用改性海泡石从锌渣酸浸液中提取回收稀散金属镓,试验结果令人满意。

Extended Terms:

meerschaum probe 海泡石探子(检铅弹)
meerschaum clay 海泡石黏土釉面砖
meerschaum sea-foam sepiolite 海泡石
meerschaum sepiolite 海泡石
modified meerschaum 改性海泡石

---

meionite [ˈmaiənait] *n.*

Definition: a silicate mineral belonging to the scapolite group 钙柱石 gài zhù shí

Formula: $Ca_4Al_6Si_6O_{24}CO_3$

Origin: from the Greek for "less", referring to its less acute pyramidal form compared with vesuvianite

Example:

Meionite was first discovered in 1801 on Mt. Somma, Vesuvius, Italy.
钙柱石最早发现于 1801 年意大利维苏威火山的索马山上。

---

melanterite [məˈlæntərait] *n.*

Definition: a widespread green, mostly fibrous mineral composed of hydrated ferrous sulphate 水绿矾 shuǐ lǜ fán

Formula: $Fe^{2+}SO_4 \cdot 7\ H_2O$

Origin: from the Greek melas, “black”

Example:

Melanterite has a Mohs hardness of 2.

水绿矾的莫氏硬度是 2。

Extended Terms:

zinc-melanterite 锌水绿矾

magnesium-melanterite 镁绿矾

zinc-copper melanterite 锌铜绿矾

mangan-melanterite 锰绿矾

iron-melanterite 铁绿矾

---

melilite ['melilait] *n.*

Definition: any mineral consisting of a solid solution of gehlenite ($Ca_2Al_2SiO_7$) and akermanite ($Ca_2MgSi_2O_7$) 黄长石 huáng cháng shí（又称方柱石）

Formula: $(Ca, Na)_2(Al, Mg, Fe^{2+})(Al, Si)_2O_7$

Origin: from a ncient Greek μελι (meli, “honey”) +-ite

Example:

Melilite has a Mohs hardness of 5-5.5.

黄长石的莫氏硬度是 5~5.5。

Extended Terms:

melilite basalt 黄长玄武岩

melilite ankaratrite 黄长橄霞玄武岩

melilite etindite 黄长白榴霞石岩

melilite italite 黄长粗白榴岩

melilite humboldtilite 黄长石

---

meliphanite [ˌməˈlifənait] *n.*

Definition: a rare calcium, sodium, beryllium, aluminum silicate 蜜黄长石 mì huáng cháng shí

Formula: $(Ca, Na)_2Be[(Si, Al)_2O_6(F, OH)]$

Origin: named from the Greek for “honey” and “to appear” in allusion to the color

Example:

Meliphanite has a Mohs hardness of 5-5.5.
蜜黄长石的莫氏硬度是 5~5.5。

---

mendipite [ˈmendipait] *n.*

Definition: a white orthorhombic mineral consisting of an oxide and chloride of lead 白氯铅矿 bái lǜ qiān kuàng

Formula: $Pb_3O_2Cl_2$

Origin: named after its locality: Mendip Hills, Somersetshire, England

Example:

Because of the lead content, the specific gravity of mendipite is high.
因为含有铅,白氯铅矿的相对密度高。

---

mercury [ˈməːkjuri] *n.*

Definition: the chemical element of atomic number 80, a heavy silvery-white metal which is liquid at ordinary temperatures 天然汞 tiān rán gǒng

Formula: Hg

Origin: Middle English: from Latin Mercurius

Example:

Mercury levels are high in marine fish, particularly in larger species, such as shark, tuna and swordfish.
海洋鱼类中的汞含量很高,尤其在体型较大的鱼类体内,如鲨鱼、金枪鱼和箭鱼。

Extended Terms:

mercury electrode 水银电极,汞电极
mercury switch 水银开关,汞开关
mercury boiler 水银锅炉,水银蒸煮器
mercury fulminate 雷酸汞,雷汞
mercury contact 水银触点,水银接点

---

merwinite [ˈməːwinait] *n.*

Definition: a monoclinic mineral, colorless to pale green 默硅钙镁石 mò guī gài měi shí (又称镁硅钙石、默硅镁、镁蔷薇辉石)

Formula: $Ca_3Mg(SiO_4)_2$

Origin: named after Herbert Eugene Merwin (1878-1963), American mineralogist and petrologist, Carnegie Institute, Washington, D.C., USA

Example:

Merwinite has a Mohs hardness of 6.
默硅钙镁石的莫氏硬度是6。

Extended Term:

manganese-merwinite 锰镁硅钙石

---

mesolite ['mesəlait] *n.*

Definition: a mineral with monoclinic crystals, of the zeolite group 中沸石 zhōng fèi shí

Formula: $Na_2Ca_2(Al_2Si_3O_{10})_3 \cdot 8H_2O$

Origin: from the Greek mesos-"middle."

Example:

Mesolite has a Mohs hardness of 5 and a specific gravity of 2.26.
中沸石的莫氏硬度为5,相对密度为2.26。

Extended Term:

turquoise mesolite 中沸石

---

mica ['maikə] *n.*

Definition: A shiny silicate mineral with a layered structure, found as minute scales in granite and other rocks, or as crystals. It is used as a thermal or electrical insulator. 云母 yún mǔ

Formula: $KAl_2[AlSi_3O_{10}][OH]_2$

Origin: early 18th century: from Latin, literally "crumb"

Example:

Electrical insulators include rubber, plastic, porcelain, and mica.
电绝缘体包括橡胶、塑料、陶瓷和云母。

Extended Terms:

mica plate 云母板
mica flake 云母片
mica cloth 云母布
lithium mica 锂云母

mica powder 云母粉

---

microcline [ˈmaikrəuklain] *n.*

Definition: a green, pink or brown crystalline mineral consisting of potassium-rich feldspar, characteristic of granite and pegmatites 微斜长石 wēi xié cháng shí（曾称钾微斜长石）

Formula: $KAlSi_3O_8$

Origin: mid 19th century: from German Microklin, from Greek mikros "small" + klinein "to lean"（because its angle of cleavage differs only slightly from 90 degrees）

Example:

Wadeite is paragenic with aegirine, nepheline, arfredsonite, microcline and apatite, and associated with natrolite, rinkite and cancrinite.

钾钙板锆石与霓石、霞石、钠铁闪石、微斜长石、磷灰石共生，与钠沸石、层硅铈钛矿、钙霞石等伴生。

Extended Terms:

microcline twinning 微斜长石变晶，微斜长石双晶

sodian microcline 钠微斜长石

microcline-perthite 微斜纹长石

microcline syenite 微斜正长岩

microcline nepheline syenite 微斜霞石正长岩

---

microlite [ˈmaikrəlait] *n.*

Definition: a basic fluoride of sodium, calcium, tantalum and niobium that is isomorphous with pyrochlore 细晶石 xì jīng shí（又称微晶）

Formula: $(Na,Ca)_2Ta_2O_6(O,OH,F)$

Origin: named from Greek mikros for "small" and lithos for "stone", 1835

Example:

Microlite is a mineral, usually including niobium, fluorine, and other impurities, occurring in cubic crystals.

微晶石是一种矿物质，通常包括铌、氟等杂质，是一种立方晶体。

Extended Term:

microlite structure 微晶结构

---

milky quartz [ˈmilki kwɔːts]

Definition: (or milk quartz) a crystalline quartz that is white and translucent to almost opaque due to numerous evenly distributed gas and/or fluid inclusions 乳石英 rǔ shí yīng

Formula: SiO_2

Origin: quartz: derived from the German word quarz, which was imported from Middle High German, twarc, which originated in Slavic (cf. Czech tvrdy ("hard"), Polish twardy ("hard"), Russian твёрдый ("hard")), from Old Church Slavonic тврьдь ("firm")

Example:

Milky quartz is white in color and cloudy.

乳石英呈白色,不透明。

---

millerite ['milərait] *n.*

Definition: a mineral consisting of nickel sulphide and typically occurring as slender needle-shaped bronze crystals 针硫镍矿 zhēn liú niè kuàng(又称针镍矿)

Formula: NiS

Origin: mid 19th century: named after William H. Miller (1801-1980), English scientist, +-ite

Example:

The Jussi ore shoot is exceptional also in that it contains considerable amounts of millerite and bornite.

洁西矿结的独特也在于它含有相当数量的针硫镍矿及斑铜矿。

Extended Term:

synthetic millerite 针镍矿

---

mimetite ['mimitait] *n.*

Definition: an arsenate mineral which forms as a secondary mineral in lead deposits, usually by the oxidation of galena and arsenopyrite 砷铅矿 shēn qiān kuàng

Formula: $Pb_5(AsO_4)_3Cl$

Origin: from the Greek mimethes "imitator", because of its resemblance to pyromorphite

Example:

Mimetite group minerals are a group of isomorphic series minerals with hexagonal system, which include mimetite, vanadinite and pyromorphite.

砷铅矿族矿物是一族六方晶系的类质同象系列矿物,包括砷铅矿、钒铅矿和磷氯铅矿。

Extended Term:

calcium barium mimetite 钙钡砷铅矿

---

minasragrite [ˌmiːnɑːsˈrɑːgrait, ˌminəsˈrægrait] *n.*

Definition: a blue, monoclinic mineral consisting of hydrated acid vanadyl sulfate; occurs in efflorescences and as aggregates or masses. 钒矾 fán fán

Formula: $VO(SO_4) \cdot 5(H_2O)$

Origin: named for the locality: Mina Ragra (Minasragra), Junin, Cerro de Pasco, Peru

Example:

Minasragrite has a Mohs hardness of 1-2.
钒矾的莫氏硬度是 1~2。

---

minium [ˈminiəm] *n.*

Definition: a tetragonal mineral, an alteration product of galena or cerussite 铅丹 qiān dān (又称红铅)

Formula: Pb_3O_4

Origin: named after its locality: the river Minius located in north-west Spain

Example:

Adding minium into the positive lead paste of VRLA battery can elevate the formation efficiency of positive plates of battery.
在阀控铅蓄电池正极铅膏中添加铅丹可以提高正极板的化成效率。

Extended Term:

alumina minium 铝红
iron minium 铁丹
ironcored minium 赭土
minium sulfate 硫酸铝

---

minnesotaite [ˌminiˈsəutəˌait] *n.*

Definition: an iron magnesium silicate of the talc group 铁滑石 tiě huá shí

Formula: $(Fe^{2+}, Mg)_3Si_4O_{10}(OH)_2$

Origin: named after its occurrence in Minnesota, USA

Example:

Minnesotaite has a Mohs hardness of 1.5-2.
铁滑石的莫氏硬度是 1.5~2。

---

mirabilite [miˈræbilait] *n.*

Definition: a saline evaporite, consisting of sodium sulphate 芒硝 mánɡ xiāo

Formula: $Na_2SO_4 \cdot 10(H_2O)$

Origin: named after the Latin, sal mirabile "miracle salt" expressing Glauber's surprise on its synthesis

Example:

Mirabilite is the highly finished crystallization of Puxiao after being boiled.
芒硝是朴硝再煮炼后所得的精制结晶。

Extended Terms:

mirabilite ore 硫酸钠矿,芒硝矿
mirabilite gypsum 芒硝石膏
mirabilite extraction 提硝
mirabilite method 芒硝法
mirabilite solution 芒硝液

---

molybdenite [mɔˈlibdənait] *n.*

Definition: A blue-grey mineral, typically occurring as hexagonal crystals. It consists of molybdenum disulphide and is the most common ore of molybdenum. 辉钼矿 huī mù kuànɡ

Formula: MoS_2

Origin: from Greek molybdos "lead"

Example:

The strong hydrophobic attractive force between molybdenite and bubble interface is the basic reason which causes the natural floatability of molybdenite.
辉钼矿与气泡之间的强疏水引力是辉钼矿具备天然可浮性的根本原因。

Extended Terms:

molybdenite concentrate 辉钼精矿
molybdenite flotation 钼浮选
nano-molybdenite 纳米辉钼矿
low grade molybdenite 低品位辉钼矿
floating separation molybdenite 选钼

---

molybdite [məˈlibdait] *n.*

Definition: (also known as molybdic ocher or molybdine) a mineral, much of which is actually ferrimolybdite 钼华 mù huá

Formula: MoO_3

Origin: named for the composition

Example:

Molybdite has a Mohs hardness of 3-4.
钼华的莫氏硬度是 3~4。

---

monalbite [məˈnɔːlˌbait] *n.*

Definition: a modification of albite with monoclinic symmetry that is stable under equilibrium conditions at temperatures (about 1000℃) near the melting point 蒙钠长石 méng nà cháng shí

Formula: $NaAlSi_3O_8$

Origin: in allusion to the crystal system, Monoclinic, and its relationship to Albite

---

monazite [ˈmɔnəzait] *n.*

Definition: a brown crystalline mineral consisting of a phosphate of cerium, lanthanum, other rare earth elements, and thorium 独居石 dú jū shí (又称磷铈镧矿)

Formula: $(Ce, La, Y, Th)[PO_4]$

Origin: mid 19th century: from German Monazit, from Greek monazein "live alone" (because of its rare occurrence)

Example:

Monazite glass-ceramics consist of both monazite and metaphosphate glass phases.
独居石微晶玻璃由偏磷酸盐玻璃和独居石两相组成。

Extended Terms:

authigenic monazite 自生独居石
monazite sand 独居石砂
monazite concentrate 独居石精矿
monazite glass-ceramics 独居石微晶玻璃
bastnaesite-monazite concentrate 氟碳铈矿—独居石混合精矿

---

monticellite [ˌmɔntiˈselait] *n.*

Definition: gray silicate minerals of the olivine group 钙镁橄榄石 gài měi gǎn lǎn shí

Formula: $CaMgSiO_4$

Origin: named after Teodoro Monticelli (1759-1846), Italian mineralogist

Example:

Monticellite has a Mohs hardness of 5.
钙镁橄榄石的莫氏硬度是5。

Extended Terms:

monticellite nephelinite 钙镁橄霞石岩
monomer monticellite 钙橄榄石
monticellite polzenite 钙镁橄黄玄岩
iron-monticellite 铁钙橄榄石
monticellite nepheline basalt 钙镁橄霞玄武岩

---

montmorillonite [ˌmɔntməˈrilənait] *n.*

Definition: an aluminium-rich clay mineral of the smectite group, containing some sodium and magnesium 蒙脱石 méng tuō shí(又称胶岭石)

Formula: $(Na,Ca)_{0.33}(Al,Mg)_2(Si_4O_{10})(OH)_2 \cdot nH_2O$

Origin: mid-19th century: from Montmorillon, the name of a town in France, +-ite

Example:

There are almost all kinds of common clay minerals in Shanghai clay, especially illite and montmorillonite which are the main contents.
上海软土中几乎含有常见的各种黏土矿物,尤以伊利石、蒙脱石为主。

Extended Terms:

activated montmorillonite 活性微晶高岭土
calcium montmorillonite 钙蒙脱石
Anji montmorillonite 安吉蒙脱石
montmorillonite dehydration 蒙脱石脱水(作用)
bentonite montmorillonite 膨润土

---

smectite [ˈsmektait] *n.*

Definition: a clay mineral (e. g. bentonite) which undergoes reversible expansion on absorbing water 蒙脱石 méng tuō shí (又称绿土)

Formula: $(Na,Ca)_{0.33}(Al,Mg)_2[(Si,Al)_4O_{10}](OH)_2 \cdot nH_2O$

Origin: early 19th century: from Greek smēktis "fuller's earth" +-ite

Example:

Smectite has obvious selectivity for the absorption of cigaret gases. It absorbs less for sweet nonpolar substances, but absorbs more for strong polar poisons substances.
蒙脱石对卷烟烟气物质吸附具有明显的选择性,对具香味的非极性物质吸附得很少,对强极性的有毒的物质吸附量很高。

Extended Terms:

smectite powder 蒙脱石散
smectite interstratified 蒙混层
dioctahedral smectite 十六角蒙脱石,双八面体蒙脱石
glauconitic smectite 海绿石相蒙皂石
smectite gel 镁皂石凝胶

---

montroydite [mɔn'trɔidait] *n.*

Definition: an orthorhombic mineral in the oxidized zone of some mercury deposits 橙红石 chéng hóng shí(又称橙汞矿、橙汞石)

Formula: HgO

Origin: named for Montroyd Sharp, an owner of the mercury deposit at Terlingua, Texas, USA

Example:

Montroydite has a Mohs hardness of 1.5-2.
橙红石的莫氏硬度是1.5~2。

---

moonstone ['muːnstəun] *n.*

Definition: a pearly white semi-precious stone, especially one consisting of alkali feldspar 月长石 yuè cháng shí(又称月光石)

Formula: $(Na,K)AlSi_3O_8$

Origin: derived from a visual effect, or sheen, caused by light reflecting internally in the moonstone from layer inclusion of different feldspars

Example:

The potash feldspar contains 5-10 per cent of blue opalescent moonstone.
正长石中约有5%~10%的发蓝色蛋白光的月光石。

Extended Terms:

albite moonstone 钠长月光石

Ceylon moonstone 锡兰月光石
Labradorite moonstone 拉长月光石
Indian moonstone 印度月光石
oriental moonstone 东方月光石

---

mordenite [ˈmɔːdənait] *n.*

Definition: a zeolite mineral, crystallizing in the form of fibrous aggregates, masses, and vertically striated prismatic crystals 发光沸石 fā guāng fèi shí（又称丝光沸石）

Formula: $(Ca, Na_2, K_2)Al_2Si_{10}O_{24} \cdot 7H_2O$

Origin: named after the small community of Morden, Nova Scotia, Canada, along the Bay of Fundy, where it was first found

Example:

Natural zeolite has been discovered more than 40 varieties in nature, but only 6 to 7 varieties have been used to industry scale; clinoptilolite and mordenite are the two most familiar varieties.
自然界天然沸石已发现 40 余种，但具有工业规模的才 6 到 7 种，以斜发沸石、丝光沸石两种最常见。

Extended Terms:

natural mordenite 天然丝光沸石
modified mordenite 改性丝光沸石
mordenite catalysts 丝光沸石催化剂
siliceous mordenite 高硅丝光沸石
mordenite membrane 丝光沸石膜

---

mullite [ˈmʌlait] *n.*

Definition: a rare silicate mineral of post-clay genesis 莫来石 mò lái shí（又称富铝红柱石、多铝红柱石、高铝红柱石）

Formula: $Al(4+2x)Si(2-2x)O(10-x)$ where $x = 0.17$ to 0.59

Origin: named after its locality: Isle of Mull, Scotland, England

Example:

Zirconia Toughened Mullite (ZTM) Ceramics is one of finest performance high temperature structure ceramics.
氧化锆增韧莫来石(ZTM)是优良的高温结构陶瓷材料。

Extended Terms:

fibre mullite 纤维状莫来石

mullite refractories 莫来石耐火材料
zirconium mullite 锆莫来石
ultrafine mullite 莫来石超细粉
nanometer mullite 纳米莫来石

muscovite ['mʌskəuvait] *n.*

Definition: a silver-grey form of mica occurring in many igneous and metamorphic rocks 白云母 bái yún mǔ

Formula: $KAl_2(AlSi_3O_{10})(OH)_2$

Origin: mid 19th century: from obsolete Muscovy glass (in the same sense) +-ite

Example:

Muscovite is usually colorless but may be light gray, brown, pale green, or rose red.
白云母通常是无色的,但也可以是浅灰色、棕色、淡绿色或玫瑰色。

Extended Terms:

muscovite mica 白云母,硅铝酸钾
muscovite-granite 白云母花岗岩,白云花冈岩
muscovite-quartzrock 白云母石英岩
muscovite-granite porphyry 白云花岗斑岩,白雪花岗斑岩
muscovite chlorite subfacies 白云绿泥分相

nacrite ['neikrait] *n.*

Definition: a clay mineral that is a polymorph (or polytype) of kaolinite, crystallizing in the monoclinic system 珍珠陶土 zhēn zhū táo tǔ(又称珍珠石)

Formula: $Al_2Si_2O_5(OH)_4$

Origin: named from nacre in reference to the mother of pearl luster of nacrite masses

Example:

Tianhuang stone is a mixture of nacrite and dickite in varying proportions.
田黄石是珍珠石与地开石呈不同比例的混合物。

nahcolite [ˈnɑːkəˌlait] *n.*

Definition: a soft, colorless or white carbonate mineral with the composition of sodium bicarbonate, also called thermokalite, crystallizing in the monoclinic system 苏打石 sū dǎ shí (又称重碳钠盐)

Formula: $NaHCO_3$

Origin: in reference to its chemical formula

Example:

Nahcolite was first described in 1928 for an occurrence in a lava tunnel at Mount Vesuvius, Italy.
苏打石最早记载于 1928 年，出现在意大利维苏威火山熔岩中。

---

native chromium [ˈneitiv ˈkrəumjəm]

Definition: a chemical element which has the symbol Cr and atomic number 24, first element in Group 6; a steely-gray, lustrous, hard metal that takes a high polish and has a high melting point, odorless, tasteless, and malleable. 自然铬 zì rán gè

Formula: Cr

Origin: derived from the Greek word "chrōma" (χρώμα), meaning "color", because many of its compounds are intensely colored

Example:

Chromium has a Mohs hardness of 7.5.
自然铬的莫氏硬度是 7.5。

Extended Terms:

chromium hydroxide 氢氧化铬
hexavalent Chromium 六价铬
chromium silicide 硅化铬
chromium nitride 氮化铬
chromium nitrate 硝酸铬

---

natrolite [ˈnætrəlait, ˈnei-] *n.*

Definition: a tectosilicate mineral species belonging to the zeolite group, being a sodium aluminosilicate 钠沸石 nà fèi shí

Formula: $Na_2Al_2Si_3O_{10} \cdot 2H_2O$

Origin: from the Greek natron, "soda," in allusion to sodium content and lithos- "stone"

Example:

Wadeite is paragenic with aegirine, nepheline, arfredsonite, microcline and apatite, and associated with natrolite, rinkite and cancrinite.
钾钙板锆石与霓石、霞石、钠铁闪石、微斜长石、磷灰石共生；与钠沸石、层硅铈钛矿、钙霞石等伴生。

Extended Terms:

iron natrolite（钠沸石与绿泥石的混合物）铁钠沸石，杂绿泥钠沸石
barium natrolite 钡钠沸石
silver natrolite 银钠沸石
lithium natrolite 锂钠沸石
thallous natrolite 铊钠沸石

---

natron ['neitrɔn;-trən] *n.*

Definition: a mineral salt found in dried lake beds, consisting of hydrated sodium carbonate
泡碱 pào jiǎn（又称苏打、天然碳酸钠）

Formula: $Na_2CO_3 \cdot 10H_2O$

Origin: late 17th century: from French, from Spanish natrón, via Arabic from Greek nitron

Example:

The government kept monopoly control of beer, salt and natron (sodium carbonate used in the preservation of mummies).
该政府垄断了啤酒、盐和泡碱(用于保存木乃伊的碳酸钠)的生产和销售。

Extended Terms:

natron-catapliite 多钠锆石
natron ceratophyre 钠质角斑岩
natron lake 泡碱湖

---

nepheline ['nefilin;-li:n] *n.*

Definition: a colorless, greenish, or brownish mineral consisting of an aluminosilicate of sodium (often with potassium) and occurring as crystals and grains in igneous rocks 霞石 xiá shí

Formula: $(Na, K)AlSiO_4$

Origin: early 19th century: from French néphéline, from Greek nephelē "cloud" (because its fragments are made cloudy on immersion in nitric acid) +-ine

Example:

The result shows that the sintering temperature of ceramic body is reduced, but the toughness is improved obviously on substitution of nepheline for feldspar.

试验结果表明,霞石替代长石不仅能降低瓷坯的烧成温度,而且瓷坯的强度也有明显改善。

Extended Terms:

nepheline syenite 霞石正长岩,霞长石

nepheline basalt 霞石玄武岩,橄榄霞岩

nepheline calcimonzonite 霞石钙质二长岩

nepheline picrite 霞石苦橄岩

nepheline melilitite 霞石黄长岩

---

nephrite [ˈnefrait] *n.*

Definition: A hard pale green or white mineral which is one of the forms of jade. It is a silicate of calcium and magnesium. 软玉 ruǎn yù(又称闪玉、闪石玉)

Formula: $Ca_2(Mg,Fe)_5Si_8O_{22}(OH)_2$

Origin: late 18th century: from German Nephrit, from Greek nephros "kidney" (with reference to its supposed efficacy in treating kidney disease)

Example:

The proportion of nephrite is close to 3, harder than most of minerals except diamond and quartz.

软玉的比重接近3,比钻石、石英以外的大多数矿石都坚硬。

Extended Terms:

landscape nephrite 风景软玉

spinach nephrite 菠菜绿玉

greenstone nephrite 软玉

nephrite jade 肾石玉

calcic nephrite syenite 钙质软玉正长岩

---

neptunite [nepˈtjuːnait] *n.*

Definition: a black mineral composed of silicate of sodium, potassium, iron, manganese, and titanium 柱星叶石 zhù xīng yè shí

Formula: $KNa_2Li(Fe^{2+}, Mn^{2+})_2Ti_2Si_8O_{24}$

Origin: Named for Neptune, the Roman god of the sea, because it was found with aegirine

Example:

Neptunite has a specific gravity of 3.19-3.23.
柱星叶石的相对密度是 3.19 至 3.23。

Extended Term:

mangan-neptunite 锰柱星叶石

---

niccolite [ˈnikəlait] *n.*

Definition: a mineral, a form of nickel arsenide, having a metallic lustre and copper-red color, that is a minor ore of nickel 红砷镍矿 hóng shēn niè kuàng

Formula: NiAs

Origin: named after its composition

Example:

Niccolite has a specific gravity of 7.8.
红砷镍矿的相对密度是 7.8。

---

nickeline [ˈnikəliːn] *n.*

Definition: a mineral, a form of nickel arsenide, having a metallic lustre and copper-red color, that is a minor ore of nickel 红砷镍矿 hóng shēn niè kuàng

Formula: NiAs

Origin: named after its composition

Example:

Nickeline has a specific gravity of 7.8.
红砷镍矿的相对密度是 7.8。

Extended Terms:

nickeline wire 镍合金线
nickeline resistance 镍克林电阻

---

niobite [ˈnaiəbait] *n.*

Definition: A black mineral group that is an ore of niobium and tantalum. It has a submetallic luster and a high density and is a niobate of iron and manganese, containing tantalate of iron. 铌铁矿 ní tiě kuàng

Formula: $(Fe, Mn)(Nb, Ta)_2O_6$

Origin: named after its content of the element niobium

Example:

The fine spar contains 40% tantalum while the tantalum niobite contains 21% tantalum.
细晶石中的钽含量占 40%,而钽铌铁矿中的钽含量占 21%。

Extended Term:

mangan-niobite 锰铌铁矿

---

niter [ˈnaitə] *n.*

Definition: Niter (US) or nitre (UK) is the mineral form of potassium nitrate, also known as saltpeter (US) or saltpetre (UK). 钾硝石 jiǎ xiāo shí (又称印度硝石、火硝)

Formula: KNO_3

Origin: derived from Herbraic neter, used in ancient times for alkaline salts extracted by water from vegetable ashes

Example:

Since the major raw material of gunpowder niter is as white as snow, Arabians called gunpowder "Chinese Snow" and "Chinese Salt".
因为制造火药的主要原料硝石洁白如雪,所以火药被阿拉伯人称为"中国雪"和"中国盐"。

Extended Terms:

soda niter 钠硝石,智利硝石
cubic niter 钠硝石,智利硝石
niter cake 硝饼
niter ball 硝石球
niter air 硝气

---

nitratine [ˈnaitrətiːnˌ-tin] *n.*

Definition: Nitratine or nitratite, also known as cubic niter (UK:nitre), soda niter or Chile saltpeter (UK: saltpetre), is a mineral, the naturally occurring form of sodium nitrate. It crystallizes in the trigonal system, but rarely occurs as well formed crystals. It is isostructural with calcite. 钠硝石 nà xiāo shí(又称智利硝石)

Formula: $NaNO_3$

Origin: named after its composition of containing nitrates

Example:

Nitratine has a Mohs hardness of 1.5-2 and a specific gravity of 2.24-2.29.

钠硝石的硬度是 1.5～2，密度是 2.24～2.29。

---

nontronite [nɔn'trəunait] *n.*

Definition: the iron(III) rich member of the smectite group of clay minerals 绿脱石 lǜ tuō shí(又称绿高岭石，囊脱石)

Formula: $Na_{0.3}Fe^{3+}2(Si,Al)_4O_{10}(OH)_2 \cdot n(H_2O)$

Origin: named after its locality：Nontrone, Dordogne, France

Example:

Some evidence suggests that microorganisms may play an important role in the formation of nontronite.

有证据显示微生物在绿脱石的形成过程中起到重要作用。

---

nosean ['nəuzən] *n.*

Definition: a rare fluorescent isometric feldspathoid mineral, forming isometric crystals of variable color：white, grey, blue, green, to brown 黝方石 yǒu fāng shí

Formula: $Na_8Al_6Si_6O_{24}(SO_4)$

Origin: named after the German mineralogist, K. W. Nose (1753-1835)

Example:

Nosean was first described in 1815 from the Rhineland in Germany.

黝方石最早记载于 1815 年，出现在德国的莱茵区。

Extended Terms:

nosean sanidinite 黝方透长岩

nosean tinguaite 黝方霓霞脉岩

nosean trachyte 黝方粗面岩

nosean tephrite 黝方碱玄岩

nosean syenite 黝方正长岩

---

nsutite ['nsjuːtait] *n.*

Definition: a dull manganese oxide mineral with a hardness of 6.5-8.5 and an average specific gravity of 4.45 六方锰矿 liù fāng měng kuàng (又称恩苏塔矿)

Formula: $(Mn_{4\,1-x}^{+}Mn_{2x}^{+}O_{2-2x}(OH)_{2x}$ where $x = 0.06-0.07$)

Origin: named for the locality：Nsuta, Ghana, where it was discovered, +-ite

Example:

Pyrolusite, cryptomelane, coronadite, psilomelane, nsutite, manganese spar, manganite and manganocalcite were found in the E'rentaolegai silver deposit.
额仁陶勒盖银矿床出现的锰矿物主要为软锰矿、隐钾锰矿、铅硬锰矿、硬锰矿、六方锰矿、菱锰矿、水锰矿及含钙菱锰矿。

oligoclase [ˈɔligəukleis; əˈli-] *n.*

Definition: a feldspar mineral common in siliceous igneous rocks, consisting of a sodium-rich plagioclase (with more calcium than albite) 奥长石 ào cháng shí (又称更长石)

Formula: $(Na,Ca)(Si,Al)_4O_8$

Origin: mid 19th century: from oligo-"relatively little" + Greek klasis "breaking" (it is thought to have a less perfect cleavage than albite.)

Example:

Latites contain plagioclase feldspar (andesine or oligoclase) as large, single crystals (phenocrysts) in a fine-grained matrix of orthoclase feldspar and augite.
安粗岩中包含斜长石(中长石或奥长石),是在正长石和普通辉石组成的细粒基质中的大单晶体(斑晶)。

Extended Terms:

oligoclase basalt 富钠长石玄武岩,奥长石玄武岩
oligoclase andesite 奥安山岩,奥长安山岩
oligoclase porphyry 奥长斑岩
corundum oligoclase pegmatite 刚玉奥长伟晶岩
epidote oligoclase pegmatite 绿帘奥长伟晶岩

---

olivenite [ˈɔlivənait; əuˈlivənait] *n.*

Definition: an orthorhombic copper arsenate mineral 橄榄铜矿 gǎn lǎn tóng kuàng

Formula: Cu_2AsO_4OH

Origin: from the German olivernerz, literally "olive ore", in allusion to its typical color

Example:

The hardness of olivenite is 3, and the specific gravity is 4.3.

橄榄铜矿的硬度是 3,相对密度是 4.3。

olivine [ˌɔliˈviːn; ˈɔliviːn] *n.*

Definition: An olive-green, grey-green, or brown mineral occurring widely in basalt, peridotite, and other basic igneous rocks. It is a silicate containing varying proportions of magnesium, iron, and other elements. 橄榄石 gǎn lǎn shí

Formula: $(Mg,Fe)_2SiO_4$

Origin: late 18th century: from Latin oliva (See olive) +-ine

Example:

Over the past eight years, the TES instrument has discovered that Martian rocks and sands are composed almost entirely of the volcanic minerals feldspar, pyroxene and olivine—the components of basalt.

在过去的八年间,TES 仪器发现火星的岩石和沙几乎全是由构成玄武岩的火山矿物——长石、辉石和橄榄石组成。

Extended Terms:

olivine gabbro 橄榄辉长岩
olivine leucitite 橄榄白榴岩
olivine nephelinite 橄榄霞石岩
olivine diabase 橄榄辉绿岩
olivine basalt 橄件武岩,橄榄玄武岩

omphacite [ˈɔmfəsait] *n.*

Definition: any of a range of green, monoclinic pyroxene minerals found in eclogites and similar rocks; they are solid solutions of jadeite and diopside. 绿辉石 lǜ huī shí

Formula: $(Ca, Na)(Mg, Fe^{2+}, Al)Si_2O_6$

Origin: derived from the Greek omphax or unripe grape for the typical green color

Example:

Feitsui can be devided into jadeit, omphacite and soda chrome pyroxene.

翡翠可分为硬玉翡翠、绿辉石翡翠及钠铬辉石翡翠。

Extended Terms:

omphacite-eclogite 绿辉榴辉岩
omphacite jade 绿辉石玉

opal [ˈəupəl] *n.*

Definition: a gemstone consisting of a quartz-like form of hydrated silica, typically semi-transparent and showing many small points of shifting color against a pale or dark ground 蛋白石 dàn bái shí

Formula: $SiO_2 \cdot nH_2O$

Origin: late 16th century: from French opale or Latin opalus, probably based on Sanskrit upala "precious stone" (having been first brought from India)

Example:

It was a round opal, red and fiery, set in a circle of tiny rubies.
那是一块圆形蛋白石,红红的,像火焰一般,上面镶着一圈红宝石。

Extended Terms:

milky opal 乳蛋白石,奶蛋白石
jasp opal 碧石蛋白石,蛋白碧玉
dendritic opal 枝状蛋白石
gem opal 宝石蛋白石
Volcanic opal 火山蛋白石

---

orpiment [ˈɔːpimənt] *n.*

Definition: a bright yellow mineral consisting of arsenic trisulphide, formerly used as a dye and artist's pigment 雌黄 cí huáng

Formula: As_2S_3

Origin: late Middle English: via Old French from Latin auripigmentum, from aurum "gold" + pigmentum "pigment"

Example:

Orpiment is a kind of reddish-yellow mineral which was used in ancient times to paint over mistakes one made when writing.
雌黄是一种黄红色的矿物,古时候人们写错了字都会用它涂抹。

Extended Terms:

gamboge orpiment 雌黄
red orpiment 雄黄,雌黄
blocky medicinal realgar/orpiment 块状药用雄(雌)黄
realgar and orpiment deposit 雄黄雌黄矿床

---

orthoclase [ˈɔːθəukleis; -kleiz] *n.*

Definition: A common rock-forming mineral occurring typically as white or pink crystals. It is a potassium-rich alkali feldspar and is used in ceramics and glass-making. 正长石 zhèng cháng shí

Formula: $KAlSi_3O_8$

Origin: mid 19th century: from ortho-"straight" + Greek klasis "breaking" (because of the characteristic of two cleavages at right angles)

Example:

The main type of the carbonatites rock includes: biotite calcite carbonatite, orthoclase biotite calcite carbonatite, pyrite calcite carbonatite and so on.
碳酸岩主要岩石类型包括:黑云母—方解石碳酸盐岩、正长石—黑云母方解石碳酸盐岩、黄铁矿—方解石碳酸盐岩等。

Extended Terms:

sodian orthoclase 钠正长石
gallium orthoclase 镓正长石
orthoclase gabbro 正长辉长石
orthoclase twin 正长石双晶
lithium orthoclase 含锂长石

---

orthopyroxene [ˌɔːθəpaiˈrɔksiːn] *n.*

Definition: a mineral of the pyroxene group crystallizing in the orthorhombic system 斜方晶辉石 xié fāng jīng huī shí (又称斜方辉石) xié fāng huī shí

Formula: $Mg_2[Si_2O_6]$ and $Fe_2[Si_2O_6]$

Example:

Both clinopyroxene and orthopyroxene show exsolution texture.
单斜辉石和斜方辉石均发育出溶结构。

Extended Term:

orthopyroxene-andesite 斜辉安山岩,直辉安山岩

---

osmium [ˈɔzmiəm; ˈɔs-] *n.*

Definition: the chemical element of atomic number 76, a hard, dense, silvery-white metal of the transition series; a hard, brittle, blue-gray or blue-black transition metal in the platinum family, and is the densest natural element. 自然锇 zì rán é

Formula: Os

Origin: early 19th century: modern Latin, from Greek osmē "smell" (from the pungent smell of its tetroxide)

Example:

Osmium has a Mohs hardness of 6.25-6.5.
自然锇的莫氏硬度为 6.25~6.5。

Extended Terms:

osmium compounds 锇化合物
osmium carbonyl 五羰基合锇
pentacarbonyl osmium 五羰基合锇
osmium tetraoxide 四氧化锇
osmium filament 锇丝

---

osumilite [ɔ'sʌməlait] *n.*

Definition: a very rare hydrate potassium-sodium-iron-magnesium-aluminum silicate mineral 大隅石 dà yú shí

Formula: $(K,Na)(Fe,Mg)_2(Al,Fe)_3(Si,Al)_{12}O_{30} \cdot H_2O$

Origin: named for the locality: historic province of Osumi in Sakkabira, Kyushu, Japan

Example:

Osumilite was first discovered in Japan.
大隅石最早发现于日本。

palygorskite [ˌpæli'gɔːskait] *n.*

Definition: (also called palygorskite or attapulgite) A magnesium aluminium phyllosilicate which occurs in a type of clay soil common to the Southeastern United States. It is one of the types of fuller's earth. 坡缕石(链状结构黏土矿物) pō lǚ shí (又称山软木,凹凸棒石)

Formula: $(Mg,Al)_2Si_4O_{10}(OH) \cdot 4(H_2O)$

Origin: named after a deposit in the Ural Mountains, Russia

Example:

There is rich palygorskite ore bed in west region in Guizhou Province.
贵州省西部地区蕴藏了大量坡缕石矿床。

Extended Terms:

modified palygorskite 改性坡缕石
palygorskite clay 坡缕石黏土
nanometer palygorskite 纳米坡缕石
magnetic palygorskite composite 磁性坡缕石复合材料
i-bearing palygorskite 含碘坡缕石

---

paragonite [pə'rægənait] *n.*

Definition: (also known as natron-glimmer) A mineral, related to muscovite. It is a common mineral in rocks metamorphosed under blueschist facies conditions along with other sodic minerals such as albite, jadeite and glaucophane. 钠云母 nà yún mǔ

Formula: $NaAl_2(Si_3Al)O_{10}(OH)_2$

Origin: named from the Greek paragon for "to mislead", in allusion to its originally having been mistaken for talc

Example:

White mica includes white mica and rare paragonite.
白云母包括白云母和少见的钠云母。

Extended Terms:

paragonite soda mica 钠云母
paragonite-schist 钠云中岩,钠云片岩

---

pargasite ['pɑːgəsit] *n.*

Definition: A complex inosilicate mineral. It occurs in high temperature regional metamorphic rocks and in the skarns within contact aureoles around igneous intrusions. It also occurs in andesite volcanic rocks and altered ultramafic rocks. 韭闪石 jiǔ shǎn shí

Formula: $NaCa_2(Mg,Fe^{2+})_4Al(Si_6Al_2)O_{22}(OH)_2$

Origin: named after its locality: Pargas, Finland

Example:

The compositions of hornblende are obviously different from pargasite in olivite enclosure in mantle.

角闪石巨晶的成分明显不同于地幔橄榄岩包体中的韭闪石。

parasite [ˈpærisait] *n.*

Definition: a rare mineral consisting of cerium, lanthanum and calcium fluoro-carbonate 氟碳钙铈矿 fú tàn gài shì kuàng（又称氟菱钙铈矿）

Formula: $Ca(Ce,La)_2(CO_3)_3F_2$

Origin: named for J. J. Paris, mine proprietor at Muzo, north of Bogota, Columbia

Example:

Since 1835 when the mineral parisite was discovered, over 150 years have passed.
氟碳钙铈矿从 1835 年发现至今已有一百五十多年的历史。

Extended Terms:

parisite-(Nd)氟碳铌钙石
parisite-(Ce)氟碳铈钙石

patronite [ˈpætrənait] *n.*

Definition: a black form of vanadium sulfide that is an ore of vanadium 绿硫钒石 lǜ liú fán shí(又称绿硫钒矿)

Formula: VS_4

Origin: named for Antenor Rizo-Patron, Peruvian engineer, discoverer of the Peruvian occurrence

Example:

Patronite was first found in 1809 in a vanadium mine near Cerro de Pasco, Peru.
绿硫钒石最早发现于 1809 年秘鲁塞罗德帕斯科附近的钒矿里。

pectolite [ˈpektəlait] *n.*

Definition: a gray-white mineral inosilicate hydroxide of calcium and sodium that crystallizes in the triclinic system 针钠钙石 zhēn nà gài shí

Formula: $NaCa_2Si_3O_8(OH)$

Origin: from the Greek pektos-“compacted” and lithos-“stone”

Example:

Pectolite has a Mohs hardness of 4.5 to 5 and a specific gravity of 2.7 to 2.9.
针钠钙石的硬度是 4.5~5,相对密度是 2.7~2.9。

Extended Terms:

magnesium pectolite 镁针钠钙石
calcium pectolite 硬硅钙石
pectolite cat's eye 针钠钙石猫眼石
calcium pectolite(xonotlite)硬硅钙石,硬矽钙石
magnesium cell pectolite 镁针钠钙石

---

pennine ['penin] *n.*, penninite ['peninait] *n.*

Definition: an emerald-green, olive-green, pale-green, or bluish mineral of the chlorite group crystallizing in the monoclinic system 叶绿泥石 yè lǜ ní shí

Formula: $(Mg,Fe)5Al(AlSi_3O_{10})(OH)_8$

Example:

Penninite has a Mohs hardness of 2-2.5 and a specific gravity of 2.6-2.85.
叶绿泥石的莫氏硬度为2~2.5,相对密度为2.6~2.85。

Extended Term:

manganese pennine 锰叶绿泥石

---

pentlandite ['pentlən,dait] *n.*

Definition: a bronze-yellow mineral which consists of a sulphide of iron and nickel and is the principal ore of nickel 镍黄铁矿 niè huáng tiě kuàng(又称硫镍铁矿)

Formula: $(Fe,Ni)9S_8$

Origin: mid 19th century: from the name of Joseph B. Pentland (1797-1873), Irish traveller, +-ite

Example:

Pyrrhotite, Pentlandite and chalcopyrite are main metallic minerals in copper-nickel sulfide deposits.
磁黄铁矿、镍黄铁矿、黄铜矿是铜镍硫化物矿床的主要金属矿物。

Extended Terms:

cobalt adamite-pentlandite 方硫钴矿
hexagonal pentlandite 六方镍黄铁矿

---

periclase [peri'kleis] *n.*

Definition: a colorless mineral consisting of magnesium oxide, occurring chiefly in marble and limestone 方镁石 fāng měi shí

Formula: MgO

Origin: mid 19th century: from modern Latin periclasia, erroneously from Greek peri "utterly" + klasis "breaking" (because it cleaves perfectly)

Example:

Normal magnesia brick is made of Alumina-Magnesia spinel, periclase, It has higher temperature strength, good thermal shock stability, and strong resistance to basic slag.
普通镁铝以镁铝尖晶石和方镁石为主要晶相,具有较高的高温强度,较好的热稳定性,抗碱性炉渣侵蚀。

Extended Terms:

periclase porcelain 方镁石瓷
periclase brick 方镁石砖
periclase refractory 方镁石耐火材料
periclase-spinel refractory 方镁石尖晶石耐火材料
corundum-periclase-carbon brick 刚玉方镁石碳砖

---

pericline [ˈperiˌklain] *n.*

Definition: a white translucent variety of albite in the form of elongated crystals 肖纳长石 xiāo nà cháng shí

Origin: from Greek periklinēs, sloping on all sides: peri-, peri-+ klīnein, to slope; see klei- in Indo-European roots

Example:

Pericline has a Mohs hardness of 8-9.
肖纳长石的莫氏硬度为 8~9。

Extended Terms:

pericline twin 肖钠(长石)双晶
pericline twinning 肖钠双晶
pericline law 肖钠长石双晶律

---

peristerite [pəˈristərait] *n.*

Definition: a gem variety of albite (An_2-An_{24}) that resembles moonstone and has a blue or bluish-white luster characterized by sharp internal reflections of blue, green, and yellow 晕长石 yùn cháng shí(又称蓝彩钠长石,鸽彩石)

Example:

Iridescent plagioclase in Peristerite area is albite oligoclase; Iridescent plagioclase in Belggild

area is andesine labradorite.
在晕长石连生区产生晕彩的斜长石为钠奥长石,在沃基尔德连生区产生晕彩的斜长石为中拉长石。

Extended Terms:

peristerite gap 晕长石间断,晕长石间隙
peristerite solvus 晕长石溶线
peristerite unmixing 晕长石不混溶
peristerite intergrowth 晕长石互生

---

perovskite [pə'rɔvzkait] *n.*

Definition: a yellow, brown, or black mineral consisting largely of calcium titanate 钙钛矿 gài tài kuàng

Formula: $CaTiO_3$

Origin: named after the Russian mineralogist, L. A. Perovski (1792-1856)

Example:

It was identified that calcium titanate of perovskite structure is a highly efficient anticorrosion pigment for paints.
结果发现具钙钛矿结构的钛酸钙是一种适用于油漆的高效防腐颜料。

Extended Terms:

double perovskite 双钙钛矿
niobium perovskite 含铌钙钛
cerium perovskite 含铈钙钛矿
organic-inorganic perovskite materials 有机—无机类钙钛矿材料
perovskite-like complex oxides 类钙钛矿复合氧化物

---

perthite ['pəːθait] *n.*

Definition: a kind of feldspar consisting of a laminated intertexture of albite and orthoclase, usually of different colors 条纹长石 tiáo wén cháng shí

Formula: $(K, Na)AlSi_3O_8$

Origin: named after Perth, Ontario, Canada

Example:

Under the pressure of 510MPa, the perthite in granite begin to disappear at 600℃, and under the 700℃, most of them vanish and turn into orthoclase.

在 510MPa 压力下，花岗岩的条纹长石于 600℃开始消失，至 700℃时它们大量消失并变为正长石。

Extended Terms:

perthite pyroxenite 条纹辉岩
braid perthite 辫状条纹长石
double perthite 复纹长石
perthite syenite 条纹正长岩
microcline-perthite 微斜纹长石

---

petalite [ˈpetəlait] *n.*

Definition: (also known as castorite) a lithium aluminium tectosilicate mineral, crystallizing in the monoclinic system 透锂长石 tòu lǐ chánɡ shí

Formula: $LiAlSi_4O_{10}$

Origin: from the Greek petalon-"leaf" in allusion to the perfect basal cleavage

Example:

Petalite has a Mohs hardness of 6-6.5 and a specific gravity of 2.4.
透锂长石的硬度是 6 至 6.5，相对密度是 2.4。

Extended Terms:

petalite-analcime 透锂长石—方沸石
lepidolite petalite bearing pegmatite 锂云母—透锂长石伟晶岩

---

phenakite [ˈfenəˌkait] *n.*

Definition: (also spelled phenacite) a fairly rare nesosilicate mineral consisting of beryllium orthosilicate, occurring as isolated crystals, which are rhombohedral with parallel-faced hemihedrism, and are either lenticular or prismatic in habit 硅铍石 ɡuī pí shí

Formula: Be_2SiO_4

Origin: from the Greek phenakos-"deceiver", in allusion to its similarity to quartz when colorless

Example:

The Mohs hardness of phenakite is high, being 7.5-8, and its specific gravity is 2.96.
硅铍石的硬度较高，为 7.5~8，其相对密度是 2.96。

Extended Terms:

synthetic phenakite 合成硅铍石
germanium phenakite 锗硅铍石

---

phengite [ˈfendʒait] *n.*

Definition: a series name for dioctahedral micas, similar to muscovite but with addition of magnesium 多硅白云母 duō guī bái yún mǔ(又称月光石)

Formula: $K(AlMg)_2(OH)_2(SiAl)_4O_{10}$

Example:

The 3T-type phengite is a variety of muscovite.
3T 型多硅白云母是白云母的一个变种。

Extended Terms:

phengite eclogite 多硅白云母榴辉岩
phengite mudstone 多硅白云母泥岩
garnet-phengite schist 石榴石多硅白云母片岩
3t/T-type phengite 3t/T 型多硅白云母
chrome phengite 铬多硅白云母

---

phillipsite [ˈfilipsait] *n.*

Definition: a mineral of the zeolite group, a hydrated potassium, calcium and aluminium silicate with monoclinic crystals 钙十字沸石 gài shí zì fèi shí(又称钙交沸石,旧称钙十字石)

Formula: $(Ca,Na_2,K_2)3Al_6Si_{10}O_32 \cdot 12H_2O$ (also with sodium replaced by calcium: $KCaAl_3Si_5O_{16} \cdot 6H_2O$)

Origin: named after William Phillips (1775-1829), English mineralogist and founder of the Geological Society of London

Extended Terms:

ba phillipsite 钡钙十字沸石
silver phillipsite 银钙十字沸石
phillipsite tepherite 钙十字碱玄岩
phillipsite tephrite 钙十字碱玄岩

---

phlogopite [ˈflɔgəpait] *n.*

Definition: a brown micaceous mineral which occurs chiefly in metamorphosed limestone and magnesium-rich igneous rocks 金云母 jīn yún mǔ

Formula: $KMg_3[AlSi_3O_{10}](F,OH)_2$

Origin: mid-19th century: from Greek phlogōpos "fiery" (from the base of phlegein "burn") + ōps, ōp-"face" +-ite

Example:

The results showed that grain amaranth could efficiently take up K from both soil and micas (biotite and phlogopite).
结果表明,籽粒苋能有效地利用土壤和云母(黑云母和金云母)中的钾。

Extended Terms:

synthetic phlogopite 人造金云母
phlogopite phlogopitum 金礞石
sodium phlogopite 钠金云母
bronze mica phlogopite 金云母
phlogopite-vermiculite 金云母蛭石

---

piedmontite ['piːdmɔntait;-mən-] *n.*

Definition: a dark red mineral occurring in metamorphic rocks: a complex hydrated silicate containing calcium, aluminium, iron, and manganese 红帘石 hóng lián shí

Formula: $Ca_2(Al,Fe,Mn)_3(SiO_4)_3OH$

Origin: named after its locality: Piedmont, Italy

Example:

Piedmontite has a Mohs hardness of 6-7.
红帘石的莫氏硬度是6~7。

Extended Term:

piedmontite schist 红帘石片岩

---

pigeonite ['pidʒinait] *n.*

Definition: a yellow-green aluminosilicate mineral of the pyroxene group, containing iron, magnesium, and calcium 易变辉石 yì biàn huī shí

Formula: $(Ca,Mg,Fe)(Mg,Fe)Si_2O_6$

Origin: early 20th century: named after its type locality on Lake Superior's shores at Pigeon Point, Cook County, Minnesota, United States, +-ite

Example:

Pigeonite has a Mohs hardness of 6 and a specific gravity of 3.17-3.46.
易变辉石的硬度是6,相对密度是3.17~3.46。

Extended Term:
pigeonite-augitite 易变辉石岩

pingguite [ˈpiŋguˌait] *n.*

Definition: a new bismuth tellurite mineral 平谷矿 píng gǔ kuàng

Formula: $Bi_6^{3+}Te_2^{4+}O_{13}$

Origin: named after its locality: at Yangjiava, Pinggu County, near Beijing, China

Example:
Pingguite has a Mohs hardness of 5.75.
平谷矿的硬度为 5.75。

pitchblende [ˈpitʃblend] *n.*

Definition: a brown to black mineral that consists of massive uraninite, has a distinctive luster, contains radium, and is the chief ore-mineral source of uranium 沥青铀矿 lì qīng yóu kuàng

Formula: UO_2

Origin: partial translation of German Pechblende, from Pech "pitch" + Blende "blend"

Example:
Pitchblende is the only primary uranium mineral, which is associated with quartz, fluorite, hematite and pyrite.
沥青铀矿是矿石中唯一的原生铀矿物,其共生矿物主要有石英、荧石、赤铁矿和黄铁矿。

plagioclase [ˈpleidʒiəkleisˌ -kleiz] *n.*

Definition: a form of feldspar consisting of aluminosilicates of sodium and/or calcium, common in igneous rocks and typically white 斜长岩 xié cháng yán (又称斜长石)

Formula: $(Na,Ca)(Si,Al)_4O_8$

Origin: mid 19th century: from plagio-"oblique" + Greek klasis "cleavage" (originally characterized as having two cleavages at an oblique angle)

Example:
Plagioclase widely spreads across the earth surface, and causes relevant geological movements such as landslide and debris flow, which become one of most broad and severe geological

catastrophes.
斜长岩在地球表面的广泛分布,导致了与之相关的地质活动如滑坡、石流等,也随之成为人类面临的分布最广泛、严重的地质灾害类型之一。

Extended Terms:

zoned plagioclase 带状斜长石
plagioclase granite 斜长花岗岩
plagioclase arkose 斜长石砂岩
acid plagioclase 酸性斜长石
plagioclase rhyolite 斜长流纹岩

---

plancheite [ˈplaːntʃiˌait] *n.*

Definition: a hydrated copper silicate mineral, closely related to shattuckite in structure and appearance 纤硅铜矿 xiān guī tóng kuàng

Formula: $Cu_8Si_8O_{22}(OH)_4 \cdot (H_2O)$

Origin: named after J. Planche who brought it from Africa

Example:

Plancheite has a Mohs hardness of 5.5.
纤硅铜矿的莫氏硬度是5.5。

---

platinum [ˈplætinəm] *n.*

Definition: A precious silvery-white metal, the chemical element of atomic number 78. It was first encountered by the Spanish in South America in the 16th century, and is used in jewellery, electrical contacts, laboratory equipment, and industrial catalysts. 自然铂 zì rán bó

Formula: Pt

Origin: early 19th century: alteration of earlier platina, from Spanish, diminutive of plata "silver"

Example:

Aqueous solution plating and metallurgy composite were main methods of preparing platinum composite electrode at present.
水溶液电镀和冶金复合是目前最常用的制备铂复合电极的方法。

Extended Terms:

platinum black 铂黑
platinum chloride 氯化铂

platinum oxide 氧化铂
platinum solder 铂焊料
platinum bronze 铂青铜

pollucite [pə'ljusait] *n.*

Definition: a colorless transparent mineral of the zeolite family consisting of hydrous cesium aluminum silicate and occurring massive or crystallizing in cubes 铯榴石 sè liú shí (又称铯沸石)

Formula: $(Cs,Na)_2Al_2Si_4O_{12}\cdot 2H_2O$

Origin: named after Pollux, a figure from Greek mythology, brother of Castor, for its common association with "castorite" (petalite)

Example:

Pollucite has a Mohs hardness of 6.5 and a specific gravity of 2.9.
铯榴石的硬度是6.5,相对密度是2.9。

Extended Term:

decomposition of pollucite by sulphuric acid 铯榴石硫酸法分解

polycrase ['pɔlikreiz] *n.*

Definition: (also known as polycrasite) a black or brown metallic complex uranium yttrium oxide mineral, crystallizing in the orthorhombic system with typically radial prismatic crystal form 复稀金矿 fù xī jīn kuàng (又称锗铀钇矿石)

Formula: $(Y,Ca,Ce,U,Th)(Ti,Nb,Ta)_2O_6$

Origin: from the Greek for "many" and "mixture", in reference to the large number of chemical elements in the formula

Example:

Polycrase has a Mohs hardness of 5 to 6 and a specific gravity of 5.
复稀金矿的硬度是5至6,相对密度是5。

polyhalite [ˌpɔli'hælait] *n.*

Definition: a mineral usually occurring in fibrous masses, of a brick-red color, being tinged with iron, and consisting chiefly of the sulphates of lime, magnesia, and soda 杂卤石 zá lǔ shí

Formula: $K_2Ca_2Mg(SO_4)_4\cdot 2(H_2O)$

Origin: from the Greek polys, meaning "much" and hals, meaning "salt", in allusion to many salt components in the formulae

Example:

The shallow polyhalite potash deposit in Sichuan are distributed in the north plunge top of Huayingshan anticline where is anhydrock of earlier Mesozoic Triassic period.
四川浅层杂卤石钾矿分布于华蓥山背斜的北倾末端，这里是早期中生代三叠纪的硬石膏岩层。

Extended Terms:

polyhalite deposit 杂卤石矿床
polyhalite ore 杂卤石矿
shallow polyhalite 浅层杂卤石
shallow polyhalite potash 浅层杂卤石，浅层杂卤石钾
shallow seated polyhalite 浅层杂卤石

---

potassium alum [pəˈtæsjəm ˈæləm] *n.*

Definition: (also known as potash alum or tawas) the potassium double sulfate of aluminium 铝钾矾 lǚ jiǎ fán（又称钾矾、钾明矾、明矾）

Formula: $KAl(SO_4)_2 \cdot 12(H_2O)$

Origin: potassium: named from the English word potash (pot ashes) and the Arabic word qalimeaning "alkali"
alum: named from the Latin alumen

Example:

Potassium alum is commonly used in water purification.
明矾常用于水的净化。

Extended Terms:

chromic potassium alum 铬钾矾
gallium potassium alum 镓钾矾
ferric potassium alum 铁钾矾
potassium indium alum 铟钾矾
potassium chrome alum 硫酸铬钾

---

potassium feldspar [pəˈtæsjəm ˈfeldspɑː]

Definition: any of several alkaline feldspars containing potassium aluminium silicate 钾长石 jiǎ cháng shí

Formula: $KAlSi_3O_8$

Origin: potassium: named from the English word potash (pot ashes) and the Arabic word "qali" meaning "alkali"

feldspar: named from Feld "field" + spath "spar, non-metallic mineral, gypsum"

Example:

The development of ceramics industry of Anhui promotes the utilization of potassium feldspar resources.

安徽陶瓷工业的发展,提供了开发钾长石资源的市场。

Extended Terms:

potassium feldspar powder 钾长石粉

dissolution of potassium feldspar 钾长石分解

---

prehnite ['preinait, 'pre-] *n.*

Definition: a mineral, a basic calcium, aluminium and iron aluminosilicate, which occurs in stalagtitic aggregates or curved crystals 葡萄石 pú táo shí

Formula: $Ca_2Al_2Si_3O_{10}(OH)_2$

Origin: named in 1789 after Hendrik Von Prehn (1733-1785), Dutch colonial governor. This was the first mineral to be named after a person.

Example:

The metamorphic grades of Minleh and Changma areas are prehnite-pumpellyite and prehnite-actinolite facies, respectively.

民乐与昌马地区变质火山岩,变质度分别为葡萄石—绿纤石相和葡萄石—阳起石相。

Extended Term:

prehnite-pumpellyite facies 葡萄石—绿纤石相

---

proustite ['pru:stait] *n.*

Definition: A sulfosalt mineral consisting of silver sulfarsenide, known also as light red silver or ruby silver ore, and an important source of the metal. It is closely allied to the corresponding sulfantimonide, pyrargyrite. 淡红银矿 dàn hóng yín kuàng (又称硫砷银矿) liú shēn yín kuàng

Formula: Ag_3AsS_3

Origin: named after French chemist Joseph L. Proust (1754-1826) in 1832

Example:

Proustite has a Mohs hardness of 2.5 and a specific gravity of 5.57.

淡红银矿的硬度是 2.5,相对密度是 5.57。

pseudobrookite [ˌsjuːdəuˈbrukait] *n.*

Definition: a rare mineral occurring in cavities in granitic, rhyolitic, basaltic and other igneous rocks 铁板钛矿 tiě bǎn tài kuàng(又称假板钛矿)

Formula: Fe_2TiO_5

Origin: from the Greek pseudo-"I mislead" and the mineral brookite

Example:

Pseudobrookite has a Mohs hardness of 6 and a specific gravity of 4.4.
铁板钛矿的硬度是 6,相对密度是 4.4。

psilomelane [psaiˈlɔməlein, psailauˈmelein] *n.*

Definition: Also known as black hematite, is a group name for hard black manganese oxides such as hollandite and romanechite. Psilomelane consists of hydrous manganese oxide with variable amounts of barium and potassium. 硬锰矿 yìng měng kuàng

Formula: $mMnO \cdot MnO_2 \cdot nH_2O$

Origin: named in 1758 from the Greek psilos-"smooth" and melas-"black"

Example:

The mineral components of manganese nodules are mainly pyrolusite, psilomelane and limonite.
锰结核的矿物成分主要是软锰矿、硬锰矿和褐铁矿。

pumpellyite [pʌmˈpeliˌait] *n.*

Definition: a group of closely related sorosilicate minerals 绿纤石 lǜ xiān shí

Formula:

pumpellyite-(Mg): $Ca_2MgAl_2[(OH)_2|SiO_4|Si_2O_7]n \cdot (H_2O)$

pumpellyite-(Fe^{2+}): $Ca_2Fe^{2+}Al_2[(OH)_2|SiO_4|Si_2O_7]\ n \cdot (H_2O)$

pumpellyite-(Fe^{3+}): $Ca_2(Fe^{3+},Mg,Fe^{2+})(Al,Fe^{3+})_2[(OH,O)_2|SiO_4|Si_2O_7]\ n. \cdot H_2O$

pumpellyite-(Mn^{2+}): $Ca_2(Mn^{2+},Mg)(Al,Mn^{3+},Fe^{3+})_2[(OH)_2|SiO_4|Si_2O_7]n \cdot (H_2O)$

pumpellyite-(Al): $Ca_2(Al,Fe^{2+},Mg)Al_2[(OH,O)_2|SiO_4|Si_2O_7]\ n \cdot H_2O$

Origin: named after US geologist, Raphael Pumpelly (1837-1923)

Example:

Pumpellyite has a Mohs hardness of 5.5 and a specific gravity of 3.2.

绿纤石的硬度是 5.5,相对密度是 3.2。

Extended Terms:

prehnite-pumpellyite facies 葡萄石—绿纤石相
prehnite-pumpellyite-metagreywacke facies 葡萄石—绿纤石—变质杂砂岩相
pumpellyite-prehnite-quartz facies 绿纤石—葡萄石—石英相
pumpellyite-(Mg)镁绿纤石
pumpellyite-(Mn)锰绿纤石

---

pyrargyrite [paiəˈrɑːdʒərait] *n.*

Definition: a dark reddish-grey mineral consisting of a sulphide of silver and antimony 硫锑银矿 liú tī yín kuàng (又称深红银矿)

Formula: Ag_3SbS_3

Origin: mid 19th century: from Greek puro-(from pur "fire") + arguros "silver" (in allusion to color and silver content)+-ite

Example:

The Mohs hardness of pyrargyrite is 2.75, and the specific gravity 5.85.
硫锑银矿的莫氏硬度是 2.75,相对密度是 5.85。

---

pyrite [ˈpairait] *n.*

Definition: the common mineral iron disulfide, of a pale brass-yellow color and brilliant metallic luster, crystallizing in the isometric system 黄铁矿 huáng tiě kuàng

Formula: FeS_2

Origin: From the Greek, pyrites lithos, "stone which strikes fire," in allusion to the sparking produced when iron is struck by a lump of pyrite.

Example:

Authigenic pyrite is the normal product of anoxic sulfate reduction in marine sediments.
自生黄铁矿是海洋沉积物缺氧硫酸盐还原过程的主要产物。

Extended Terms:

pyrite cinder 黄铁矿烬滓
magnetic pyrite 磁黄铁矿
spear pyrite 茅白铁矿
white pyrite 白铁矿
pyrite furnace 黄铁矿炉

---

pyrochlore (=pyrochlorite) [ˈpaiərəukləː]

Definition: a mineral whose composition is that of a mixed niobate mostly of sodium, calcium and cerium 烧绿石 shāo lǜ shí (又称焦绿石)

Formula: $(Na,Ca)_2Nb_2O_6(OH,F)$

Origin: From the Greek πῦρ, "fire", and χλωρὸς, "green" because it typically turns green on ignition in classic blowpipe analysis.

Example:

When the Ca content increases, the crystal structures of the samples transform from monoclinic pyrochlore to cubic pyrochlore.

当钙含量增加时,样品的晶体结构由单斜烧绿石相转变为立方烧绿石。

Extended Terms:

uranium pyrochlore 铀烧绿石
cubic pyrochlore 立方焦绿石
pyrochlore structure 烧绿石结构
pyrochlore phase 焦绿石相
pyrochlore-rich 富烧绿石

---

pyrolusite [ˌpaiərəˈljuːsait; paiˈrɔlju-] *n.*

Definition: a black or dark grey mineral with a metallic lustre, consisting of manganese dioxide 软锰矿 ruǎn měng kuàng

Formula: MnO_2

Origin: early 19th century: from pyro-"fire, heat" + Greek lousis "washing" (because of the mineral's use in decolorizing glass)

Example:

The main silver-bearing minerals were psilomelane, pyrolusite, limonite, galena, pyrite and sphalerite.

主要载银矿物有硬锰矿、软锰矿、褐铁矿、方铅矿、黄铁矿和闪锌矿。

Extended Terms:

pyrolusite powder 软锰粉
pyrolusite ore 软锰砂
pyrolusite pulp 软锰矿浆
pyrolusite slurry 软锰矿浆
low-grade pyrolusite 低品位软锰矿

---

pyromorphite [ˌpaiərəˈmɔːfait] *n.*

Definition: a mineral consisting of a chloride and phosphate of lead, typically occurring as green, yellow, or brown crystals in the oxidized zones of lead deposits 磷氯铅矿 lín lǜ qiān kuàng

Formula: $Pb_5(PO_4)_3Cl$

Origin: early 19th century: from pyro-"fire, heat" + Greek morphē "form" +-ite

Example:

Mimetite group minerals are a group of isomorphic series minerals with hexagonal system, which include mimetite, vanadinite and pyromorphite.
砷铅矿族矿物是一族六方晶系的类质同象系列矿物,包括砷铅矿、钒铅矿和磷氯铅矿。

Extended Terms:

calcium pyromorphite 钙磷氯铅矿
brom pyromorphite 磷溴铅矿
fluor pyromorphite 氟磷氯铅矿
hydroxyl pyromorphite 羟磷铅石
silicate pyromorphite 硅磷氯铅矿

---

pyrope [ˈpaiərəup] *n.*

Definition: a deep red variety of garnet 镁铝榴石 měi lǚ liú shí

Formula: $Mg_3Al_2(SiO_4)_3$

Origin: from the Greek pyropos, "fiery-eyed", in allusion to the red hue

Example:

Red pyrope is one of the most important gem resources in Yunnan.
红色的镁铝榴石是云南最重要的宝石资源之一。

Extended Terms:

Arizona pyrope 亚利桑那镁铝榴石
pyrope-garnet 红榴石
pyrope-spessartine 镁铝—锰铝榴石
pyrope-almandine 镁铝—铁铝榴石
pyrope-gamet 红榴石—石榴石

---

pyrophyllite [ˌpaiərəuˈfilait] *n.*

Definition: a phyllosilicate mineral composed of aluminium silicate hydroxide, occurring in two more or less distinct varieties, namely, as crystalline folia and as compact masses 叶蜡石 yè là shí

Formula: $Al_2Si_4O_{10}(OH)_2$

Origin: from the Greek pyros "fire" and phyllos "a leaf", in allusion to its tendency to exfoliate into fan shapes when heated

Example:

The reaction activity of kaolinite, illite and pyrophyllite is obviously different and the reaction activity of pyrophyllite is the highest.
高岭石、伊利石和叶蜡石的反应活性存在较明显的不同,其中叶蜡石的反应活性最大。

Extended Terms:

pyrophyllite ceramics 叶蜡石陶瓷
pyrophyllite gasket 叶蜡石垫圈
pyrophyllite tube 叶蜡石管
zircon-pyrophyllite brick 锆英石—叶蜡石砖
pyrophyllite anhydride 偏叶蜡石

---

pyroxene [ˈpaiərɔksiːn; paiəˈrɔ-] *n.*

Definition: any of a large class of rock-forming silicate minerals, generally containing calcium, magnesium, and iron and typically occurring as prismatic crystals 辉石 huī shí

Formula: Pyroxenes have the general formula $XY(Si, Al)_2O_6$ (where X represents calcium, sodium, $iron^{+2}$ and magnesium and more rarely zinc, manganese and lithium and Y represents ions of smaller size, such as chromium, aluminium, $iron^{+3}$, magnesium, manganese, scandium, titanium, vanadium and even $iron^{+2}$).

Origin: early 19th century: from pyro-"fire" + Greek xenos "stranger" (because the mineral group was supposed alien to igneous rocks)

Example:

Jadeite jade consists of minerals of pyroxene, amphibole and feldspar groups.
翡翠由辉石族、闪石族、长石族矿物组成。

Extended Terms:

manganese pyroxene 含锰辉石
monoclinic pyroxene 单斜辉石,斜辉石
pyroxene andesite 辉石安山岩
pyroxene minette 辉石云煌岩
pyroxene amphibolite 辉石闪岩

---

pyrrhotite ['pirətait] *n.*

Definition: a reddish-bronze mineral consisting of iron sulphide, typically forming massive or granular deposits 磁黄铁矿 cí huáng tiě kuàng

Formula: $Fe_{(1-x)}S$ (x=0 to 0.2)

Origin: mid 19th century: from Greek purrhotēs "redness" +-ite

Example:

Pyrrhotite is the most common and abundant iron sulphide mineral in mine wastes worldwide.
磁黄铁矿是矿山尾矿堆中最为常见且分布很广的一种黄铁矿矿物。

Extended Terms:

pyrite pyrrhotite 硫化铁,黄铁矿
pyrrhotite peridotite 磁黄铁橄榄岩
micritic pyrrhotite 微晶磁黄铁矿
nickel pyrrhotite 镍磁黄铁矿
hexagonal pyrrhotite 六方磁黄铁矿

qitianlingite ['tʃitiæliŋˌait] *n.*

Definition: a black orthorhombic mineral 骑田岭矿 qí tián lǐng kuàng

Formula: $(Fe,Mn)_2(Nb,Ta)_2WO_{10}$

Origin: named for the locality: Qitianling, Hunan Province, China

Example:

Qitianlingite has a Mohs hardness of 5.5.
骑田岭矿的莫氏硬度是 5.5。

---

quartz [kwɔːts] *n.*

Definition: A hard mineral consisting of silica, found widely in igneous and metamorphic rocks and typically occurring as colorless or white hexagonal prisms. It is often colored by impurities (as in amethyst, citrine, and cairngorm) 石英 shí yīng

Formula: SiO_2

Origin: mid 18th century: from German Quarz, from Polish dialect kwardy, corresponding to Czech tvrdý "hard"

Example:

Quartz has much potential to be developed as device because of its material properties including piezoelectricity, isolation, transparent, high hardness, and high thermal stability.
石英的材料特性包括压电性、绝缘性、透光性、高硬度、热稳定性高等,是一种极具发展为元件潜力的材料。

Extended Terms:

quartz plate 石英片,水晶片
bastard quartz 白玻璃石英,块状白石英
quartz porphyry 石英斑岩
star quartz 星彩水星,星彩石英
quartz fibre 石英纤维,石英丝

rankinite [ˈræŋkinait] *n.*

Definition: a monoclinic mineral composed of calcium silicate 硅钙石 guī gài shí

Formula: $Ca_3Si_2O_7$

Origin: named for George Atwater Rankin (1884-?), physical chemist of the Geophysical Laboratory, Washington, D.C., USA

Example:

Rankinite has a Mohs hardness of 5.5.
硅钙石的莫氏硬度是5.5。

---

realgar [riˈælgə] *n.*

Definition: a soft reddish mineral consisting of arsenic sulphide, formerly used as a pigment and in fireworks 鸡冠石 jī guān shí(又称雄黄)

Formula: AsS

Origin: late Middle English: via medieval Latin from Arabic rahj al-gār "arsenic", literally "dust of the cave"

Example:

Realgar has a Mohs hardness of 1.5-2.
鸡冠石的莫氏硬度是 1.5～2。

Extended Terms:

realgar powder 雄黄散
realgar deposit 雄黄矿床
realgar tailings 雄黄尾矿
realgar ore district 雄黄矿区
cinnabar and realgar 朱砂雄黄

reinerite [riː'inərit]

Definition: a rare arsenite (arsenate (Ⅲ)) mineral, known especially from the polymetallic Tsumeb deposit 砷锌矿 shēn xīn kuàng

Formula: $Zn_3(AsO_3)_2$

Origin: named for Willy Reiner (1895-1965), Senior Chemist, Tsumeb Corporation, Tsumeb, Namibia, who analyzed this material

Example:

Reinerite has a Mohs hardness of 5-5.5.
砷锌矿的莫氏硬度是 5～5.5。

rhodizite ['rəudizait] *n.*

Definition: a rare potassium cesium beryllium aluminum borate mineral 硼锂铍矿 péng lǐ pí kuàng

Formula: $(K,Cs)Al_4Be_4(B,Be)12O_28$

Origin: named from the Greek "rose-colored" because it tinges the blowpipe flame red

Example:

Rhodizite has a Mohs hardness of 8.5.
硼锂铍矿的莫氏硬度是 8.5。

rhodochrosite [ˌrəudəu'krəusait] *n.*

Definition: a mineral consisting of manganese carbonate, typically occurring as pink, brown, or grey rhombohedral crystals 菱锰矿 líng měng kuàng

Formula: $MnCO_3$

Origin: mid 19th century: from Greek rhodokhrōs "rose-coloured" +-ite

Example: The rhodochrosite pellets can be used as the burden of medium and small blast furnace and electric furnace.
菱锰矿球团可作为中、小型高炉和电炉冶炼的配料。

Extended Terms:

rhodochrosite ore 菱锰矿,碳酸锰矿石
granular rhodochrosite 球粒菱锰矿
massive rhodochrosite 块状菱锰矿
fine rhodochrosite 微粒菱锰矿
phosphorus-rich rhodochrosite 高磷贫碳酸锰矿石

---

rhodonite [ˈrəudənait; ˈrɔ-] *n.*

Definition: a brownish or rose-pink mineral consisting of a silicate of manganese and other elements 蔷薇灰石 qiáng wēi huī shí

Formula: $(Mn, Fe, Ca)_5[Si_5O_{15}]$

Origin: early 19th century: from Greek rhodon "rose" +-ite

Example: The rhodonite deposit was formed mainly through two stages, sedimentation and regional metamorphism.
蔷薇辉石矿床的形成主要经历了沉积期和区域变质期两大阶段。

---

richterite [ˈritʃtərait] *n.*

Definition: a sodium calcium magnesium silicate mineral belonging to the amphibole group 碱镁闪石 jiǎn měi shǎn shí (又称锰闪石)

Formula: $Na(Ca, Na)Mg_5Si_8O_{22}(OH)_2$

Origin: named in 1865 for the German mineralogist Theodore Richter

Example:

Richterite has a Mohs hardness of 5.0-6.0 and a specific gravity of 3.0-3.5.
碱镁闪石的莫氏硬度是 5.0 至 6.0,相对密度是 3.0 至 3.5。

Extended Terms:

soda richterite 富钠透闪石,含锰镁钠钙闪石,钠锰闪石

iron richterite 铁镁钠钙闪石，铁钠透闪石
fluor richterite 氟钠透闪石，氟钠钙镁闪石
ferri richterite 锰亚铁钠闪石
potash richterite 镁钠钙闪石，钠透闪石

R

riebeckite [ˈriːbekait] *n.*

Definition: a dark blue or black mineral of the amphibole group, occurring chiefly in alkaline igneous rocks or as blue asbestos (crocidolite) 钠闪石 nà shǎn shí

Formula: $Na_2(Fe,Mg)_5Si_8O_{22}(OH)_2$

Origin: late 19th century: from the name of Emil Riebeck (1853-1885), German explorer, +-ite

Example:

The Mohs hardness of riebeckite is 5.0-6.0, and its specific gravity is 3.0-3.4.
钠闪石的莫氏硬度是 5.0 至 6.0，其相对密度是 3.0 至 3.4。

Extended Terms:

riebeckite rhyolite 钠闪流纹岩
riebeckite aplite 钠闪细晶岩
riebeckite syenite 钠闪正长岩
riebeckite granite 钠闪花冈岩
riebeckite trachyte 钠闪粗面岩

rock crystal [rɔk ˈkristəl]

Definition: transparent quartz, typically in the form of colorless hexagonal crystals 水晶 shuǐ jīng

Formula: SiO_2

Example:

Many stunning art decoration jewelry designs featured the black and white quartz combination of rock crystal and onyx.
许多令人震撼的艺术装饰首饰设计以岩石水晶和石华的黑和白色石英组合为特色。

Extended Terms:

rock crystal growth 水晶生长
rock crystal deposit 水晶矿床
natural rock crystal 天然水晶
synthetic rock crystal 合成水晶
rock crystal sphere 水晶球

roemerite [ˌreməˈrait] *n.*

Definition: a rust-brown to yellow mineral composed of hydrous ferric and ferrous iron sulfate 粒铁矾 lì tiě fán

Formula: $FeFe_2(SO_4)_4 \cdot 14H_2O$

Origin: named for Friedrich Adolph Römer (1809-1869), German geologist

Example:

Lishizhenite was found in a crack or cavity of anhydrite associated with roemerite, copiapite, sulphur, gypsum, pyrite and quartz.
李时珍石产于硬石膏裂隙或孔洞中，与粒铁矾、叶绿矾、自然硫、石膏、黄铁矿、石英等共生。

Extended Term:

zinc roemerite 锌粒铁矾，锌亚铁铁矾

---

rosasite [ˈrəuzəsait] *n.*

Definition: a secondary mineral found in the oxidization zone of copper-zinc deposits 纤维绿铜锌矿 xiān wéi lǜ tóng xīn kuàng（又称锌孔雀石、斜方绿铜锌矿）

Formula: $(Cu,Zn)_2(CO_3)(OH)_2$

Origin: named after its locality: Rosas mine, Narcao, Cagliari, Sardegna (Sardinia), Italy

Example:

Rosasite has a Mohs hardness of 4.
纤维绿铜锌矿的莫氏硬度是4。

---

roscoelite [ˈrɔskəulait] *n.*

Definition: a green mineral from the mica group that contains vanadium 钒云母 fán yún mǔ

Formula: $K(V_3^+, Al, Mg)_2AlSi_3O_{10}(OH)_2$

Origin: named for Henry Enfield Roscoe (1833-1915) of Manchester, England who first prepared pure vanadium

Example:

Roscoelite has been found in numerous places in USA, Australia, Japan, New Guinea and Czech Republic.
钒云母被发现于很多地方，如美国、澳大利亚、日本、新几内亚和捷克共和国。

---

roselite [ˈrəuzəlait] *n.*

Definition: a beautifully colored mineral and varies from rose-red to pink 砷钴钙石 shēn gǔ gài shí(又称玫瑰砷酸钙石)

Formula: $Ca_2(Co_2^+, Mg)[AsO_4]\ n \cdot 2^* H_2O$

Origin: named after Gustave Rose (1798-1873), professor on mineralogy at the University of Berlin, Germany

Example:

Roselite has a Mohs hardness of 3.5.
砷钴钙石的莫氏硬度是3.5。

Extended Term:

beta roselite β 砷钴钙石

---

rose quartz [rəuz kwɔːts]

Definition: a translucent pink variety of quartz 蔷薇石英 qiáng wēi shí yīng

Formula: SiO_2

Example:

Rose quartz is more often carved into figures such as hearts.
蔷薇石英经常被雕刻成心形等形状。

---

ruby [ˈruːbi] *n.*

Definition: a precious stone consisting of corundum in colour varieties varying from deep crimson or purple to pale rose 红宝石 hóng bǎo shí

Formula: Al_2O_3

Origin: Middle English: from Old French rubi, from medieval Latin rubinus, from the base of Latin rubeus "red"

Example:

Rubies have a hardness of 9.0 on the Mohs scale of mineral hardness.
红宝石的莫氏矿物硬度为9.0。

Extended Terms:

Brazilian ruby 巴西红宝石
natural ruby 天然红宝石

ruby crystal 红宝石晶体
ruby laser 红宝石激光
Bohemian ruby 波希米亚红宝石

ruthenium [ruːˈθiːniəm] *n.*

Definition: the chemical element of atomic number 44, a hard silvery-white metal of the transition series; a rare transition metal of the platinum group of the periodic table; and like the other metals of the platinum group, ruthenium is inert to most other chemicals 自然钌 zì rán liǎo

Formula: Ru

Origin: mid 19th century: modern Latin, from medieval Latin Ruthenia (See Ruthenia), so named because it was discovered in ores from the Urals

Example:

Iron, nickel, copper, silver, gold, iridium, or ruthenium may be included as co-catalyst metals.
铁、镍、铜,银、金,铱或钌,可用作复合催化剂金属。

Extended Terms:

ruthenium oxide 氧化钌
ruthenium catalyst 钌催化剂
ruthenium tetrachloride 四氯化钌
ruthenium hydroxide 三氢氧化钌,三羟化钌
ruthenium chloride 氯化钌

rutherfordine [ˈrʌðəˌfɔːdiːn] *n.*

Definition: a mineral containing almost pure uranium carbonate 菱铀矿 líng yóu kuàng (又称纤碳铀矿)

Formula: UO_2CO_3

Origin: named for Ernest Rutherford (1871-1937) British atomic physicist and Nobel Laureate

Example:

Rutherfordine is found primarily in the Morogoro Region of Tanzania in Africa.
菱铀矿最初发现于非洲坦桑尼亚的莫罗戈罗区。

rutile [ˈruːtail; -tiːl] *n.*

Definition: a black or reddish-brown mineral consisting of titanium dioxide, typically occurring as needle-like crystals 金红石 jīn hóng shí

Formula: TiO_2

Origin: early 19th century: from Latin rutilus "reddish", in reference to the deep red color observed in some specimens when viewed by transmitted light

Example:

The antibacterial activity of anatase TiO_2 is better than that of rutile TiO_2, and its antibacterial rate can get 80%.

锐钛矿型 TiO_2 的抗菌活性优于金红石型 TiO_2,其在棉织物上的抑菌率可达 80%。

Extended Terms:

artificial rutile 人造金红石
rutile porcelain 金红石瓷
rutile ceramic 金红石瓷
rutile prism 金红石棱镜
rutile ilmenitite 金红钛铁岩

sal ammoniac [sæl əˈməuniæk]

Definition: (also known as ammonium chloride) a rare mineral composed of ammonium chloride found around volcanic fumaroles and guano deposits 氯化氨 lǜ huà ān (又称卤砂)

Formula: NH_4Cl

Origin: Middle English: from Latin sal ammoniacus "salt of Ammon"

Example:

Sal ammoniac has a Mohs hardness of 1.5 to 2, and a low specific gravity of 1.5.

氯化氨的莫氏硬度是 1.5 至 2,相对密度较低,为 1.5。

Extended Terms:

sal ammoniac spirit 盐精
sal ammoniac lozenge 硇砂锭剂

sal ammoniac cell 氯化氨电池
sal ammoniac battery 氯化铵电池

samarskite [sə'mɑːskait] *n.*

Definition: a shiny black radioactive mineral that contains a complex mixture of rare earth oxides 铌钇矿 ní yǐ kuàng

Formula: $(Y,U,\ldots)(Nb,Ta)_2O_6$

samarskite-(Y):$(Y,Fe^{3+},U)(Nb,Ta)_5O_4$

samarskite-(Yb):$(Yb,Y,REE,U,Th,Ca,Fe^{2+})(Nb,Ta,Ti)O_4$

Origin: named for Vasilii Yefrafovich von Samarski-Bykhovets (1803-1870), Chief of Staff of the Russian Corps of Mining Engineers

Example:

Samarskite was first described in 1847 for an occurrence in Miass, Ilmen Mountains, Southern Ural Mountains of Russia.

铌钇矿最早记载于 1847 年,出现在俄罗斯南乌拉尔山脉伊尔门山的米阿斯市。

sanidine ['sænidiːn] *n.*

Definition: a glassy mineral of the alkali feldspar group, typically occurring as tabular crystals 透长石 tòu cháng shí(又称玻璃长石)

Formula: $(K,Na)(Si,Al)_4O_8$

Origin: early 19th century:from Greek sanis, sanid-"board" +-ine

Example:

Sanidine has a Mohs hardness of 6.

透长石的莫氏硬度是 6。

Extended Terms:

sanidine phonolite 透长响岩
sanidine nephelinite 透长霞石岩
sodian sanidine 钠透长石
sanidine carbonatite 透长碳酸岩
sanidine rock 透长岩

saponite ['sæpənait] *n.*

Definition: A monoclinic mineral of the montmorillonite group. It is soft, massive, and plastic, and exists in veins and cavities in serpentinite and basalt. 皂石 zào shí

Formula: $(1/2\ Ca,Na)_{0.33}(Mg,Fe^{2+})_3(Si,Al)_4O_{10}(OH)_2 \cdot 4H_2O$

Origin: from the Latin, sapo meaning "soap"

Example:

Compared with the alumina pillared montmorillonite, the pillared saponite exhibited higher cracking activity and better thermal stability.

相对于层柱蒙脱土,层柱皂石显示了更高的催化裂解性能和热稳定性。

Extended Terms:

zinc-saponite 锌蒙脱石,锌皂石

pillared saponite 交联皂石,层柱皂石

nickel saponite 镍皂石

magnesium-saponite 镁皂石

ga-substituted saponite 镓皂石

---

sapphirine [ˈsæfəriːn, sæˈfaiərin] *n.*

Definition: a transparent precious stone, typically blue, which is a variety of corundum (aluminium oxide) 假蓝宝石 jiǎ lán bǎo shí

Formula: $(Mg,Al)_8(Al,Si)_6O_{20}$

Origin: Middle English: from Old French safir, via Latin from Greek sappheiros, probably denoting "lapis lazuli"

Example:

Sapphirine is relatively hard (7.5 on Mohs scale), usually transparent to translucent, with a vitreous lustre.

假蓝宝石相对较硬(莫氏硬度为 7.5),经常呈透明状或半透明状,带有玻璃光泽。

---

sapphire [ˈsæfaiə] *n.*

Definition: a gemstone variety of the mineral corundum, an aluminium oxide, when it is a color other than red or dark pink, in which case the gem would instead be called a ruby, considered to be a different gemstone 蓝宝石 lán bǎo shí

Formula: aluminium oxide, Al_2O_3

Origin: from Ancient Greek σάπφειρος "sappheiros"

Example:

Sapphire is one of the best insulators.

蓝宝石是最好的绝缘材料之一。

Extended Terms:

Cashmere sapphire 克什米尔蓝宝石
Brazilian sapphire 巴西蓝宝石,蓝碧硒
leuco sapphire 白蓝宝石
Siam sapphire 暹罗蓝宝石
Rospoli Sapphire 罗斯波里蓝宝石

---

scapolite [ˈskæpəlait] *n.*

Definition: any of several mixed sodium and calcium aluminosilicates which also contain chloride, carbonate and sulfate and are found in metamorphic rocks 方柱石 fāng zhù shí

Formula: $(Na,Ca)_4[Al_3Si_9O_{24}]$ n. Cl ; The variations in composition of the different members of the group may be expressed by the isomorphous mixture of the following: $Ca_4(Si, Al)_{12}O_{24}(CO_3, SO_4)$ and $Na_4(Al, Si)l_2O_{24}Cl$, which are referred to as the meionite (Me) and marialite (Ma) endmembers respectively.

Origin: named in 1800 from the Greek skapos-"rod" and lithos-"stone"

Example:

The hardness of scapolite is 5-6, and the specific gravity varies with the chemical composition between 2.7 (meionite) and 2.5 (marialite).
方柱石的硬度是5至6,相对密度根据其化学成分的不同而变化,介于2.7(钙柱石)和2.5(钠柱石)之间。

Extended Terms:

Brazilian scapolite 巴西方柱石
Burmese scapolite 缅甸方柱石
scapolite cats-eye 方柱石猫眼
scapolite-belugite 方柱中长辉长岩
scapolite-gabbro 方柱辉长岩

---

scheelite [ˈʃeilait, ˈʃiː-] *n.*

Definition: a fluorescent mineral, white when pure, which consists of calcium tungstate and is an important ore of tungsten 白钨矿 bái wū kuàng(又称钙钨矿、钨酸钙矿)

Formula: $CaWO_4$

Origin: mid 19th century: from the name of Karl W. Scheele (1742-1786), Swedish chemist, +-ite

Example:

The scheelite orebody of the Yangjingou mining area occurred in the contact zone of the lower Paleozoic Wudaogou Group strata and the Indo-Sinian adamellite.
杨金沟矿区白钨矿体产于下古生界五道沟群地层与印支期二长花岗岩内外接触带中。

Extended Terms:

synthetic scheelite 合成白钨
scheelite concentrate 白钨精矿
scheelite middling 白钨中矿
artificial scheelite 人造白钨
scheelite flotation 白钨矿浮选

---

scholzite [ˈʃɔltsait] *n.*

Definition: a minature of the hydrated calcium zinc phosphate 磷钙锌矿 lín gài xīn kuàng

Formula: $CaZn_2(PO_4)_2 \cdot 2(H_2O)$

Origin: named for Adolph Scholz (1894-1950), mineral collector and chemist, Regensburg, Germany

Example:

Scholzite has a Mohs hardness of 4.
磷钙锌矿的莫氏硬度是4。

---

schorl [ˈʃɔːl] *n.*

Definition: a black iron-rich variety of tourmaline 黑电气石 hēi diàn qì shí

Formula: $NaFe^{2+}_3Al_6Si_6O_{18}(BO_3)_3(OH)_4$

Origin: late 18th century: from German Schörl, of unknown origin

Example:

The most common species of tourmaline is schorl.
最常见的电气石是黑电气石。

Extended Terms:

titanic schorl 金红石
white schorl 钠长石
precious schorl 贵电气石
glass schorl 斧石
red schorl 金红石,碧硒

---

scolecite [ˈskɔləsait, ˈskəu-] *n.*

Definition: A tectosilicate mineral belonging to the zeolite group. It is a hydrated calcium silicate. 钙沸石 gài fèi shí

Formula: $CaAl_2Si_3O_{10} \cdot 3H_2O$

Origin: from the Greek skolec "worm" in reference to the mineral's reaction to the blowpipe flame

Example:

Scolecite is usually colorless or white, but can also be pink, salmon, red or green.
钙沸石常常是无色或白色,但也有粉色、鲜肉色、红色或绿色。

---

scorodite [ˈskɔrədait, ˈskɔː-] *n.*

Definition: a common hydrated iron arsenate mineral, found in hydrothermal deposits and as a secondary mineral in gossans worldwide 臭葱石 chòu cōng shí

Formula: $FeAsO_4 \cdot 2H_2O$

Origin: from the Greek skorodon "garlic" alluding to the arsenic odor when heated

Example:

Beudanite occurred in the gossan of the oxidized zone of sulfide ore deposit, always associated with scorodite, mimetesite, garminite, bayldonite.
砷铅铁矾一般产出于含砷硫化物矿床氧化带铁帽中,常有臭葱石、砷铅石、砷铅铁石、乳砷铅铜石等矿物与之共生。

---

selenium [siˈliːniəm] *n.*

Definition: the chemical element of atomic number 34, a grey crystalline non-metal with semiconducting properties; a nonmetal, chemically related to sulfur and tellurium, and rarely occurs in its elemental state in nature 自然硒 zì rán xī

Formula: Se

Origin: early 19th century: modern Latin, from Greek selēnē "moon"

Example:

Selenium is one of the most important metalloid element existing in biotic environment.
硒是生物环境中存在的一种重要的类金属元素。

Extended Terms:

selenium tetrachloride 四氯化硒

selenium polymers 晒化物聚体
selenium layer 硒层
selenium analyzer 硒含量分析仪
selenium steel 加硒钢

senarmontite [ˌsenəˈmɔntait] *n.*

Definition: an oxide mineral of antimony 方锑矿 fāng tī kuàng

Formula: Sb_2O_3

Origin: named after the French mineralogist, H. Hureau de Senarmot (1808-1862).

Example:

Antimony trioxide is found in nature as the minerals valentinite and senarmontite.
在自然界中三氧化锑被发现于锑华和方锑矿中。

sepiolite [ˈsiːpiəlait] *n.*

Definition: a hydrated magnesium silicate, claylike mineral used for carving into decorative articles and smoking pipes 海泡石 hǎi pào shí

Formula: $Mg_4Si_6O_{15}(OH)_2 \cdot 6H_2O$

Origin: from the Greek, sepia- "cuttlefish" and lithos- "stone." From the German, meerschaum meaning "sea froth"

Example:

Sepiolite is a fibrous natural hydrated magnesium silicate with good absorptive property.
海泡石是一种纤维状含水的镁硅酸盐矿石,具有良好的吸附性能。

Extended Terms:

antibacterial sepiolite 抗菌海泡石
sepiolite adsorbent 海泡石吸附剂
sepiolite catalyst 海泡石催化剂
sepiolite clay 铝海泡石
sepiolite fiber 海泡石纤维

sericite [ˈserisait] *n.*

Definition: a fine-grained fibrous variety of muscovite formed by the alteration of feldspar, found chiefly in schist and in hydrothermally altered rock 绢云母 juàn yún mǔ

Formula: $K_{0.5-1}(Al, Fe, Mg)_2(SiAl)_4O_{10}(OH)_2nH_2O$

Origin: mid 19th century: from Latin sericum "silk" +-ite

Example:

Sericite is an important layered silicate mineral; its excellent nano-layered unit has the potential of making nano-modifying agent.
绢云母是重要的层状硅酸盐矿物,优良的纳米片层单元展示了以其制备绢云母纳米改性剂材料的潜在优势。

Extended Terms:

sericite schist 绢云母片岩
quartz-sericite schist 石英绢云片岩
ferri-sericite 铁绢云母
sericite-phyllite 绢云千枚岩
chlorite-sericite schist 绿泥绢云母片岩

---

serpentine [ˈsəːpəntain, -tiːn] *n.*

Definition: a dark green mineral consisting of hydrated magnesium silicate, sometimes mottled or spotted like a snake's skin 蛇纹石 shé wén shí

Formula: $(Mg, Fe)_3Si_2O_5(OH)_4$

Origin: late Middle English: via Old French from late Latin serpentinus

Example:

This stone is quite rare and difficult to cut, so in many cases other types of rock were used as substitutes, for example green feldspar, basalt, and serpentine.
这个石头实际上相当稀罕和难以切割,因此在许多情形里其他类型岩石被当作代用品使用,例如绿长石、玄武岩和蛇纹石。

Extended Terms:

serpentine marble 蛇纹石大理石
serpentine asbestos 蛇纹石石棉
resinous serpentine 脂光蛇纹石
serpentine rock 蛇纹石,蛇根碱石
serpentine jade 蛇纹石玉

---

serpentine asbestos [ˈsəːpəntain æzˈbestɔs]

Definition: 蛇纹石石棉 shé wén shí shí mián

serpentine: a group of common rock-forming hydrous magnesium iron phyllosilicate minerals 蛇纹石 shé wén shí

asbestos: a set of six naturally occurring silicate minerals exploited commercially for their desirable physical properties 石棉 shí mián

Formula: $(Mg, Fe)_3Si_2O_5(OH)_4$

Example:

Nephrite cat's eye and serpentine cat's eye were newly found in the serpentine asbestos deposit in Shimian town, Sichuan Province, China.
在中国四川省石棉县的蛇纹石石棉矿区新发现软玉猫眼和蛇纹石猫眼。

---

siderite [ˈsaidərait, ˈsi-] *n.*

Definition: a brown mineral consisting of ferrous carbonate, occurring as the main component of some kinds of ironstone or as rhombohedral crystals in mineral veins 菱铁矿 líng tiě kuàng (又称陨铁)

Formula: $FeCO_3$

Origin: late 16th century (denoting lodestone): from Greek sidēros "iron" +-ite

Example:

Siderite ore is one of the complex and refractory ores, from which it is difficult to obtain high grade iron concentrate.
菱铁矿为复杂难选矿种之一,采用常规选矿方法难以得到较高铁品位的铁精矿。

Extended Terms:

aerosiderite siderite 陨铁
siderite concentrate 菱铁精矿
magnesian siderite 镁菱铁矿
gypsum-siderite deposit 石膏菱铁矿矿床
copper-bearing siderite deposits 含铜菱铁矿床

---

siderophyllite [ˌsidərəˈfilait] *n.*

Definition: a rare member of the mica group of silicate minerals 针叶云母 zhēn yè yún mǔ (又称铁叶云母)

Formula: $KFe^{2+}Al(Al_2Si_2)O_{10}(F,OH)_2$

Origin: named from the Greek sideros, "iron", and phyllon, "leaf", in reference to its iron rich composition and perfect basal cleavage

Example:

Siderophyllite was first described in 1880 for an occurrence near Pikes Peak, Colorado, USA.
针叶云母最早记载于1880年,出现在美国科罗拉多州的派克斯峰。

siegenite [ˈsiːgənait] *n.*

Definition: a cobalt nickel sulfide mineral found at the Browns deposit, Batchelor, Northern Territory, Australia 块硫镍钴矿 kuài liú niè gǔ kuàng (又称硫镍钴矿)

Formula: $(Ni,Co)_3S_4$

Origin: named after its ore-bearing locality, Siegen, Germany

Example:

The cobalt in ore is in the form of independent minerals, such as siegenite, cobaltite, skutterudite and co-bearing pyrite, etc.
矿石中钴呈独立矿物硫钴镍矿、辉钴矿、方钴矿和含钴黄铁矿等产出。

Extended Term:

selenio-siegenite 硒硫镍钴矿,硒碲镍钴矿

silicon-oxygen tetrahedron [ˈsilikən ˈɔksidʒən ˌtetrəˈhiːdrən]

Definition: a four-sided geometric form created by the tight bonding of four oxygen atoms to each other, and also to a single silicon atom that lies in the middle of the form 硅氧四面体 guī yǎng sì miàn tǐ

Formula: SiO_4

Example:

The structure of silicon-oxygen tetrahedron in the raw materials may influence the properties of geopolymer to a certain extent.
原材料的硅氧四面体结构在很大程度上影响着地质聚合物的性能。

sillimanite [ˈsiliməˌnait] *n.*

Definition: (also known as fibrolite) an aluminosilicate mineral typically occurring as fibrous masses, commonly in schist or gneiss 硅线石 guī xiàn shí (又称夕线石)

Formula: $Al(AlSiO_5)$

Origin: mid 19th century: from the name of Benjamin Silliman (1779-1864), American chemist +-ite

Example:

Heilongjiang Province is China's most important sillimanite origin, where Jixi sillimanite has a natural quality of the best.
黑龙江省是中国最主要的硅线石产地,其中以鸡西硅线石所具有的天然品质最好。

Extended Terms:

sillimanite zone 硅线石带,矽线石带
sillimanite brick 硅线石砖
sillimanite crucible 硅线石坩埚
sillimanite-gneiss 硅线片麻岩,矽线片麻岩
kyanite-sillimanite type facies series 蓝晶石—硅线石型相系

---

silver [ˈsilvə] *n.*

Definition: a precious shiny greyish-white metal, the chemical element of atomic number 47
自然银 zì rán yín

Formula: Ag

Origin: Old English seolfor, of Germanic origin; related to Dutch zilver and German silber

Example:

Gold and silver are tractable metals.
金和银是容易加工的金属。

Extended Terms:

nickel silver 镍银,德银,洋铜,镍黄铜
silver alloy 银合金
nano silver 纳米银
sterling silver 标准纯银,纯银
silver oxide 氧化银,酸化银

---

skutterudite [ˈskʌtərʌdait] *n.*

Definition: a grey metallic mineral, typically forming cubic or octahedral crystals, consisting chiefly of an arsenide of cobalt and nickel 方钴矿 fāng gǔ kuàng (又称方砷钴矿)

Formula: $CoAs_3$

Origin: mid 19th century: from Skutterud (now Skotterud), a village in SE Norway, +-ite

Example:

The paper introduces several methods reported recently which are intended for reducing the

thermal conductivity of skutterudite related materials with phonon scattering mechanism.
论文结合声子散射机制介绍了近年来报道的几种降低方钴矿类热电材料热导率的途径。

Extended Terms:

nickel-skutterudite 镍砷钴矿
iron-skutterudite 铁方钴矿，方砷铁矿
filled skutterudite 填充式方钴矿
skutterudite compound 方钴矿化合物
nanobulk skutterudite 纳米块体方钴矿

---

smaltite ['smɔːltait] *n.*

Definition: a grey metallic mineral consisting chiefly of cobalt arsenide, typically occurring as cubic or octahedral crystals 砷钴矿 shēn gǔ kuàng(又称少砷方钴矿)

Formula: (Co,Fe,Ni)As_2

Origin: given by F. S. Beudant in 1832 because the mineral was used in the preparation of smalt for producing a blue color in porcelain and glass

Example:

The hardness of smaltite is 5.5 and the specific gravity is 6.5.
砷钴矿的硬度是5.5，相对密度是6.5。

Extended Terms:

nickel-smaltite 镍砷钴矿
high smaltite 高砷钴矿

---

smithsonite ['smiθsənait] *n.*

Definition: a yellow, grey, or green mineral consisting of zinc carbonate typically occurring as crusts or rounded masses 菱锌矿 líng xīn kuàng

Formula: $ZnCO_3$

Origin: named for English chemist and mineralogist, James Smithson (1754-1829), whose estate financed the Smithsonian Institution.

Example:

Smithsonite has a Mohs hardness of 4.5 and a specific gravity of 4.4 to 4.5.
菱锌矿的莫氏硬度是4.5，相对密度是4.4至4.5。

Extended Terms:

smithsonite ore 炉甘石
smithsonite deposit 菱锌矿床

calamine smithsonite 菱锌矿
smithsonite extract 菱锌矿提取物
cobalt smithsonite 钴菱锌矿

S

smoky quartz ['sməuki kwɔːts]

Definition: (also spelled smokey quartz) a semi-precious variety of quartz ranging in colour from light greyish brown to nearly black 烟晶 yān jīng (又称烟水晶、墨晶、茶晶)

Formula: SiO_2

Extended Term:

varieties of smoky quartz

soda niter ['səudə 'naitə], nitratine ['naitrətiːn, -tin] *n.*

Definition: Nitratine or nitratite, also known as cubic niter (UK: nitre), soda niter or Chile saltpeter (UK: saltpetre), is a mineral, the naturally occurring form of sodium nitrate. It crystallizes in the trigonal system, but rarely occurs as well-formed crystals. It is isostructural with calcite. 钠硝石 nà xiāo shí

Formula: $NaNO_3$

Origin: named after its composition of containing nitrates

Example:

Soda niter has a Mohs hardness of 1.5 to 2 and a specific gravity of 2.24 to 2.29.
钠硝石的莫氏硬度是1.5至2,相对密度是2.24至2.29。

sodalite ['səudəlait] *n.*

Definition: a blue mineral consisting chiefly of an aluminosilicate and chloride of sodium, occurring chiefly in alkaline igneous rocks 方钠石 fāng nà shí

Formula: $Na_8[AlSiO_4]_6Cl_2$

Origin: early 19th century: from soda +-lite

Example:

The Shiling sodalite syenite, Guangdong Province, is the only Early Cretaceous SiO_2 undersaturated alkali syenite known in the Nanling area, southeastern China.
广东省石岭方钠石正长岩是南岭地区唯一已知的早白垩世硅酸不饱和碱性岩。

Extended Terms:

sodalite phonelite 方钠响岩

transparent sodalite 透明方钠石
sodalite tephrite 方钠碱玄岩
sodalite nephelinite 方钠霞石岩
sodalite sanidinite 方钠透长石岩

---

suolunite [ˈsuɔːlənait] *n.*

Definition: a calcium silicate mineral 索伦石 suǒ lún shí

Formula: $Ca_2Si_2O_5(OH)_2 \cdot (H_2O)$

Origin: named for the locality: Suolun, Inner Mongolia, China

Example:

Suolunite has a Mohs hardness of 3.5.
索伦石的莫氏硬度是 3.5。

---

specularite [ˈspekjulərait] *n.*, specular iron ore [ˈspekjulə ˈaiən ɔː]

Definition: (also known as gray hematite; iron glance; specular iron) a black or gray variety of hematite with brilliant metallic luster, occurring in micaceous or foliated masses, or in tabular or disklike crystals 镜铁矿 jìng tiě kuàng

Formula: Fe_2O_3

Origin: in allusion to the specular appearance of this type of hematite

Example:

The iron oxides in rock crystals are specularite, hematite and lepidocrocite.
晶体中铁氧化物包裹体为镜铁矿、赤铁矿、纤铁矿。

Extended Terms:

specularite concentrate 镜铁矿精粉
Brazilian specularite 巴西镜铁矿
specularite and siderite ore 镜铁矿—菱铁矿矿石
pure specularite 纯镜铁矿

---

sperrylite [ˈsperilait] *n.*

Definition: A platinum arsenide mineral, an opaque metallic tin white mineral which crystallizes in the isometric system with the pyrite group structure. It forms cubic, octahedral or pyritohedral crystals in addition to massive and reniform habits. 砷铂矿 shēn bó kuàng

Formula: $PtAs_2$

Origin: named after its discoverer, the American chemist, Francis L. Sperry of Tallmadge, Ohio, USA

Example:

Sperrylite has a Mohs hardness of 6-7 and a very high specific gravity of 10.6.
砷铂矿的莫氏硬度是 6 至 7,相对密度很高,为 10.6。

S

---

spessartine [speˈsɑːtain] *n.*

Definition: a form of garnet containing manganese and aluminium, occurring as orange-red to dark brown crystals 锰铝榴石 měng lǚ liú shí(又称斜煌岩)

Formula: $Mn_3Al_2(SiO_4)_3$

Origin: mid 19th century: from French, from Spessart, the name of a district in NW Bavaria, +-ine

Example:

The end member of garnet is dominated by andradite, with minor grossular and spessartine.
石榴石的端元主要以钙铁榴石为主,伴以少量钙铝榴石和锰铝榴石。

Extended Terms:

calc spessartine 钙锰铝榴石
spessartine garnet 锰铝榴石
spessartine-gamet 锰铝榴石—石榴石
spessartine pyrope 锰铝—镁铝榴石

---

spessartite [ˈspesətait] *n.*

Definition: (previously named spessartite), a nesosilicate, manganese aluminium garnet species 锰铝榴石 měng lǚ liú shí (又称斜煌岩,闪斜煌斑岩)

Formula: $Mn(II)_3Al_2(SiO_4)_3$

Origin: named after its locality: Aschaffenburg, Spessart Mountains, Bavaria, Germany

Example:

Spessartite has a Mohs hardness of 6.5-7.5.
锰铝榴石的莫氏硬度是 6.5~7.5。

Extended Terms:

proterobase spessartite 角闪斜煌斑岩,次闪斜煌斑岩
diabase spessartite 辉绿闪斜煌斑岩,辉绿闪斜煌岩

---

sphalerite [ˈsfæləraɪt, ˈsfeɪl-] *n.*

Definition: a shiny mineral, yellow to dark brown or black in color, consisting of zinc sulphide 闪锌矿 shǎn xīn kuàng

Formula: ZnS

Origin: mid 19th century:from Greek sphaleros "deceptive" +-ite

Example:

The principal minerals of lead-zinc deposit in Zhenxun are sphalerite, galena, pyrite, quartz and barite.

闪锌矿、方铅矿、黄铁矿、石英及重晶石是镇旬铅锌矿田的主要矿物。

Extended Terms:

spherical sphalerite 球状闪锌矿
skeletal sphalerite 闪锌矿斑点
sphalerite flotation 闪锌矿浮选
sphalerite concentrate 锌精矿
high iron sphalerite 高铁闪锌矿

---

sphene [sfiːn] *n.*

Definition: a greenish-yellow or brown mineral consisting of a silicate of calcium and titanium, occurring in granitic and metamorphic rocks in wedge-shaped crystals 榍石 xiè shí

Formula: $CaTiSiO_5$

Origin: early 19th century:from French sphène, from Greek sphēn "wedge"

Example:

The sphene is a kind of stable mineral which is widely used in solidifying actinides separated from high level radioactive waste.

榍石是一种稳定矿物,是人造岩石固化高放射性废弃物较理想的基材之一。

Extended Term:

gem grade sphene 宝石级榍石

---

spinel [spiˈnel, ˈspɪnəl] *n.*

Definition: a hard glassy mineral occurring as octahedral crystals of variable color and consisting chiefly of magnesium and aluminium oxides 尖晶石 jiān jīng shí

Formula: $MgAl_2O_4$

Origin: early 16th century: from French spinelle, from Italian spinella, diminutive of spina "thorn"

Example:

The alumina rich spinel and magnesium rich spinel have different growth mechanisms.
富铝尖晶石和富镁尖晶石具有不同的生长机理。

Extended Terms:

ruby spinel 红尖晶石,红宝石尖晶石
ferromagnetic spinel 铁磁性尖晶石
synthetic spinel 合成尖晶石,人造尖晶石
iron spinel 铁尖晶石
normal spinel 正尖晶石

---

spodumene [ˈspɔdjumiːn] *n.*

Definition: a translucent, typically greyish-white aluminosilicate mineral which is an important source of lithium 锂辉石 lǐ huī shí

Formula: $LiAlSi_2O_6$

Origin: early 19th century: from French spodumène, from Greek spodoumenos "burning to ashes", present participle of spodousthai, from spodos "ashes"

Example:

Spodumene is the principle source of lithium, which is the lightest of all metals.
锂辉石是富含锂的矿物,锂在各种金属中最为闪亮耀眼。

Extended Terms:

spodumene deposit 锂辉石矿床
emerald spodumene 翠锂辉石,祖母绿色锂辉石
spodumene glaze 锂辉石釉
spodumene honeycomb ceramics 锂辉石质蜂窝陶瓷
β-spodumene β-锂辉石

---

spurrite [ˈspəːrait, ˈspʌ-] *n.*

Definition: a granular mineral composed of a mixed silicate and carbonate of calcium 灰硅钙石 huī guī gài shí

Formula: $Ca_5(SiO_4)_2CO_3$

Origin: named for Josiah Edward Spurr (1870-1950), American geologist

Example:

Spurrite has a Mohs hardness of 5 and a specific gravity of 3.
灰硅钙石的莫氏硬度是 5,相对密度是 3。

---

stannite [ˈstænait] *n.*

Definition: a dark-gray lustrous mineral, a mixed sulphide of copper, iron and tin, used as an ore of tin 黄锡矿 huáng xī kuàng (又称黝锡矿)

Formula: Cu_2FeSnS_4

Origin: from the Latin stannum-"tin"

Example:

In addition to cassiterite and stannite, galena also contains a high amount of tin.
除锡石和黄锡矿外,方铅矿也含有较高的锡。

---

staurolite [ˈstɔːrəlait] *n.*

Definition: A brown glassy mineral that occurs as hexagonal prisms often twinned in the shape of a cross. It consists of a silicate of aluminium and iron.十字石 shí zì shí

Formula: $(Fe,Mg,Zn)_2Al_9(Si,Al)_4O_{20}(OH)_4$

Origin: early 19th century: from Greek stauros "cross" +-lite

Example:

The staurolite schist and cordierite schist have been discovered in northern Xiangshan area for the first time.
在相山北部首次发现十字石片岩及堇青石片岩。

Extended Terms:

staurolite zone 十字石带
staurolite kyanite subfacies 十字蓝晶分相
staurolite schist 十字石片岩

---

stibarsen [stiˈbɑːsin] *n.*

Definition: a mineral of the compound SbAs of antimony and arsenic 砷锑矿 shēn tī kuàng

Formula: SbAs

Origin: named for its composition of antimony (Latin = stibium) and arsenic

Example:

Stibarsen has a Mohs hardness of 3-4.
砷锑矿的莫氏硬度是 3~4。

S

---

stibnite ['stibnait] *n.*

Definition: a lead-grey mineral, typically occurring as striated prismatic crystals, which consists of antimony sulphide and is the chief ore of antimony 辉锑矿 huī tī kuàng

Formula: Sb_2S_3

Origin: mid 19th century: from Latin stibium "black antimony" +-ite

Example:

In W-Sb-As-Au-ore deposits of central and western Hunan, stibnite is an important mineral carrier of gold.
在湖南中西部钨锑砷金矿床中,辉锑矿是金的重要载体矿物之一。

Extended Terms:

stibnite deposit 锑矿
stibnite concentrate 硫化锑精矿
stibnite veins 辉锑矿脉
Se-stibnite 硒辉锑矿
Te-rich stibnite 富碲辉锑矿

---

stilbite ['stilbait] *n.*

Definition: a tectosilicate zeolite mineral consisting of hydrated calcium aluminium silicate, common in volcanic rocks 辉沸石 huī fèi shí(又称束沸石)

Formula: $(Ca,Na_2,K_2)(Al_2Si_7O_{18}) \cdot 7H_2O$

Origin: from the Greek stilbe-"luster" in allusion to the pearly to vitreous luster

Example:

Stilbite has a Mohs hardness of 3.5-4.
辉沸石的莫氏硬度是 3.5~4。

---

stilpnomelane [ˌstilpnəuˈmelein] *n.*

Definition: a metamorphic phyllosilicate mineral of the mica group 黑硬绿泥石 hēi yìng lǜ ní shí

Formula: $K(Fe^{2+},Mg,Fe^{3+})_8(Si,Al)_{12}(O,OH)_{27}$

Origin: from the Greek stilpnos for "shining", and melanos for "black", in allusion to its luster and color

Example:

Stilpnomelane was first described in 1827 for an occurrence in Moravia in the Czech Republic.
黑硬绿泥石最早记载于1827年,出现在捷克共和国的摩拉维亚。

stolzite [ˈstəulzait] *n.*

Definition: a mineral, a lead tungstate 钨铅矿 wū qiān kuàng

Formula: $PbWO_4$

Origin: named after Joseph Alexis Stolz (1803-1896) from Teplice, Czechoslovakia, 1845

Example:

Although stolzite of Yaogangxian, Hunan Province, China, was discovered in 1948, no formal report concerned with this mineral has been published.
湖南瑶岗仙钨铅矿发现于1948年,但一直未作正式报道。

strengite [stredˈʒait, ˈstreŋait] *n.*

Definition: a pale-red mineral crystallizing in the orthorhombic system, isomorphous with variscite and dimorphous with phosphosiderite 红磷铁矿 hóng lín tiě kuàng(又称粉红磷铁矿)

Formula: $FePO_4 \cdot 2H_2O$

Origin: named after Johann August Streng (1830-1897), German mineralogist, University of Giessen, Germany

Example:

Strengite has a Mohs hardness of 3-4 and a specific gravity of approximately 2.87.
红磷铁矿的莫氏硬度是3~4,相对密度约为2.87。

Extended Term:

manganiferous strengite 锰红磷铁矿

strontianite [ˈstrɔnʃiənait] *n.*

Definition: a rare pale greenish-yellow or white mineral consisting of strontium carbonate 菱

锶矿 líng sī kuàng（又称碳酸锶矿）

Formula: $SrCO_3$

Origin: named after the village of Strontian, Lochaber, Scotland, where it was first discovered / named after its containing the element Strontium

Example:

Strontianite has a Mohs hardness of 3.5.
菱锶矿的莫氏硬度是 3.5。

Extended Term:

calcium strontianite 钙菱锶矿，钙碳锶矿

---

struvite ['struːvait] *n.*

Definition: a phosphate mineral, crystallizing in the orthorhombic system as white to yellowish or brownish-white pyramidal crystals or in platey mica-like forms 鸟粪石 niǎo fèn shí

Formula: $(NH_4)MgPO_4 \cdot 6H_2O$

Origin: named after the Russian diplomat, H. G. von Struve (1772-1851)

Example:

Phosphorus can be recovered by way of phosphate sedimentation. Recovery of phosphorus in the forms of struvite and calcium phosphate is considered to be the most promising way.
磷可以以磷酸盐沉淀的形式回收，其中以鸟粪石和磷酸钙的形式回收磷被认为是最具有前景的磷回收途径。

Extended Terms:

struvite precipitation 鸟粪石沉淀法
struvite calculus 鸟粪石
struvite stone 鸟粪石
struvite crystals 鸟粪石

---

sudoite [sjuː'dɔit] *n.*

Definition: a dioctahedral chlorite 须藤石 xū téng shí（又称铝绿泥石）

Formula: $Mg_2(Al,Fe^{3+})_3Si_3AlO_{10}(OH)_8$

Origin: named for Toshio Sudo (1911-　), University of Tokyo, Japan

Example:

Sudoite has a Mohs hardness of 2.5-3.5.
须藤石的莫氏硬度是 2.5~3.5。

---

sulphur [ˈsʌlfə] *n.*

Definition: the chemical element of atomic number 16; a yellow combustible non-metal 自然硫 zì rán liú

Formula: S

Origin: Middle English: from Anglo-Norman French sulfre, from Latin sulfur, sulphur

Example:

Sulphur can be used to make gunpowder.
硫磺可以用来制造火药。

Extended Terms:

sulphur residue 硫磺废渣
sulphur cycling 硫循环
sulphur solidification 固硫
sulphur furnace 焚硫炉
sulphur monochloride 一氯化硫

---

sunstone [ˈsʌnstəun] *n.*

Definition: a chatoyant gem consisting of feldspar, with a red or gold luster 日长石 rì cháng shí(又称日光石、太阳石、琥珀、猫睛石)

Formula: $(Ca,Na)((AI,Si)_2Si_2O_8)$

Example:

The sunstone theory was first proposed in 1966 by the Danish archaeologist Thorkild Ramskou.
关于(使用)日长石(导航的)理论,最初是由丹麦考古学家托基尔德·拉姆斯库于 1966 年提出的。

Extended Terms:

labradorite sunstone 拉长日光石
oregon sunstone 俄勒冈日光石

---

sylvite [ˈsilvait] *n.*

Definition: a colorless or white mineral consisting of potassium chloride, occurring typically as cubic crystals 钾盐 jiǎ yán

Formula: KCl

Origin: mid 19th century: from modern Latin (sal digestivus) Sylvii, the old name of this salt, +-ite

Example:

There are acute shortages of diamonds, platinum, chromite and sylvite.
金刚石、铂、铬铁矿、钾盐等矿产资源供需缺口较大。

Extended Terms:

sylvite nitrate 硝酸钾盐
organic sylvite 有机酸钾
sylvite area 钾盐矿区
sylvite supplementary treatment 补钾治疗
nitrate-type sylvite deposit 硝酸盐型钾盐矿床

---

sylvine [sil'vain] *n.*

Definition: a colorless or white mineral consisting of potassium chloride, occurring typically as cubic crystals 钾盐 jiǎ yán

Formula: KCl

Origin: mid 19th century: from modern Latin (sal digestivus) Sylvii, the old name of this salt, +-ine

Example:

There is a sylvine deposit in Yunnan Province of China.
中国云南有一个钾石盐矿藏。

Extended Terms:

sylvine resources 钾盐资源
crude sylvine 粗钾盐

---

syngenite ['sindʒənait] *n.*

Definition: a saline evaporite, consisting of a mixed potassium and calcium sulphate 钾石膏 jiǎ shí gāo

Formula: $K_2Ca(SO_4)_2 \cdot (H_2O)$

Origin: named from the Greek for related, for the chemical resemblance to polyhalite

Example:

Syngenite has a Mohs hardness of 2.5.

钾石膏的莫氏硬度是 2.5。

Extended Term:

Ca-rich syngenite 多钙钾石膏

szaibelyite [seiˈbelait] *n.*

Definition: a fibrous mineral which contains magnesium and boron 硼镁石 péng měi shí

Formula: $MgBO_2(OH)$

Origin: named for Stephan Szaibely (1777-1855), Hungarian mine surveyor who first collected the mineral

Example:

Szaibelyite has a Mohs hardness of 3-3.5.
硼镁石的莫氏硬度是 3~3.5。

Extended Terms:

hydroxyl szaibelyite 羟硼镁石
szaibelyite deposit 硼镁石矿床

talc [tælk] *n.*

Definition: a white, grey, or pale green soft mineral with a greasy feel, occurring as translucent masses or laminae and consisting of hydrated magnesium silicate 滑石 huá shí

Formula: $Mg_3(Si_4O_{10})(OH)_2$

Origin: late 16th century (denoting the mineral): from medieval Latin talcum

Example:

Talc, quartz and wollastonite had better reinforcing effect than others did in eight kinds of minerals selected.
滑石、石英和硅灰石对硅橡胶基体具有较好的增强作用。

Extended Terms:

grey talc 灰滑石
white talc 白滑石

micro talc 超微粒滑石粉
disseminated talc 浸染状滑石
fired talc 焙烧滑石

tangeite [ˈtæŋait, ˈtænait] (=tangueite) *n.*

Definition: (also known as calciovolborthite) a green, yellow, or gray mineral consisting of a basic vanadate of calcium and copper 矾钙铜矿 fán gài tóng kuàng

Formula: $CaCu(VO_4)(OH)$

Origin: named for the locality: Tange Gorge, Fergana Valley, Kyrgystan

Example:

Tangeite has a Mohs hardness of 3.5.
矾钙铜矿的莫氏硬度是 3.5。

tantalite [ˈtæntəlait] *n.*

Definition: a rare, dense black mineral consisting of a mixed oxide of iron and tantalum, of which it is the principal source 钽铁矿 tǎn tiě kuàng

Formula: $(Fe,Mn)(Ta,Nb)_2O_6$

Origin: early 19th century: from tantalum +-ite

Example:

We want to buy tantalite, Ta_2O_5, coltan, columbite in Uganda and East African region.
我们想在乌干达和东非地区购买钽铁矿、五氧化二钽、钽矿石、钶铁矿。

Extended Terms:

tantalite powder 钽粉
ferro tantalite 低铁钽矿
columbite-tantalite 铌钽铁矿
lithum tantalite single crystal 钽酸锂单晶
strontium bismuth tantalite 钽酸锶铋

tarbuttite [ˈtɑːbəˌtait] *n.*

Definition: a triclinic mineral of varying color, consisting of basic zinc phosphate 三斜磷锌矿 sān xié lín xīn kuàng

Formula: $Zn_2(PO_4)(OH)$

Origin: named for Percy Coventry Tarbutt, director of the Broken Hill Explaration Co., Zambia

Example:

Tarbuttite has a Mohs hardness of 4.
三斜磷锌矿的莫氏硬度是 4。

---

tengchongite [ˈtəntʃɔnait] *n.*

Definition: a new mineral of hydrated calcium uranyl molybdate 腾冲铀矿 téng chōng yóu kuàng

Formula: $CaU_6^{6+}Mo_2^{6+}O_{25} \cdot 12H_2O$

Origin: named for the locality: Tongbiguan village, Tengchong Co., Yunnan Province, China

Example:

Tengchongite has a Mohs hardness of 2-2.5.
腾冲铀矿的莫氏硬度是 2~2.5。

---

tenorite [ˈtenərait] *n.*

Definition: a copper oxide mineral 黑铜矿 hēi tóng kuàng

Formula: CuO

Origin: named for the Italian botanist Michele Tenore (1780-1861)

Example:

Tenorite is an oxide of cupper which has been studied in detail on its structure, force constants and spectrum.
黑铜矿为铜的氧化矿物,前人对其结构、力常数和光谱做过较详细的研究。

---

tetrahedrite [ˌtetrəˈhiːdrait, -ˈhe-] *n.*

Definition: a grey mineral consisting of a sulphide of antimony, iron, and copper, typically occurring as tetrahedral crystals 黝铜矿 yǒu tóng kuàng

Formula: $(Cu, Fe)_{12}Sb_4S_{13}$

Origin: named for its distinctive tetrahedron shaped cubic crystals

Example:

Tetrahedrite minerals in different types of gold deposits have different characteristic parameters.

在不同类型金矿中,黝铜矿具有不同的特征参数。

Extended Terms:

silver tetrahedrite 银黝铜矿
mercurial tetrahedrite 汞黝铜矿
tetrahedrite family 黝铜矿族
tetrahedrite group 黝铜矿族
tetrahedrite concetsate 黝铜矿精矿

---

thenardite [θiˈnɑːdait, ti-] *n.*

Definition: a white to brownish translucent crystalline mineral occurring in evaporated salt lakes, consisting of anhydrous sodium sulphate 无水芒硝 wú shuǐ máng xiāo

Formula: $Na_2(SO_4)$

Origin: mid 19th century: from the name of Baron Louis-Jacques Thénard (1777-1857), French chemist, +-ite

Example:

Thenardite salt rock is a kind of soft rock, the strength of which is low and the deformation is distinct.
无水芒硝岩盐是一种软岩,强度较低,变形较大。

Extended Term:

thenardite salt rock 无水芒硝岩盐

---

thermonatrite [ˈθəːməuˌneitrait] *n.*

Definition: a naturally occurring evaporite mineral form of sodium carbonate 水碱 shuǐ jiǎn (又称一水碳酸钠)

Formula: $Na_2CO_3 \cdot H_2O$

Origin: named from the Greek for heat and natron, as the dehydration product from heating natron

Example:

Thermonatrite has a Mohs hardness of 1.
水碱的莫氏硬度是 1。

---

thomsonite [ˈtɔmsənait] *n.*

Definition: the name of a series of tecto-silicate minerals of the zeolite group, with the mineral species being named thomsonite-Ca and thomsonite-Sr 杆沸石 gǎn fèi shí

Formula:

thomsonite-Ca: $NaCa_2Al_5Si_5O_{20} \cdot 6H_2O$
thomsonite-Sr: $(Sr,Ca)_2Na[Al_5Si_5O_{20}]\ n \cdot 7(H_2O)$

Origin: named after the Scottish chemist, Thomas Thomson (1773-1852)

Example:

The Mohs hardness of thomsonite is 5 to 5.5.
杆沸石的莫氏硬度是 5~5.5。

Extended Terms:

silver thomsonite 银杆沸石
strontium thomsonite 锶镁杆沸石

---

thorianite [ˈθɔːriənait, ˈθəu-] *n.*

Definition: a radioactive mineral containing thorium dioxide plus oxides of uranium, lanthanum, cerium neodymium and praseodymium in smaller quantities 方钍石(含放射能) fāng tǔ shí

Formula: ThO_2

Origin: named for its chemical composition containing thorium

Example:

Thorianite has a Mohs hardness of 6.
方钍石的莫氏硬度是 6。

Extended Terms:

alpha thorianite 阿伐方钍矿
gamma-thorianite 加马方钍矿
thorianite sand 方钍石

---

thuringite [ˈθuːrinˌdʒait] *n.*

Definition: a variety of the chlorite mineral chamosite, a hydrous iron and aluminium silicate mineral 鳞绿泥石 lín lǜ ní shí

Formula: $(Mg,Fe)_3(Fe^{3+},Fe^{2+})_3[Al_2Si_2O_{10}](OH)_8,FeO_{19}$

Origin: named after the German state of Thuringia

---

titanite [ˈtaitəˌnait] *n.*

Definition: a greenish-yellow or brown mineral consisting of a silicate of calcium and titanium, occurring in granitic and metamorphic rocks in wedge-shaped crystals 榍石 xiè shí (同 sphene)

Formula: $CaTiSiO_5$

Origin: late 18th century: from titanium +-ite

Example:

Garnet, zircon, rutile, apatite, tourmaline and titanite are six common clastic heavy minerals.
石榴石、锆石、金红石、磷灰石、电气石和榍石是六种常见的稳定重砂矿物。

Extended Terms:

zinc titanite 钛酸锌
zirconium titanite 钛酸锆
zirconium stannic titanite 钛酸锆锡
bismuth titanite 钛酸铋
titanite jacupirangite 榍石钛铁霰辉岩

---

tongbaite [tʌŋ'beit]

Definition: a greyish mineral, having the chemical formula Cr_3C_2, or chromium carbide 桐柏矿 tóng bǎi kuàng

Formula: Cr_3C_2

Origin: named after its discovery locality, Liu village, Tongbai county, Henan Province, China

Example:

Tongbaite has a Mohs hardness of 8.5.
桐柏矿的莫氏硬度是 8.5。

---

topaz ['təupæz] *n.*

Definition: a precious stone, typically colorless, yellow, or pale blue, consisting of a fluorine-containing aluminium silicate 黄晶 huáng jīng (又称黄水晶、黄玉)

Formula: $Al_2SiO_4(F,OH)_2$

Origin: Middle English (denoting a yellow sapphire): from Old French topace, via Latin from Greek topazos

Example:

The Yashan topaz lepidolite granite was emplaced at the latest stage of the composite batholith.
雅山黄玉锂云母花岗岩是雅山复式岩体的最晚阶段岩体。

Extended Terms:

asteriated topaz 星彩黄宝石
quartz topaz 石英黄宝石，黄水晶
California topaz 加利福尼亚黄宝石
Uralian topaz 乌拉尔黄宝石
Siberian topaz 西伯利亚黄宝石

---

torbernite [ˈtɔːbənait] *n.*

Definition: a radioactive, hydrated green copper uranyl phosphate mineral, found in granites and other uranium-bearing deposits as a secondary mineral 铜铀云母 tóng yóu yún mǔ

Formula: $Cu(UO_2)_2(PO_4)_2 \cdot 12\ H_2O$

Origin: named after the Swedish chemist, Tornbern Bergmann (1735-1784)

Example:

Torbernite has a Mohs hardness of 2-2.5.
铜铀云母的莫氏硬度是 2~2.5。

Extended Terms:

ortho-torbernite 正铜铀云母
meta torbernite 偏铜铀云母，变铜铀云母

---

tourmaline [ˈtuəməliːn, -lin] *n.*

Definition: A brittle gray or black mineral that occurs as prismatic crystals in granitic and other rocks. It consists of a boron aluminosilicate and has pyroelectric and polarizing properties, and is used in electrical and optical instruments and as a gemstone. 电气石 diàn qì shí (又称碧玺)

Formula: $(Ca,K,Na,n.)(Al,Fe,Li,Mg,Mn)_3(Al,Cr,Fe,V)_6(BO_3)_3(Si,Al,B)_6O_{18}(OH,F)_4$

Origin: mid 18th century: from French, based on Sinhalese tōramalli "carnelian"

Example:

Tourmaline is a semi-precious stone and mineral that emits negative ions and far-infrared rays.
电气石在矿产中是一种次级的宝石，它可以放射负离子和远红外线。

Extended Terms:

aquamarine tourmaline 海蓝宝石碧玺
Siberian tourmaline 西伯利亚碧玺

Pala tourmaline 帕拉碧玺
Paraibatourmaline 帕拉依巴电气石;帕拉依巴碧玺
liddicoatite tourmaline 钙锂电气石

T

tremolite [ˈtreməlait] *n.*

Definition: a white to grey amphibole mineral which occurs widely in igneous rocks and is characteristic of metamorphosed dolomitic limestones 透闪石 tòu shǎn shí

Formula: $Ca_2Mg_5Si_8O_{22}(OH)_2$

Origin: late 18th century: from Tremola Valley, Switzerland, +-ite

Example:

Laozhuang Tremolite-diopside Deposit is a superior large deposit firstly found and evaluated in Henan Province.
南召老庄透闪石—透辉石矿是河南省首次发现与评价的一个特大型矿床。

Extended Terms:

asbestos tremolite 阳起石
soda tremolite 钠透闪石
manganoan tremolite 含锰透闪石
fibrous tremolite 纤维状透闪石
tremolite talc 透闪石滑石

tridymite [ˈtridimait] *n.*

Definition: a high-temperature form of quartz found as thin hexagonal crystals in some igneous rocks and stony meteorites 鳞石英 lín shí yīng

Formula: SiO_2

Origin: mid 19th century: from German Tridymit, from Greek tridumos "threefold", from tri-"three" + -dumos (as in didumos "twin"), because of its occurrence in groups of three crystals

Example:

The order of toxicity is found to be: tridymite >α-quartz>silica gel.
毒性作用顺序为:鳞石英>α-石英>硅胶。

Extended Terms:

tridymite alboranite 鳞英辉玄岩,鳞英苏玄岩
tridymite latite 鳞英二长安岩
tridymite peralboranite 鳞英淡苏安玄岩

tridymite-trachyte 鳞英粗面岩,鳞矽粗面岩,鳞硅粗面岩
low tridymite 低温鳞石英

trona [ˈtrəunə] *n.*

Definition: a grey mineral which occurs as an evaporite in salt deposits and consists of a hydrated carbonate and bicarbonate of sodium 天然碱 tiān rán jiǎn

Formula: $Na_3H(CO_3)_2 \cdot 2H_2O$

Origin: late 18th century: from Swedish, from Arabic natrūn (See natron)

Example:

The calcined trona can replace soda ash for use in the flotation operation.
经过煅烧的天然碱可以在浮选作业中代替纯碱使用。

Extended Terms:

trona deposit 天然碱矿床
trona rock 天然碱岩
trona ore 碳酸钠矿,天然碱矿
trona potash 氯钾天然碱
trona-water geothermometer 天然碱—水氢同位素地热温标

tschermakite [ˈtʃəːmækait] *n.*

Definition: a complex silicate mineral (a kind of hornblende) containing calcium, magnesium, aluminium and ferric iron found in metamorphic rocks 钙镁闪石 gài měi shǎn shí

Formula: $(Ca_2(Mg,Fe^{2+})_3Al_2(Si_6Al_2)O_{22}(OH)_2)$

Origin: named for Gustav Tschermak von Sessenegg (1836-1927), Austrian mineralogist

Example:

On the Mohs hardness scale, zoisite and hornblende (or tschermakite) have hardnesses about 6 ½, whereas corundum has a hardness of 9.
黝帘石和角闪石(或者钙镁闪石)的摩氏硬度为6.5,而刚玉则达到9。

Extended Terms:

alumino tschermakite 铝镁钙闪石
ferro alumino tschermakite 铁铝镁钙闪石

tungstite [ˈtʌŋstait] *n.*

Definition: a yellow mineral consisting of hydrated tungsten oxide, typically occurring as a powdery coating on tungsten ores 钨华 wū huá

Formula: $WO_3 \cdot H_2O$

Origin: mid 19th century: from tungsten (Swedish, tung sten = "heavy stone") +-ite

Example:

Tungstite has a Mohs hardness of 2.5 and a specific gravity of 5.5.
钨华的莫氏硬度是 2.5，相对密度是 5.5。

---

turquoise [ˈtəːkwɔiz, -kwaːz] *n.*

Definition: a semi-precious stone, typically opaque and of a greenish-blue or sky-blue color, consisting of a hydrated phosphate of copper and aluminium 绿松石 lǜ sōng shí (又称土耳其玉)

Formula: $CuAl_6(PO_4)_4(OH)_8 \cdot 4H_2O$

Origin: late Middle English: from Old French turqueise "Turkish" (stone)

Example:

Situated in the east margin of the south Qinling orogen, Yungaishi area is well-known throughout the world as one of the high quality turquoise occurrences.
云盖寺地区位于南秦岭造山带东缘，以盛产优质绿松石而闻名。

Extended Terms:

tooth turquoise 齿绿松石
ferri turquoise 铁绿松石
porcelain turquoise 瓷绿松石
Alexandrian turquoise 亚历山大绿松石
Nevada turquoise 内华达绿松石

---

tychite [ˈtaikait] *n.*

Definition: a carbonate and sulphate of magnesium and sodium which occurs in colorless octahedral crystals at Borax Lake, San Bernardino County, California 杂芒硝 zá máng xiāo

Formula: $Na_6Mg_2(CO_3)_4(SO_4)$

Origin: named from the Greek for "luck or chance" because out of 5,000 northupite crystals, the last 10 were tychite

Example:

Tychite has a Mohs hardness of 3.5.

杂芒硝的莫氏硬度是 3.5。

tyuyamunite [ˌtjuːjəˈmuːnait] *n.*

Definition: A yellowish earthy mineral which is an ore of uranium. It consists of a hydrated vanadate of calcium and uranium. 钙矾铀矿 gài fán yóu kuàng(又称钒钙铀矿)

Formula: $Ca(UO_2)_2(VO_4)_2 \cdot 5\text{-}8H_2O$

Origin: named from its locality: the Tyuya-Muyun (Tuja Mujun) hill, a northern spur of the Alai Mountains, Ferghana, in Turkestan, +-ite

Example:

Uranium minerals mainly are uranophane, tyuyamunite and carnotite.
含铀矿物主要是硅钙铀矿、钒钙铀矿、钒钾铀矿等。

Extended Term:

meta tyuyamunite 偏钒钙铀矿

ulexite [ˈjuːleksait] *n.*

Definition: A mineral occurring on alkali flats as rounded masses of small white crystals. It is a hydrated borate of sodium and calcium. 钠硼解石 nà péng jiě shí (又称硼钠方解石)

Formula: $NaCaB_5O_6(OH)_6 \cdot 5(H_2O)$

Origin: mid 19th century: from George L. Ulex (died 1883), German chemist, +-ite

Example:

The calcium borate product was prepared with natural ulexite powder, through hydrothermal depolymerization and phase inversion.
以钠硼解石天然矿粉为原料,硼酸钙产品经水热解聚和相转化制备而出。

Extended Term:

borate ulexite 钠硼解石

ulvospinel [ʌlˈvɔspainel] *n.*

Definition: an iron titanium oxide mineral, belonging to the spinel group of minerals 钛尖晶

石 tài jiān jīng shí

Origin: named after the locality (Sodra Ulvön island, Angermanland, Sweden) and the spinel group of minerals

Example:

Ulvospinel has a Mohs hardness of 5.5 to 6.
钛尖晶石的莫氏硬度是 5.5 至 6。

Extended Term:

chromian ulvospinel 铬钛铁晶石

---

uraninite [ˌjuəˈræninait] *n.*

Definition: a black, grey, or brown mineral which consists mainly of uranium dioxide and is the chief ore of uranium 沥青铀矿 lì qīng yóu kuàng (又称晶质铀矿)

Formula: UO_2

Origin: late 19th century: from urano- +-ite

Example:

In the older rocks below these glacial deposits are detrital uraninite and pyrite, two minerals considered evidence for very low levels of atmospheric oxygen.
这些冰河沉积物下方的较老岩层,是沥青铀矿与黄铁矿碎屑;这两种矿物被认定为大气中氧气极为稀少的证据。

Extended Terms:

thorian uraninite 钍晶质铀矿
uraninite group 沥青铀矿族
ulrichite uraninite 沥青方铀矿
thor uraninite 钍铀矿
metamict uraninite 沥青铀矿,晶质铀矿

---

uvarovite [juːˈvɑːrəvait] *n.*

Definition: an emerald green variety of garnet, containing chromium 钙铬榴石 gài gè liú shí(又称钙铬石榴子石)

Formula: $Ca_3Cr_2(SiO_4)_3$

Origin: mid 19th century: from the name of Count Sergei S. Uvarov (1785-1855), Russian statesman, +-ite

Example:

Uvarovite from the eastern Tibet is coexisted with quartz and calcite.

西藏东部的钙铬榴石与石英和方解石存在共生关系。

valentinite [ˈvæləntinait] *n.*

Definition: an antimony oxide mineral, crystallizing in the orthorhombic system and typically forms as radiating clusters of euhedral crystals or as fibrous masses 锑华 tī huá

Formula: Sb_2O_3

Origin: named after the German alchemist, Basilius Valentinus (16th or 17th centuries)

Example:

Valentinite has a Mohs hardness of 2.5 to 3 and a specific gravity of 5.67.
锑华的莫氏硬度是 2.5 至 3，相对密度是 5.67。

---

vanadinite [vəˈnædinait] *n.*

Definition: a rare reddish-brown mineral consisting of a vanadate and chloride of lead, typically occurring as an oxidation product of lead ores 钒铅矿 fán qiān kuàng（曾称褐铅矿）

Formula: $Pb_5(VO_4)_3Cl$

Origin: mid 19th century: from vanadium +-ite

Example:

Mimetite group minerals are a group of isomorphic series minerals with hexagonal system, which include mimetite, vanadinite and pyromorphite.
砷铅矿族矿物是一族六方晶系的类质同象系列矿物，包括砷铅矿、钒铅矿和磷氯铅矿。

---

variscite [ˈværisait] *n.*

Definition: a hydrated aluminium phosphate mineral, a relatively rare phosphate mineral, a secondary mineral formed by direct deposition from phosphate-bearing water that has reacted with aluminium-rich rocks in a near-surface environment 磷铝石 lín lǚ shí

Formula: $AlPO_4 \cdot 2H_2O$

Origin: named after Variscia, the historical name of Vogtland, Germany

Example:

Sulfate-variscite occurs in the oxidation zone of a sedimentary pho-sphatic deposit at Muqiongkudouke, Weili County, Bayinguoleng Mongolia Autonomous Prefecture.
硫磷铝石产于巴音郭楞蒙古自治州尉犁县木穷库都克沉积型磷矿氧化带中。

Extended Terms:

meta-variscite 变磷铝石
variscite-matrix 带脉石磷铝石

---

vermiculite [vəː'mikjulait] *n.*

Definition: a yellow or brown mineral found as an alteration product of mica and other minerals, and for insulation or as a moisture-retentive medium for growing plants 蛭石 zhì shí

Formula: $(MgFe,Al)_3(Al,Si)_4O_{10}(OH)_2 \cdot 4H_2O$

Origin: early 19th century: from Latin vermiculari "be full of worms" (because on expansion due to heat, it shoots out forms resembling small worms) +-ite

Example:

Vermiculite was first described in 1824 for an occurrence in Millbury, Worcester County, Massachusetts, USA.
蛭石最早记载于1824年,出现在美国马萨诸塞州伍斯特市的米尔伯里。

Extended Terms:

vermiculite mortar 蛭石灰浆
vermiculite gypsum 蛭石石膏板
vermiculite concrete 蛭石混凝土
vermiculite aggregate 蛭石骨料
cement vermiculite 水泥蛭石

---

vesuvianite [vi'sjuːviənait] *n.*

Definition: (also known as idocrase) a green, brown, yellow, or blue silicate mineral, occurring as tetragonal crystals in skarn deposits and limestones that have been subjected to contact metamorphism 符山石 fú shān shí

Formula: $Ca_{10}(Mg, Fe)_2Al_4(SiO_4)_5(Si_2O_7)_2(OH,F)_4$

Origin: late 19th century: from Mount Vesuvius (in Italy), +-ite

Example:

The skarn type vesuvianite occurred in Huilong region, Tongbai, Henan Province.
这种矽卡岩型符山石玉产于河南桐柏的回龙地区。

Extended Terms:

beryllium vesuvianite 铍符山石
mangan vesuvianite 锰符山石

---

violarite [vaiˈəulərait] *n.*

Definition: a supergene sulfide mineral associated with the weathering and oxidation of primary pentlandite nickel sulfide ore minerals 紫硫镍矿 zǐ liú niè kuàng (又称紫硫镍铁矿)

Formula: $Fe^{2+}Ni^{3+}2S_4$

Origin: from the Latin for "purple", its color in a polished section

Example:

Primary violarite has comparatively big particle size, and it is considered the easiest flotation copper-nickle-sulphide ore.
紫硫镍铁矿原生粒度较大,被认为是最易浮选的硫化铜镍矿物。

---

vivianite [ˈviviənait] *n.*

Definition: A mineral consisting of a phosphate of iron which occurs as a secondary mineral in ore deposits. It is colorless when fresh but becomes blue or green with oxidization. 蓝铁矿 lán tiě kuàng

Formula: $Fe_3(PO_4)_2 \cdot 8(H_2O)$

Origin: early 19th century: named after John H. Vivian (1785-1855), British mineralogist, +-ite

Example:

Vivianite was first described in 1817 and named after John Henry Vivian, who first discovered crystals of the mineral in Cornwall, England.
蓝铁矿最早记载于 1817 年,(英文名)以 John Henry Vivian 的名字命名,此人在英国的康沃尔郡第一次发现这种矿物的晶体。

Extended Term:

vivianite nodule 蓝铁矿结核

wavellite [ˈweivəlait] *n.*

Definition: a phosphate mineral, normally translucent green, found in fractures in aluminous metamorphic rock, in hydrothermal regions and in phosphate rock deposits 银星石 yín xīng shí

Formula: $Al_3(PO_4)_2(OH,F)_3 \cdot 5H_2O$

Origin: named after William Wavell (? -1829) of England who discovered the mineral in a quarry in Devon, England in 1805

Example:

Some phosphates-vivianite, turquoise, wavellite, etc., are discovered in the oxidation zone.
在氧化带中发现有蓝铁矿、绿松石和银星石等磷酸盐矿物。

Extended Term:

lime wavellite 纤磷钙铝石

---

weloganite [ˈweləugənait] *n.*

Definition: a rare carbonate mineral, crystallizing in the triclinic system and shows pseudo-hexagonal crystal forms 水碳锆锶石 shuǐ tàn gào sī shí

Formula: $Na_2(Sr,Ca)_3Zr(CO_3)_6 \cdot 3H_2O$

Origin: named for William E. Logan (1798-1875), first director of the Canadian Geologic Survey

Example:

Weloganite has a Mohs hardness of 3.5.
水碳锆锶石的莫氏硬度是 3.5。

---

willemite [ˈwiləmait, ˈvi-] *n.*

Definition: a mineral, typically greenish-yellow and fluorescent, consisting of a silicate of zinc 硅锌矿 guī xīn kuàng

Formula: Zn_2SiO_4

Origin: mid 19th century: from the name of Willem I (1772-1843), king of the Netherlands, +-ite

Example:

In the glaze, willemite possesses spherulitic nature.
在釉中,硅锌矿具有球粒体性质。

Extended Terms:

pseudo willemite 假硅锌矿
green willemite phosphor 绿光硅酸锌荧光体

---

winchite ['wintʃait] *n.*

Definition: a rare amphibole group mineral 蓝透闪石 lán tòu shǎn shí

Formula: $(CaNa)Mg_4(Al,Fe^{3+})Si_8O_22(OH)_2$

Origin: named for its composition and for Howard J. Winch, Geological survey of India.

Example:

Mineral constituents consist mainly of jadeite, aegirine-augite, magnesio-riebeckite, winchite, magnesio-hornblende, actinolite, quartz, albite, rutile and sphene.
主要组成矿物为硬玉、霓辉石、镁纳闪石、蓝透闪石、镁角闪石、阳起石、石英、钠长石、金红石和榍石。

Extended Terms:

ferri winchite 高铁蓝透闪石
alumino winchite 铝蓝透闪石
ferro alumino winchite 铁铝蓝透闪石

---

witherite ['wiðərait] *n.*

Definition: a rare white mineral consisting of barium carbonate, occurring especially in veins of galena 毒重石 dú zhòng shí(又称碳钡矿、碳酸钡矿)

Formula: $BaCO_3$

Origin: late 18th century: from the name of William Withering (1741-1799), the English physician and scientist who first described it, +-ite

Example:

More than 40 witherite and barite deposits (or points) locate in northern Daba Mountains, which is a large and unique witherite metallogenic belt in the world.
北大巴山地区已发现毒重石—重晶石矿床(点)达40余处,是世界范围内独一无二的大型毒重石成矿带。

Extended Terms:

witherite deposit 毒重石矿床
witherite mineralized belt 毒重石成矿带

---

wolframite ['wulfrəmait, 'vɔːl-] *n.*

Definition: A black or brown mineral which is the chief ore of tungsten. It consists of a tungstate of iron and manganese. 黑钨矿 hēi wū kuàng(又称钨锰铁矿)

Formula: $(Fe, Mn)WO_4$

Origin: from German Wolfram for tungsten and providing the chemical symbol, W, for tungsten

Example:

Hukeng tungsten, located in Wugongshan metallogenic belt in central part of Jiangxi Province, is one large scale quartz vein type wolframite deposit.
浒坑钨矿床是位于江西省中部武功山成矿带的大型石英脉型黑钨矿床。

Extended Terms:

domangano wolframite 多锰黑钨矿
ferromangano wolframite 铁锰黑钨矿
fine wolframite 细粒黑钨矿
quartz type wolframite 石英脉型黑钨矿
veined wolframite deposit 脉钨矿床

---

wollastonite ['wuləstənait] *n.*

Definition: A white or greyish mineral typically occurring in tabular masses in metamorphosed limestone. It is a silicate of calcium and is used as a source of rock wool.硅灰石 guī huī shí(又称硅酸钙岩矿、钙硅石)

Formula: $CaSiO_3$

Origin: early 19th century: from the name of W. H. Wollaston (see Wollaston) +-ite

Example:

There are two main constituents that form the mineral wollastonite: CaO and SiO_2.
硅灰石矿石由两种主要成分组成:氧化钙和二氧化硅。

Extended Terms:

modified wollastonite 改性硅灰石

wollastonite whisker 硅灰石晶须
acicular wollastonite 针状硅灰石
magnesium wollastonite 镁硅灰石
wollastonite pegmatite 硅灰伟晶岩

---

wulfenite [ˈwulfənait] *n.*

Definition: an orange-yellow mineral consisting of a molybdate of lead, typically occurring as tabular crystals 钼铅矿 mù qiān kuàng（又称彩钼铅矿）

Formula: $PbMoO_4$

Origin: mid 19th century: from the name of F. X. von Wulfen (1728-1805), Austrian scientist, +-ite

Example:

Wulfenite is not hard enough to be classified as agemstone.
钼铅矿的硬度不足以使它归为宝石之列。

Extended Terms:

color wulfenite 彩钼铅矿
wulfenite type 钼铅矿晶组

---

wurtzite [ˈwəːtsait] *n.*

Definition: a mineral consisting of zinc sulphide, typically occurring as brownish-black pyramidal crystals 纤锌矿 xiān xīn kuàng(又称纤维锌矿)

Formula: ZnS

Origin: mid 19th century: from the name of Charles A. Wurtz (1817-1884), French chemist, +-ite

Example:

The simulation showed that wurtzite boron nitride would withstand 18% more stress than diamond.
模拟实验显示,纤锌矿型氮化硼的抗压能力比钻石高 18%。

Extended Terms:

wurtzite structure 纤维锌结构
hexagonal wurtzite 六方纤锌矿结构
wurtzite phase 纤锌矿相,纤锌矿晶型
wurtzite boron nitride 纤锌矿型氮化硼

wurtzite type boron nitride 纤锌矿型氮化硼

wustite ['wuːstait] *n.*

Definition: a mineral form of iron (II) oxide found with meteorites and native iron, crystallizing in the isometric-hexoctahedral crystal system in opaque to translucent metallic grains 方铁矿 fāng tiě kuàng

Formula: FeO

Origin: named for Ewald Wüst (1875-1934), German metallurgist and Director for Iron Research of the Kaiser Wilhelm Institute, Dusseldorf, Germany who first synthesized the compound

Example:

There are a lot of domain structures in magnetite and they may be divided into basal and nonbasal domains in which the coexistence of wustite is sometimes found.
磁铁矿还含有丰富的畴结构，畴界面分基面和非基面两类，有方铁矿与畴结构共生。

Extended Terms:

wustite iron 方铁体铁，维氏体铁
wustite iron phase 维氏体铁相
porous wustite 多孔方铁体，多孔维氏体
magnesio wustite 镁方铁矿
wustite formation stage 维氏体阶段

xenolith ['zenəliθ, 'ziː-] *n.*

Definition: a piece of rock within an igneous mass which is not derived from the original magma but has been introduced from elsewhere, especially the surrounding country rock 捕虏体，捕虏岩 bǔ lǔ tǐ, bǔ lǔ yán（指火成岩中与其无成因关系的包体）

Example:

The mafic granulite xenolith was first reported to have occurred in Early Mesozoic diorite in Harkin region, Eastern Inner Mongolia Autonomous Region, China.
在内蒙古东部喀喇沁地区早中生代大营子闪长岩中首次发现基性麻粒岩捕虏体。

Extended Terms:

cognate xenolith 同源捕虏体
accidental xenolith 外源捕虏体
basic xenolith 基性捕虏体
granulite xenolith 麻粒岩捕虏体
peridotite xenolith 橄榄岩捕虏体

---

xenotime [ˈzenətaim, ˈziː-] *n.*

Definition: a yellowish-brown mineral which occurs in some igneous rocks and consists of a phosphate of yttrium and other rare-earth elements 磷钇矿 lín yǐ kuàng

Formula: YPO_4

Origin: mid 19th century: from xeno-, apparently erroneously for Greek kenos "vain, empty", + timē "honour" (because it was wrongly supposed to contain a new metal)

Example:

Xenotime was first described for an occurrence in Vest-Agder, Norway in 1832.
据最早记载,磷钇矿出现在1832年的挪威西阿格德尔郡。

---

xiangjiangite [zjəŋdʒiənd'ʒait] *n.*

Definition: a yellow mineral composed of aluminium, hydrogen, iron, oxygen, phosphorus, sulphur, and uranium 湘江铀矿 xiāng jiāng yóu kuàng

Formula: $(Fe_3^{+}, Al)(UO_2)_4(PO_4)_2(SO_4)_2(OH) \cdot 22H_2O$

Origin: named after the locality: Xiangjiang River, Hunan, China

Example:

Xiangjiangite has a Mohs hardness of 1-2.
湘江铀矿的莫氏硬度是1~2。

---

xifengite ['zjəfend'ʒait] *n.*

Definition: a metallic mineral, occurring as steel gray inclusions within other meteorite derived nickel iron mineral phases 喜峰矿 xǐ fēng kuàng

Formula: Fe_5Si_3

Origin: named for the eastern passageway, Xifengkou, of the Great Wall of China

Example:

Xifengite has a specific gravity of 6.45 and a Mohs hardness of 5.5.
喜峰矿的相对密度是 6.45,莫氏硬度是 5.5。

---

xilingolite [ˌzjəlingəˈlait] *n.*

Definition: a grey mineral with metallic lustre 锡林郭勒矿 xī lín guō lè kuàng

Formula: $Pb_3Bi_2S_6$

Origin: named for the locality: Chaobuleng district, Xilingola League, Inner Mongolia, China

Example:

Xilingolite has a Mohs hardness of 3.
锡林郭勒矿的莫氏硬度是 3。

Ximengite [ˌzimenˈgait] *n.*

Definition: a trigonal mineral 西盟石 xī méng shí

Formula: $BiPO_4$

Origin: named for the locality: undefined locality in Ximeng county, about 420 km southwest of Kunming, Yunnan Province, China

Example:

Ximengite has a Mohs hardness of 4.5.
西盟石的莫氏硬度是 4.5。

---

xitieshanite [ziˈtiːʃənait] *n.*

Definition: a conchoidal green or yellow mineral 锡铁山石 xī tiě shān shí

Formula: $Fe^{3+}(SO_4)(OH) \cdot 7H_2O$

Origin: named for the locality: Xitieshan lead-zinc mine, Qaidam basin, Qinghai, China

Example:

Xitieshanite has a Mohs hardness of 3.
锡铁山石的莫氏硬度是 3。

yoderite [ˈjəudərait] *n.*

Definition: a hydrous magnesium iron alumino-silicate 紫硅铝镁石 zǐ guī lǚ měi shí

Formula: $Mg_2(Al,Fe^{3+})_6Si_4O_{18}(OH)_2$

Origin: named for Hatten Schyuler Yoder, Jr. (1921-), petrologist of the Geophysical Laboratory, Washington D. C. USA.

Example:

Yoderite has a Mohs hardness of 6.
紫硅铝镁石的莫氏硬度是 6。

---

yugawaralite [juˈgæwərəlait] *n.*

Definition: a zeolite mineral consisting of hydrous calcium aluminum silicate 汤河原沸石 tāng hé yuán fèi shí

Formula: $CaAl_2Si_6O_{16} \cdot 4H_2O$

Origin: named for the locality: Yugawara Hot Springs, Kanagawa Prefecture, Honshu, Japan.

Example:

Yugawaralite has a Mohs hardness of 4.5.
汤河原沸石的莫氏硬度是 4.5。

zabuyelite [zəˈbaiəlait]

Definition: the natural mineral form of lithium carbonate, forming colorless vitreous monoclinic crystals 扎布耶石 zhā bù yē shí

Formula: Li_2CO_3

Origin: named for the locality: Zabuye Salt Lake, Nagri, Tibet, China

Example:

Zabuyelite has a Mohs hardness of 3.
扎布耶石的莫氏硬度是3。

---

zeolite ['ziːəlait] *n.*

Definition: Any of a large group of minerals consisting of hydrated aluminosilicates of sodium, potassium, calcium, and barium. They can be readily dehydrated and rehydrated, and are used as cation exchangers and molecular sieves. 沸石 fèi shí

Origin: late 18th century: from Swedish and German zeolit, from Greek zein "to boil" +-lite (from their characteristic swelling when heated in the laboratory)

Example:

Zeolite is a natural and cheap mineral, and its application in purification of organic waste water is a popular research subject of waste water treatment because of its better adsorption.
沸石是一种天然廉价矿物,对有机污染物具有良好的吸附作用,将其用于有机废水的处理是目前污水处理研究的热点。

Extended Terms:

organic zeolite 有机沸石
modification zeolite 改性沸石
version zeolite 改型沸石
acidic zeolite 酸性沸石
hydrophobic zeolite 疏水性沸石

---

zeunerite ['zuːnərait] *n.*

Definition: a green copper uranium arsenate mineral, a member of the autunite group 翠砷铜铀矿 cuì shēn tóng yóu kuàng(又称铜砷铀云母)

Formula: $Cu(UO_2)_2(AsO_4)_2 \cdot 10\text{-}16(H_2O)$.

Origin: named after Gustav Anton Zeuner (1828-1907), German physicist, director, of School of Mines, Freiberg, Germany

Example:

Zeunerite has a Mohs hardness of 2.5.
翠砷铜铀矿的莫氏硬度是2.5。

Extended Term:

meta zeunerite 偏水砷铜铀矿,变水砷铜铀矿

zhanghengite [æŋhin'gait] *n.*

Definition: a mineral consisting of 80% copper and zinc, 10% iron with the balance made up of chromium and aluminium 张衡矿 zhāng héng kuàng

Formula: CuZn

Origin: named for Zhang Heng (78-139), famous astronomer in ancient Ghina

Example:

The color of zhanghengite is golden yellow.
张衡矿的颜色是金黄色。

zinc blende [ziŋk blend]

Definition: a shiny mineral, yellow to dark brown or black in color, consisting of zinc sulphide 闪锌矿 shǎn xīn kuàng

Formula: ZnS

Example:

Indium phosphide is a direct band gap Ⅲ-Ⅴ compound semiconductor, with zinc blende structure, and its band gap is about 1.34eV.
磷化铟是直接跃迁带隙、闪锌矿结构的化合物半导体,禁带宽度约为1.34eV。

Extended Terms:

zinc blende lattice 闪锌矿晶格,晶格
zinc blende structure 闪锌矿型,闪锌矿结构
zinc blende roaster 闪锌矿焙烧炉
zinc blende lattice structure 闪锌矿晶格结构
zinc-blende semiconductor 闪锌矿半导体

zincite ['ziŋkait] *n.*

Definition: a rare deep red or orange-yellow mineral consisting chiefly of zinc oxide, occurring typically as granular or foliated masses 红锌矿 hóng xīn kuàng

Formula: ZnO

Origin: mid 19th century: from zinc +-ite

Example:

The zincite found at Franklin Furnace is red-colored (mostly due toiron and manganese) and associated with willemite and franklinite.
在富兰克林弗尼斯找到的红锌矿呈红色(主要由铁和锰造成),并伴有硅锌矿和锌铁矿。

Extended Terms:

synthetic zincite 合成红锌矿
zincite detector 红锌矿检波器
zincite micro crystallites 氧化锌微晶
zincite type 红锌矿晶组
zincite structure 红锌矿结构

---

zincobotryogen [ˌziŋkəuˈbɔtriəˌdʒən] *n.*

Definition: a soft hydrous sulfate mineral that forms bright orange-red monoclinic prismatic crystals with a vitreous to greasy lustre 锌赤铁矾 xīn chì tiě fán

Formula: $(Zn,Mg,Mn)Fe^{3+}(SO_4)2(OH) \cdot 7H_2O$

Origin: named for its zinc content and its relationship to botryogen

Example:

Zincobotryogen is a sulphate mineral containing Zn^{2+} and Fe^{3+}, which belongs to the family of botryogen.
锌赤铁矾是一种含 Zn^{2+} 和 Fe^{3+} 的硫酸盐矿物,属赤铁矾族。

---

zinnwaldite [ˈtsinvɑːldait] *n.*

Definition: a potassium lithium iron aluminium silicate hydroxide fluoride silicate mineral in the mica group 铁锂云母 tiě lǐ yún mǔ

Formula: $KLiFeAl(AlSi_3)O_{10}(OH,F)_2$

Origin: named after its locality: Erzgebirge of Saxony at Zinnwald (now Cinovec)—"Ore Mountains", Czechoslovakia

Example:

Qitianlingite is associated with quartz, K-feldspar, albite, zinnwaldite, muscovite, cassiterite, wolframite, wolframoixolite, etc.
骑田岭矿与石英、钾长石、钠长石、铁锂云母、白云母、锡石、黑钨矿、铌黑钨矿等矿物共生。

Extended Term:

protolith zinnwaldite 黑鳞铁锂云母

---

zircon [ˈzəːkɔn] *n.*

Definition: A mineral occurring as prismatic crystals, typically brown but sometimes in translucent forms of gem quality. It consists of zirconium silicate and is the chief ore of zirconium. 锆石 gào shí

Formula: $ZrSiO_4$

Origin: named from its containing the element, zirconium; which was named from the Persian, zar "gold" and gun "like"

Example:

The oldest known zircons are from Western Australia, with an age of 4.4 billion years.
已知的最古老的锆石出产于西澳大利亚,已有 44 亿年的历史。

Extended Terms:

zircon syenite 锆石正长岩
low zircon 低型锆石
high zircon 高型锆石
intermediate zircon 中型锆石
hydrothermal zircon 热液锆石

---

zoisite [ˈzɔisait] *n.*

Definition: a greyish-white or greyish-green crystalline mineral consisting of a basic silicate of calcium and aluminium 黝帘石 yǒu lián shí

Formula: $Ca_2Al_3(SiO_4)(Si_2O_7)O(OH)$

Origin: early 19th century: from the name of Baron S. von Edelstein Zois (1747-1819), Austrian scholar, +-ite

Example:

Sources of zoisite include Tanzania (tanzanite), Kenya (anyolite), Norway (thulite), Switzerland, Austria, India, Pakistan, and Washington in the USA.
黝帘石的产地包括坦桑尼亚(出产坦桑黝帘石)、肯尼亚(出产红宝黝帘石)、挪威(出产锰黝帘石)、瑞士、奥地利、印度、巴基斯坦和美国华盛顿。

Extended Terms:

tanzanite 坦桑黝帘石
thulite 锰黝帘石
anyolite 红宝黝帘石
manganese salicylate zoisite 锰黝帘石
chrome zoisite 铬黝帘石

第三部分
矿物学词汇

abundance [əˈbʌndəns] *n.*

Definition: the ratio of the total mass of an element in the earth's crust to the total mass of the earth's crust; expressed as a percentage or in parts per million 丰度 fēng dù

Origin: mid-14 century, from Old French abundance (Modern French abondance), from Latin abundantia "fullness", noun of state from abundans (genitive abundantis), present partiliple of abundare "to overflow"

Example:

It is shown that the higher the D value, the larger the oil-gas abundance (oil-bearingarea or output of oil, gas and water) in the subregions.

与油田已知油气分布特征对比后,发现并证实油田内子区域分维值大,油田丰度(含油面积或流体产能)就大。

Extended Terms:

cosmic abundance 宇宙丰度

isotopic abundance 同位素丰度

D abundance 氘丰度

elemental abundance 元素丰度

abundance pattern 丰度模式

---

accessory [əkˈsesəri] *adj.*

Definition: (of a mineral) being present in small amounts in a rock and having no bearing on the classification of the rock, as zircon in granite. contributing to a general effect; supplementary; subsidiary 副的 fù de,附属的 fù shǔ de

Origin: 1400-1450; late Middle English accessorie (from Middle French) from Medieval Latin accessōrius

Example:

As a kind of accessory mineral, chrome spinel contained in ultrabasic rocks is complicated in chemical composition, stable in physical property, and strong in antial-teration ability.

产于超基性岩中的副矿物铬尖晶石化学成分复杂,物理性质稳定,抗蚀变能力强。

Extended Terms:

accessory mineral 副矿物
accessory structure 附属结构
accessory devices 附属配件
accessory chromosome 副染色体
accessory equipment 辅助设备
accessory ingredient 助剂;配合剂

---

achondrite [ei'kɔndrait] *n.*

Definition: An achondrite is a stony meteorite that does not contain chondrules. 无球粒陨石 wú qiú lì yǔn shí

Origin: from the prefix a-(privative a) and the word chondrite

Examples:

Achondrites are the only known samples of volcanic rocks originating outside the Earth-Moon system.
无球粒陨石是唯一已知的起源于地球—月球系之外的火山岩实例。
GR99027 meteorite from Blue Ice Area in Antarctica is an achondrite.
来自南极格罗夫山蓝冰地区的 GR99027 陨石为一无球粒陨石。

Extended Terms:

olivine-bronzite chondrite 橄榄古铜球粒陨石
calcium-poor achondrite 贫钙无球粒陨石
calcium-rich achondrite 富钙无球粒陨石

---

acicular [ə'sikjulə] (needle) *adj.*

Definition: needle-shaped 针状的 zhēn zhuàng de, 针尖状的 zhēn jiān zhuàng de

Origin: 1785-1795; acicul(a) + -ar

Examples:

The microstructure of deposit is mainly acicular ferrite when Si content is low, and block ferrite when Si content is high.
低硅含量时,熔敷金属显微组织以针状铁素体为主,高硅含量时,则以块状铁素体为主。
In the process of wollastonite fining, we can forge and then cool it to protect its crystal structure which is short acicular fiber.
硅灰石微粉碎过程中,为保护硅灰石的短纤维针状结构,可在粉碎前对硅灰石进行煅烧空冷处理。

Extended Terms:

acicular crystal 针状晶体
acicularstucture 针状组织
acicular ferrite 针状铁素体
acicular bismuth 针硫铝矿
acicular powder 针状粉末

acline [ə'klain] *adj.*

Definition: 水平的 shuǐ píng de,无倾斜的 wú qīng xié de

Origin: a+cline, prefix meaning "not," from Greek a-, an-"not," from Greek klinein (to lean)

Example:

This kind of converters can realize uninterruptible power supply, power factor correction and output voltage's regulation, and it has the advantages, such as simple topology, high power density, high conversion efficiency, and high power factor, uninterruptible power supply, electrical isolation between battery and AC line, and can be applied to the computer, telecom and medical industry.
这类变换器实现了不间断供电、功率因数校正、输出直流电压的稳定与调节,具有拓扑简洁、变换效率高、功率密度高、不间断供电、功率因数高、蓄电池与交流电网电气隔离等优点。在计算机、通信、医疗等电子设备中有广泛的应用前景。

Extended Term:

acline twins 底面双晶(水平双晶)

adamantine [ædə'mæntain] *adj.*

Definition: having the hardness or luster of a diamond 金刚石质的 jīn gāng shí zhì de,坚硬的 jiān yìng de

Origin: Middle English, 13th century from Latin adamantinus, from Greek adamantinos

Example:

Special fat-exploding adamantine ceramic iron can deeply heat subcutaneous fat for 60℃ to remove oil, with rapid effect.
特殊的脂肪爆破金刚陶瓷烫斗,皮下深层脂肪升温 60℃膨胀走油。

Extended Terms:

adamantine drill 金刚石钻头
adamantine clinker 硬钢砖金刚石钻

adamantine luster(lustre)金刚光泽
adamantine spar 黑褐色的刚玉
adamantine luster 金刚光泽

adularia [ˌædjuˈlɛəriə] *n.*

Definition: a transparent or translucent orthoclase 冰长石 bīng cháng shí

Origin: Italian, from French adulaire, from Adula, Swiss mountain group

Example:

The adularia associated with Au, Ag mineralization in Bitian gold deposit is identified by Electron Microprobe, Scanning Electron Microscope, and X-ray Diffractometer.
电子探针、扫描电子显微镜、X 射线衍射等测试方法的应用,揭示了碧田金矿床冰长石的存在及其矿物学特征。

Extended Terms:

adularia moonston 冰长月光石
sodian adularia 钠冰长石; 钠冰长石
adularia-ablite series 冰钠长石系; 冰钠长石系
sericite-adularia type 绢云母-冰长石型
adularia-calcite assemblage 冰长石方解石组合

ageing [ˈeidʒiŋ] *n.*

Definition: to make old; cause to grow or seem old 陈化 chén huà, 老化 lǎo huà

Origin: 1225-1275; (n.) Middle English, from Anglo-French, Old French. aage, eage, equivalent. to aé (from Latin aetātem of ae (vi) tās age; aev (um) time, lifetime + -itās -ity) + -age; (v.) Middle English agen, derived of the noun.

Example:

Prevent electric equipment ageing to cause the step with the mostsignificant fire, it is seasonable examination and accurate decide ageing degree tries to update.
防止电气设备老化引起火灾最有效的措施,就是及时检查与正确判定老化程度并加以更新。

Extended Terms:

column ageing 柱老化
ageing condition 老化条件
ageing treatment 老化处理
plant ageing 设备老化
ageing oven 老化炉

ageing stability 老化稳定性

aggregate [ˈægrigət, ˈægrigeit] *n.*

Definition: (1) a rock, such as granite, consisting of a mixture of minerals 集合体 jí hé tǐ; (2) the mineral materials, such as sand or stone, used in making concrete 集料 jí liào, 骨料 gǔ liào, 粒料 lì liào

Origin: from Latin aggregatus "associated," literal "united in a flock," past participle of aggregare "add to," from ad-"to" + gregare "herd", so "to lead to a flock"

Example:

Selection of materials includes characteristics of cement, particle size, intrinsic strength, shape and texture of aggregate, chemical admixtures and mineral admixtures, etc.

材料的选择包括水泥的品质、集料粒径、集料强度、颗粒形状和质地、化学外加剂、矿物掺合料等。

Extended Terms:

dendritic aggregate 树枝状集合体
earthy aggregate 土状集合体
fibrous aggregate 纤维状集合体
filmy aggregate 被膜状集合体
massive aggregate 块状集合体
oolitic aggregate 鲕状集合体
packet aggregate 束状集合体
phanerocrystalline aggregate 显晶集合体
pisolitic aggregate 豆状集合体
powdery aggregate 粉末状集合体
radiating aggregate 放射状集合体
reniform aggregate 肾状集合体
stalactic aggregate 钟乳状集合体
blotter aggregate 吸油集料
calcareous aggregate 石灰骨料
ceramic aggregate 陶瓷骨料
spheroidal aggregate 球状聚集体
synthetic aggregate 合成集料
tailor-made aggregate 特制集料
ungraded aggregate 无级配骨料
aggregate of soil 土壤团块

aggregated [ˈægrigətid, ˈægrigeitid] *adj.*

Definition: 聚合的 jù hé de,聚集的 jù jí de

Origin: from aggregate+d

Example:

The result shows that the main faults in this area have good sealing characteristics in the Xujiahe group and can seal the hydrocarbon effectively thus to be aggregated and preserved in reservoirs.
结果表明,该区主要断层在许家河组地层,具有良好的封堵性,能有效地封闭油气,有利于许家河组气藏的聚集与保存。

Extended Terms:

aggregated elemen 聚集元素
aggregated loss 累积损失
aggregate limit 累积限额
aggregated data 合计数据
aggregated model 总体模式

---

albitization [ˈælbitiˈzeiʃən] *n.*

Definition: 钠长石化 nà chánɡ shí huà

Examples:

The hydration-hydrolysis and dissolution rock fragments and feldspar and albitization of feldspar make many ions free from their framework and go into pore-water.
长石的钠长石化作用和长石、岩屑的水化水解作用、溶解作用,导致了不稳定组分的转变以及大量离子游离于孔隙水中,同时导致大量的自生矿物沉淀。
The distinctness of rock for deep clastic gas layers in Songliao basin is laumontitization, sericitization and water-bearing property of illite, and extensive albitization.
松辽盆地深层碎屑岩储气层在岩石矿物方面的特殊性在于浊沸石化、伊利石的绢云母化及其含水性和广泛的钠长石化。

Extended Terms:

glaucophane-albitization 蓝闪钠长石化

---

alkaline [ˈælkəlain] *adj.*

Definition:

1.of or like an alkali.
2.containing an alkali.

3.having the properties of an alkali.
4.having a PH value greater than 7. Compareacid.
5.any of various bases, the hydroxides of the alkali metals and of ammonium, that neutralize acids to form salts and turn red litmus paper blue 碱性的 jiǎn xìng de,碱的 jiǎn de

Origin: 1670s, from Modern Latin alcalinus (from alkali) + chemical suffix-ine. of soils, attested from 1850

Example:

Improving the quenching ability of the alkaline solution, prolonging the using life of the alkaline solution, and cutting down the cost of heat-treat are important to heat-treat worker.
提出几种防止淬火剂老化及再生的方法,提高碱液的淬火能力,延长碱液使用时间,降低成本成为热处理工作者所关心的问题。

Extended Terms:

alkaline earths 碱土
alkaline soil 碱性土壤
alkaline metal 碱性金属
alkaline battery 碱性电池
aikalinesolution 碱性溶液
alkaline amylase 碱性淀粉酶
alkaline basalt 碱性玄武岩
ocean island alkaline basalt 洋岛碱性玄武岩
alkaline olivine basalt 碱性橄榄玄武岩

---

alkalis ['ælkəlaiz] *n.*

Definition: a substance having highly basic properties; a strong base 碱质 jiǎn zhì

Origin: 1670s, from Modern Latin alcalinus (from alkali) + chemical suffix-ine. of soils, attested from 1850

Example:

Their oxides are frequently amphoteric, and react with both acids and alkalis.
他们的氧化物通常是两性的,与酸、碱都能反应。

Extended Terms:

underground natural alkalis 地下天然碱
dissolution by alkalis 碱溶作用
huanglian alkalis 黄连生物碱
color fastness to alkalis 耐碱牢度
available alkalis 有效碱

---

allochromatic [ˌæləkrəˈmætik] *adj.*

Definition: (of a mineral) having no color in itself but bearing colored impurities(矿)无色但含有色杂质之矿物的 wú sè dàn hán yǒu sè zá zhì zhī kuàng wù de

Origin: 1875-1880; allo-+ chromatic

Example:

We prepare some allochromatic BTF by irradiation, then study its stability and compatibility with contact materials such as PVB by VST, DTA, DSC et al. The reaction mechanism between BTF and PVB by TG IR is discused by the way.
通过光照制备了一定量的变色苯并三呋咱氧化物（BTF），再分别采用差热（DTA）、差示扫描量热（DSC）、真空安定性试验（VST）等热分析方法，对变色前后的BTF进行了热安定性研究及与聚乙烯醇缩丁醛（PVB）等接触材料的相容性研究，并通过热重红外等仪器分析手段，初步探讨了BTF与PVB之间的作用机理。

Extended Terms:

allochromatic color 他色
allochromatic crystal 义质色晶体
allochromatic photoconductor 义质色光电导体

---

allotype [ˈæləutaip] *n.*

Definition: 1. A biological specimen that is the opposite sex of a holotype.
2. Immunology. Any of the genetically determined variants in the constant region of a given subclass of immunoglobulin that is detectable as an antigen by members of the same species having a different constant region. 他型 tā xíng(异型 yì xíng)

Origin: 1915-1920; allo-+ type; in immunological sense, probably back formation from French allotypie, coined in 1956

Example:

The lading polishing device with recycled polishing agent and special effect to allotype materials is designed with reference to the principle of stretching lubricating pressure die.
依据拉伸润滑压力模的原理，设计了抛光剂可以循环使用、对异型材特别有效的载线抛光装置，使用该装置，提高了有色金属线材特别是钛合金线材的表面光洁度。

Extended Terms:

allotype spout 异型喷头
allotype profile 异型材
alltype suppression 同种异型抑制
complex allotype 复合同种异型

---

aluminosilicate [əˌljuːminəuˈsilikət] *n.*

Definition: 铝硅酸盐 lǚ guī suān yán

Origin: alumino-+ silicate; Latin alumin-, alumen + -o-+ International Scientific Vocabulary Silicate

Example:

Zeolites are hydrated aluminosilicate mineral combinations which have a unique "open" micro-porous molecular structure often described as a honeycomb. There are many forms of zeolite.
沸石是水化铝硅酸盐矿物结合,具有独特的"开放式"微孔分子结构,即人们所说的"蜂窝"结构。

Extended Terms:

crystalline aluminosilicate 结晶硅酸铝
aluminosilicate glass 铝硅酸盐玻璃
sodium aluminosilicate 铝硅酸钠
calcium aluminosilicate 铝硅酸钙

---

amorphous [əˈmɔːfəs] *adj.*

Definition: lacking definite form; having no specific shape; formless: the amorphous clouds 非晶态的 fēi jīng tài de,无定形的 wú dìng xíng de

Origin: 1725-1735; ámorphos (shapeless)

Example:

Crystalline silicon must be made 200 microns thick to absorb a sufficient amount of sunlight for efficient energy conversion, whereas only 1 micron of the proper amorphous materials is necessary.
必须要使晶体硅达到 200 微米厚,才能吸收足够的阳光进行有效的能量转换,而合适的非结晶体材料只要 1 微米厚就够了。

Extended Terms:

amorphous form 非晶态
amorphous silicon 非晶硅
amorphous ferromagnets 非晶态铁磁体
amorphous hydroxide 无定形氢氧化物
amorphous ceramic 非晶陶瓷
amorphous structure 非晶态结构,无定形结构

---

amygdaloid [əˈmigdəlɔid] *n.*

Definition: A volcanic rock in which rounded cavities formed by the expansion of gas or steam have later become filled with deposits of various minerals. 杏仁体 xìng rén tǐ, 杏仁岩 xìng rén yán

Origin: Greek amygdaloeidēs, from amygdalē; 1785-1795; from Latin amygdal (a) almond + suffix + -oid

Example:

To understand the relationship of neurotransmitter between the striatum and limbic system such as amygdaloid nucleus and bed nucleus of the stria terminalis, 30 male Sprague Dawley rats were used.
为了解大鼠纹状体内的神经活性物质与边缘系统重要结构杏仁核和终纹床核内相应神经活性物质之间的关系，选用了30只成年斯普拉·道来大鼠。

Extended Terms:

amygdaloid nucleus 杏仁核
amygdaloid complex 杏仁核复合体
amygdaloid body 杏仁体
nuclei amygdaloid 类杏仁核
anterior amygdaloid area 杏仁前区

---

anabolism [əˈnæbəulizəm] *n.*

Definition: The phase of metabolism in which simple substances are synthesized into the complex materials of living tissue. 合成作用 hé chéng zuò yòng (同化作用 tóng huà zuò yòng)

Origin: international scientific vocabulary; ana-+ (meta) bolism, 1885-1890

Example:

The phase of metabolism, in which large, complex molecules are built up from smaller, simpler ones, is called anabolism.
代谢作用的这一方面，即由小的简单分子形成大的复杂分子的过程被称为同化作用。

Extended Terms:

assimilation anabolism 同化作用
facilitated anabolism 促进合成作用
energy anabolism 能量组成代谢
anabolism-promoting factor 促合成代谢因子
substance anabolism 物质组成代谢

---

anion [ˈænaiən] *n.*

Definition: 1. a negatively charged ion, as one attracted to the anode in electrolysis.

2. any negatively charged atom or group of atoms (opposed to cation) 阴离子 yīn lí zǐ

Origin: Greek, neuter of aniōn, present participle of anienai to go up, from ana-+ ienai

Example: In the air includes the cation and the anion, but between the positive and negative ion must obtain the certain balance.
空气中含有正离子与负离子,而正负离子之间必须取得一定的平衡。

Extended Terms:

anion beam 负离子束,阴离子束
anion interference 阴离子干扰
anion activator 阴离子活化剂
anion detergent 释义:阴离子去污剂
radical anion 自由基负离子
enolate anion 烯醇阴离子
anion channel 阴离子通道

---

antiferromagnetism [ˈæntiˌferəuˈmægnitizəm] *n.*

Definition: type of magnetism in solids such as manganese oxide (MnO) in which adjacent ions that behave as tiny magnets (in this case manganese ions, Mn^{2+}) spontaneously align themselves at relatively low temperatures into opposite, or antiparallel, arrangements throughout the material so that it exhibits almost no gross external magnetism. In antiferromagnetic materials, which include certain metals and alloys in addition to some ionic solids, the magnetism from magnetic atoms or ions oriented in one direction is canceled out by the set of magnetic atoms or ions that are aligned in the reverse direction. 反铁磁性 fǎn tiě cí xìng

Origin: 1935-1940; anti-+ ferromagnetic

Example:

According to the results of the temperature dependencies of the intensity, linewidth and resonance field, we conclude that the peak at T (subscript M_2) is a transition of antiferromagnetism.
通过对铁磁共振线的强度、线宽和共振场与温度依赖关系的讨论,我们得出 T(下标 M_2)附近的峰对应着体系的反铁磁转变。

Extended Terms:

antiferromagnetism axis 反铁磁性轴
landau antiferromagnetism 朗道反磁性

---

antiperthite [ˌæntiˈpəːθait] *n.*

Definition: 反条纹长石 fǎn tiáo wén cháng shí

Origin: anti- opposite + peth + -ite stone

Example:

These rocks are composed of ortho- and clinopyroxenes, fayalite, perthite, antiperthite, and quartz, and are characterized by positive ε Nd values (+4.8 ~ +5.9).
紫苏花岗岩由斜方辉石、单斜辉石、铁橄榄石、反条纹长石、石英组成,以正 ε Nd 为特征(εNd = +4.8 ~ +5.9)。

Extended Terms:

antiperthite pegmatite 反条长石伟晶岩
antiperthitic texture 反纹结构

---

antiphase ['ænti,feis] *n.*

Definition: [电]反相 fǎn xiàng;逆相位 nì xiàng wèi

Origin: anti- opposite + phase

Example:

In accordance with situation of Meng No. 1 combination station Huabei oilfield, the auther develops BH-1 antiphase demulsifying agent.
华北油田蒙一联的污水处理一直是一个老大难问题,污水组成复杂,性质特殊,针对其特性研制出了 BH-1 反相破乳剂。

Extended Terms:

antiphase domains 反相畴
antiphase boundary 反相边界
antiphase reflection 反相反射
antiphase defects 反相缺陷
antiphase domain 逆相区

---

antistress ['æntistres] *adj.*

Definition: the action on a body of any system of balanced forces whereby strain or deformation results 反应力的 fǎn yìng lì de

Examples:

Ingredient:Golden lotus oil is good for antistress, relief the exhaustion and balancing your mind.
成分:其作用是提高反应力,减轻精疲力竭和平衡大脑。
The recent study indicated that chromium had special function inantistress and improving immune function.
近年来研究发现铬在抗应激和提高免疫机能上有特殊的作用。

Extended Terms:

antistress mineral 反应力矿物
antistress action 抗应激作用
antistress function 抗应激作用
antistress agent 抗应激剂
antistress capability 抗应激能力

---

arsenate [ˈɑːsənit] *n.*

Definition: a salt or ester of arsenic acid 砷酸盐 shēn suān yán

Origin: 1790-1800; arsen-+-ate

Examples:

As the thermal treatment temperature of Ti-MMT increased, the adsorption of arsenate in water obviously decreased.
随着对钛柱撑蒙脱石热处理温度升高,材料对水体中砷酸根的吸附能力呈明显下降趋势。
They may help to define the industrial type of primary deposit and prospecting by the association type of arsenate minerals.
根据砷酸盐矿物的组合类型可较快速地确定原生矿床(体)的工业类型和找矿方向。

Extended Terms:

mercuric arsenate 砷酸汞;砷酸汞
arsenate dicitrophenol 砷酸二硝基酚
copper arsenate 砷酸铜
magnesium arsenate 砷酸镁
arsenate mineral 砷酸盐矿物

---

artificial [ˌɑːtiˈfiʃəl] *adj.*

Definition: made by human skill; produced by humans (opposed to natural)人造的 rén zhào de

Origin: 1350-1400; Middle English, from Latin artificiālis contrived by art

Example:

At present it has become an urgent task to assess the artificial mineral deposit by means of geology technique and to utilize such great potential resources.
对人工矿床进行地质工艺评价,开发利用这一巨大的潜在资源,是现阶段矿业生产发展的一项迫切的新任务。

Extended Terms:

artificial mineral 人造矿物,人工矿物

artificial respiration 人工呼吸
artificial viscosity 人工黏度,类黏度
artificial radioactivity 人工放射
artificial stones 人造宝石

---

associate [ə'səuʃieit, ə'səuʃiət,-eit] *adj.*

Definition: 有关联的 yǒu guān lián de,伴随的 bàn suí de,伴生的 bàn shēng de

Origin: Middle English, associat associated, from Latin associatus, past participle of associare to unite, from ad-+ sociare to join, from socius companion

Example:

With the increasing of temperature, the degree of ion associate increases, the number of water nearest the ion decreases, and the multinuclear ions structure appears.
随着温度升高,溶液的结构发生了变化,离子的缔合度增加,内层配位水的数目减少,出现了离子的多核簇组成。

Extended Terms:

associate species 伴生种
associate structure 伴生构造
associate cumulation 结合累积
associate neuron 联络神经元
associate bundle 配从

---

associated [ə'səuʃieitid] *adj.*

Definition: anything usually accompanying or associated with another; an accompaniment or concomitant;allied; concomitant 伴随的 bàn suí de,伴生的 bàn shēng de

Origin: 1400-1450; late Middle English from Latin associātus joined to, united with (past participle of associāre), equivalent. to as-as-+ soci- + -ātus -ate1 ; comparative Anglo-French associer (v.), associé (n.)
Middle English *associat* associated, from Latin *associatus*, past participle of *associare* to unite, from *ad*-+ sociare to join, from *socius* companion.

Example:

Pitchblende is the only primary uranium mineral, which is associated with quartz, fluorite, hematite and pyrite.
沥青铀矿是矿石中唯一的原生铀矿物,其共生矿物主要有石英、萤石、赤铁矿和黄铁矿。

Extended Terms:

associated mineral 伴生矿物

associated corpuscular emission 伴生微粒辐射
associated matrix 关联矩阵
associated metal 伴生金属
associated dimension 关联维数

asterism [ˈæstərizəm] *n.*

Definition: a property of some crystallized minerals of showing a starlike luminous figure in transmitted light or, in a cabochon-cut stone, by reflected light 星彩 xīng cǎi

Origin: 1590-1600; from Greek asterism (ós) a marking with stars. See asterisk,-ism

Examples:

The asterism of star sapphire are due to the existence of three groups of mutually intercrossed (60°) needle-like particles in the surface of sapphire.
星光宝石中的星线是由于在宝石基平面中存在着互成60°的针状沉淀物。
Based on these models, physical principles behind imaging effects of defocused blur, motion blur and asterism were analyzed thoroughly, and methods to simulate these effects were proposed as well.
基于这些模型,深入分析了星空环境下成像过程中背景离焦模糊、运动模糊和星芒这三种效果的形成原因并提出了针对性的模拟方法。

Extended Terms:

asterism constellation 星群
multiple groups of asterism 多组星光效应

atmophile [ˈætəmɔfail] *adj.*

Definition: (of a chemical element in the earth) having an affinity for the atmosphere, as neon or helium 亲大气的 qīn dà qì de

Origin: atmo- + -phile

Example:

Elements that concentrate into the gaseous atmosphere are termed atmophile (from the Greek *atmos* = vapour).
从而提出亲氧(石)、亲硫(铜)、亲铁和亲气元素的概念

Extended Term:

atmophile element 亲气元素

attachment [əˈtætʃmənt] *n.*

Definition: an act of attaching or the state of being attached 依附 yī fù,附着的 fù zhuó de

Origin: 1400-1450; late Middle English attachement seizure from Auglo-French. See attach, -ment

Example:

This paper is about the Effects of triphenyltin on feeding, attachment and survival of bay scallop Argopecten irradians.

该论文是关于三苯基氯化锡对海湾扇贝摄食、附着和存活的影响。

Extended Terms:

attachment kinetics 附着能(吸附能)
bracket attachment 腋板连接
attachment face 安装面
attachment protein 吸附蛋白
drill attachment 钻床配件
attachment frame 连装附件框架

---

authigenic [ˌɔːθiˈdʒenik] *adj.*

Definition: (of a constituent of a rock) formed in the rock where it is found. Compare allogenic 自生的 zì shēng de

Example:

As the main flow manner of formation water in deep reservoir, circulated convection current controlled authigenic mineral location and secondary pore distribution.

循环热对流是深部储层流体的主要流动方式,它控制着自生矿物在砂层中的析出位置和次生孔隙的分布特征。

Extended Terms:

authigenic mineral 自生矿物
authigenic element 自生元素
authigenic mineralizations 自生矿化作用
authigenic kaolinite 自生高岭石
marine authigenic mineral 海相自生矿物
authigenic sediment 自生沉积

---

autocatalysis [ˌɔːtəukəˈtælisis] *n.*

Definition: catalysis caused by a catalytic agent formed during a reaction 自催化作用 zì cuī huà zuò yòng

Origin: 1890-1895; auto- + catalysis

Examples:

autocatalysis: catalysis of a chemical reaction by one of the products of the reaction
自身催化:指通过某一反应的产物来催化该化学反应。

Adsotption-autocatalysis Theory and Ionic Diffusion Theory are the two dominating theories on the iron ore reduction mechanism.
吸附-自动催化和离子扩散论理是铁矿石还原机理中两个基本的理论。

Extended Terms:

magmatic autocatalysis 岩浆自催化作用
monomolecular autocatalysis 单分子自动催化
autocatalysis reaction 自动催化反应
asymmetric autocatalysis 不对称自动催化
chemicomechanical autocatalysis 化学力学自催化

---

automorphic [ˌɔːtəuˈmɔːfik] *n.*

Definition: 自型 zì xíng

Origin: 1870-1875; auto- + -morphic

Example:

The automorphic, hypidiomorphic and xenomorphic granular textures are the main ore textures and vein and disseminated structures are the main ore structure.
矿石多具白形、半自形、他形粒状结构,以脉状、浸染状构造为主。

Extended Terms:

automorphic granulartexture 自形粒状结构
automorphic granular 自然粒状
automorphic soil 自型土
automorphic function 自守函数
automorphic crystal 自形晶体
automorphic extension 自守扩张

band [bænd] *n.*

Definition: a flat, thin strip or loop of material, typically one used to fasten things together, to reinforce something, or as decoration 带 dài

Origin: 1480-1490; from Middle French bande from banda; from Late Lalin bandum from Germanic; akin to Gothic bandwa standard, band, bend, bond

Example:

The heat exchanger outlet temperature is 80℃, which is now above the P-band, and the sensor now signals the controller to shut down the steam valve.
热交换器的出口温度是 80℃,这个温度已经超过了 P-带了,传感器将给控制器一个信号关闭蒸汽阀门。

Extended Terms:

band gap 带隙
band theory 能带理论
sawsvertical band 立式带锯
band spectrum 带状光谱
detection band 检测带

---

barium ['bɛəriəm] *n.*

Definition: a soft silvery-white metallic element of the alkaline earth group. It is used in bearing alloys and compounds are used as pigments. Symbol: Ba; atomic no: 56; atomic wt: 137.327; valency: 2; relative density: 3.5; melting point 729℃; boiling point 1805℃ 钡 bèi

Origin: [C19: from bar (yta) + -ium] 1808, Mod.L., from Gk. barys "heavy;" so called by its discoverer, English chemist Sir Humphrey Davy (1778-1829), because it was present in the mineral barytes "heavy spar," from Greek barys "heavy" (see grave (adj.)).

Examples:

Common ceramic elements are barium titanate and lead zirconate-titanate.
普通的陶瓷元件分别是钛酸钡和锆钛酸铅。
The bright blues are created by copper compounds, while the green colours areproduced by barium nitrate.
明亮的蓝色是铜的化合物产生的,而绿颜色是由硝酸钡所产生。

Extended Terms:

barium anorthitite 钡斜长岩
barium feldspar 钡长石
barium glass 钡玻璃
barium permanganate 过锰酸钡
barium oxide 氧化钡

---

barrier [ˈbæriə] *n.*

Definition: an antarctic ice shelf or ice front 堡坝 bǎo bà

Drigin: 1275-1325; Middle English, from Middle French *barriere* (barre bar +-iere from Latin -āria -ary); from Middle English barrere from Anglo-French from Medieval Latin barrera

Examples:

Catalysts reduce the activation energy Barrier between reactants and products. 催化剂会减少反应物和产物之间的活化能障碍。

It is one of seven barrier islands that, along with some mainland areas of Mississippi and Florida, make up the Gulf Islands National Seashore.

它是在密西西比和佛罗里达的一些大陆地区沿岸,组成海湾群岛国家海滨的七座屏障岛屿中的一座。

Extended Terms:

barrier reef 堡礁
barrier spring 堡坝泉
ice barrier 冰障
protective barrier 防护栏障
diffusion barrier 扩散势垒

---

basal [ˈbeisəl] *adj.*

Definition: of or constituting a foundation or basis; fundamental; essential 基础的 jī chǔ de,基本的 jī běn de

Origin: 1820-1830; base + -al

Example:

Leaves all basal, simple, palmate, ternate, or pinnate, sometimes reduced and scale like, rarely absent.全部基生的、单一的、掌状的、三出的或羽状的箔金属片有时会退化如鳞片状,但不会完全消失。

Extended Terms:

basal complex 基底杂岩
basal conglomerate 底砾岩
basal granule 基粒
basal sandstone 基底砂岩
basal structure 基底构造

---

base [beis] *n.*

Definition: the line or surface forming the part of a figure that is most nearly horizontal or on which it is supposed to stand 底面 dǐ miàn ;盐基 yán jī

Examples:

He could see no gap in it. He swam down to its base.
他看不到岩石上有任何缝隙,于是游到了它的底部。

The upgrade bonuses have been added to the base Possessed Marine unit. 额外奖金已经被增加到各基本海事单位。

Extended Terms:

base address 基础地址
base centered lattice 底心晶格
base desaturation 脱碱酌
base exchange 盐基代换
base exchange capacity 盐基交换容量
base flow 底流基本水流
base level 基准面
base level of erosion 侵蚀基准面
base line 基线
base of crude oil 原油的基类
base plate 座板
base runoff 底流基本水流
base temperature 基底温度

---

basement [ˈbeismənt] *n.*

Definition: a story of a building, partly or wholly underground 基底 jī dǐ

Origin: 1720-1730; base suffix + -ment

Example:

Basement fractures are closely related with environment geochemistry landscape and distribution features of soil erosion in the Ordos Basin. 鄂尔多斯盆地的环境地球化学景观与水土流失分布特征与基底断裂关系密切。

Extended Terms:

basement membrane 基底膜,基膜
basement folding 基底褶皱作用
basement complex 基底杂岩
basement fold 基底褶曲
basement nappe 基底推复体

batholith [ˈbæθəliθ] *n.*

Definition: a very large irregular-shaped mass of igneous rock, especially granite, formed from an intrusion of magma at great depth, especially one exposed after erosion of less resistant overlying rocks 岩基 yán jī

Origin: 1903, from German. (1892), coined by German. geologist Eduard Suess (1831-1914) from Greek bathos "depth" + -lith, from lithos "stone"

Examples:

The forming mechanism of the magma is partly melting of deep material, multipleactivity of magma, emplacing along regional great fault forming E-W trendinglinear batholith.
岩浆形成机制主要是深部物质部分熔融,岩浆多幕性活动,沿区域性大断裂侵位,形成东西向展布的线状岩基。

Lincang granite intrusive body is a composite batholith formed by polyphasicintrusion, whose main rock type is the biotite adamellite.
临沧花岗岩侵入体是一个多期次侵入活动所形成的复式岩基,其主要岩石类型为黑云母二长花岗岩。

Extended Terms:

concordant batholith 整合岩基
batholithic joint 岩基节理
anatectic batholith 深源岩基
subautochthonous batholith 半原地岩基
composite batholith 复合岩基
discordant batholith 不整合岩基

---

bed [bed] *n.*

Definition: a layer of rock; a stratum 层 céng

Example:

The United States is technically advanced in coal bed methane drilling. 美国钻煤层气井技术较成熟。

Extended Terms:

coal bed 煤层
bed load 推移质
bed moisture 内在潮湿
bed roughness 层粗糙度
bed succesion 层序

bed vein 板状脉

bedded [ˈbedid] *adj.*

Definition: of or pertaining to rocks that exhibit bedding 层状的 céng zhuàng de

Origin: 1820-1830; bed + -ed

Example:

The drilled result shows that beneficial thin-bedded sand bodies will be found out in the deltal-fan front sheet sand. 钻探结果表明,在扇三角洲前缘席状砂地带,可寻找具有圈闭条件的薄层砂体。

Extended Terms:

bedded rockill 成层填石
bedded vein 板状脉
bedded rock 层状岩
bedded deposit 层状矿床
bedded formation 层状建造,层状岩系

bedding [ˈbediŋ] *n.*

Definition: arrangement of sedimentary rocks in strata 层理 céng lǐ

Origin: before 1000; Middle English; Old English bedd; cognate with Old Frisian, Dutch bed, Old Saxon bed (de), Old High German betti (German Bett), Gothic badi; akin to Latin fodere to dig. OCS+Lith bedù

Example:

Now you can remove the compost and fill the empty space with new bedding.
这时你就可以把堆肥取走,再用新的垫料补充空位。

Extended Terms:

bedding fault 层面断层
bedding fissility 层面裂开性
bedding foliation 顺层面理
bedding joint 层状节理
bedding thrust 层面冲断层

bench [bentʃ] *n.*

Definition: a shelflike area of rock with steep slopes above and below 海蚀平台 hǎi shí píng tái

Origin: before 1000; Middle English, Old English benc; cognate with Old Frisian benk, Old Saken, Dutch, Old High German bank,

Example:

Proceeding from reasonable utilization of explosive energy, the paper discusses the relation between the ratio of stemming height/bench height, and loading height.
本文从合理利用炸药能量出发,论述高台阶爆破时填塞高度与装药高度比随台阶高度变化的关系。

Extended Terms:

bench mark 水准点
bench mining 阶梯式开采
bench placer 阶地砂矿
bench stoping 阶梯式开采
bench terrace 阶地段丘

---

behaviour [biˈheivjə] *n.*

Definition: the action or reaction of any material under given circumstances: *the behavior of tin under heat* 性态 xìng tài,性能 xìng néng

Origin: 1375-1425; behave + affix -ior (on model of havior, variation of havor from Middle French avoir (Latin habēre to have); from. late Middle English behavoure, behaver

Example:

Human-factors engineering: or human engineering or human factors engineering, profession of designing machines, tools, and work environments to Best accommodate human performance and Behaviour.
人体工学:亦称人类工程学或人机工程学,设计机器、工具和工作环境以最好地适应人们的操作和行为的专业。

Extended Terms:

creep behaviour 蠕变性
binary eutectic 二元共晶
aseismic behaviour 抗震性能
deformation behaviour 变形性能
diffusion behaviour 扩散行为

---

biogeochemical [ˈbaiəuˌdʒiːəuˈkemikəl] *adj.*

Definition: the scientific discipline that involves the study of the chemical, physical, geological, and biological processes and reactions that govern the composition of the natural environment (including the biosphere, the hydrosphere, the pedosphere, the atmosphere, and the lithosphere) 生物地球化学的 shēng wù dì qiú huà xué de

Origin: 1935-1940; bio-+ geochemistry

Example:

The interaction between microorganisms and minerals can facilitate the process of exogenic biogeochemical reaction, which is one of the important research contents in exogenic geochemistry.

微生物—矿物相互作用可以促进许多表生生物地球化学反应过程,是表生地球化学研究的重要内容。

Extended Terms:

biogeochemical cycles 生物地球化学循环
biogeochemical reaction 生物地球化学反应
biogeochemical agent 生物地球化学因子
biogeochemical exploration 生物地球化学勘查
biogeochemical element 生物地球化学元素

---

biomineralization [ˈbaiəuˌminərəlaiˈzeiʃən] *n.*

Definition: 生物矿化作用 shēng wù kuàng huà zuò yòng

Examples:

The new strategies for synthesis of functional biomaterials can be derived from mechanism of biomineralization.

生物矿化过程机理为制备具有特殊功能的仿生材料提供了新的思路。

Their affinities are still unclear, but they may provide a critical new dimension to the record of the initial phase of animal evolution and biomineralization.

这些化石的亲缘关系仍不清楚,但它们为我们提供了动物演化和矿化初始阶段的化石记录。

Extended Terms:

abnormal biomineralization 病例矿化
biomimetic biomineralization 仿生矿化
biomineralization synthesis 生物矿化合成
bone biomineralization 骨矿化
in-situ biomineralization 原位矿化

---

borate [ˈbɔːreit] *n.*

Definition: a salt or ester of boric acid

(loosely) a salt or ester of any acid containing boron 硼酸盐 péng suān yán

Origin: 1810-1820, bor-+ -ate

Examples:

Yellow transparent diamond crystals were readily obtained when the solution of boric acid or ammonium borate is added into graphite powders.
添加硼酸和硼酸氨的溶液时,易于合成出黄色透明的金刚石。
This paper presents laboratory research results of cement solidification of borate acid liquid wastes and concentrates arising from PWR.
本文介绍了压水堆核电站产生的硼酸废液和浓缩废液水泥固化的实验室研究结果。

Extended Terms:

borate minerals 硼酸盐矿物
calcium borate 硼酸钙
ammonium borate 硼酸铵
borate glass 硼酸玻璃
cupric borate 硼酸铜

---

botryoidal [ˌbɔtriˈɔidəl, ˈbɔtriɔidl] *n*

Definition: having the form of a bunch of grapes: botryoidal hematite 一串葡萄状的 yī chuàn pú táo zhuàng de,也作 botryose

Origin: 1810-1820; from Greek botryoeid (ēs) shaped like a bunch of grapes (bótry(s) bunch of grapes + -oeidēs -oid) + -al[1]

Example:

studying on the origin of botryoidal lace in dolomintite of dengying formation, from Ziyang Sichuan
四川资阳地区震旦系灯影组白云岩中葡萄花边的成因研究

Extended Terms:

botryoidal aggregate 葡萄状集合体
botryoidal structure 葡萄状构造
botryoidal grape hyacinth 葡萄风信子,蓝壶花

---

brittleness [ˈbritlnis] *n.*

Definition: having hardness and rigidity but little tensile strength; breaking readily with a comparatively smooth fracture, as glass 脆性 cuì xìng,矿物脆性 kuàng wù cuì xìng

Origin: 1350-1400; Middle English britel, equivalent. to brit-(akin to Old English brysten fragment) +-el adj. suffix-ness

Example: Experimental results show that there are ductile removal and brittleness removal simultaneously, and the model can analyze qualitatively the material removal ratio in dual-lapping of sapphire.
结果表明,蓝宝石双面研磨中同时存在延性去除和脆性去除,该模型可以定性地描述双面研磨加工材料的去除率。

Extended Terms:

brittleness distortion 脆性变形
brittleness rupture 脆性断裂
brittleness crackle 脆性裂纹
brittleness level 脆值平台
surface brittleness 表面脆性

capacity [kəˈpæsəti] *n.*

Definition: the ability to receive or contain:This hotel has a large capacity. 容量 róng liàng, 能力 néng lì

Origin: 1375-1425; late Middle English capacite from Middle French, from Latin capācitāt-(of capācitās), equivalent to capāci-, s. of capāx roomy (cap to hold +-āci-adj. suffix) +-tāt-ty

Example:

Capacity Performance and Radio Resource Management Algorithms in Multi-service CDMA Cellular Communication Systems
多业务 CDMA 蜂窝通信系统容量性能和无线资源管理算法研究

Extended Terms:

heatcapacity 热容
capacity expansion 容量扩展
capacity definition 容量定义
transportation capacity 交通容量
supply capacity 供应能力
processing capacity 处理能力

capillary [kə'piləri, 'kæpi-] *adj.*

Definition: resembling a strand of hair; hairlike 毛发状的 máo fà zhuàng de,似毛的 sì máo de

Origin: 1570-1580; capill (ar) [obs., from Latin capillāris pertaining to hair, equivalent. to capill (us) hair +-āris-ar] + -ary

Example:

Because of reduced accuracy for flow applications, the use of diaphragm seals including size of diaphragm and capillary should be carefully evaluated.
由于对在流量的应用会降低精度,膜片密封的使用包括膜片和毛细管的尺寸宜仔细计算。

Extended Terms:

capillary aggregate 毛发状集合体
capillary effect 毛细作用
capillary electrophoresi 毛细管电泳
capillary impregnation 毛细渗透
capillary driving force 毛管力

---

carbonate ['kaːbəneit] *n.*

Definition: In chemistry, a carbonate is a salt of carbonic acid, characterized by the presence of the carbonate ion. 碳酸盐 tàn suān yán

Origin: from New Latin carbonatem, from Latin carbo

Example:

The effect of sodium carbonate on the corrosion inhibition of aluminum alloy in sodium chloride solution was investigated by SEM and electrochemical means.
应用电化学实验方法和扫描电子显微镜(SEM)研究了碳酸钠在氯化钠溶液中时铝合金的缓蚀作用。

The karst surface collapse in Wuhan happened at the intersection between the NE first terrace binary-structured deposits of Yangtze River and carbonate in the core of E-W syncline.
武汉市岩溶地面塌陷发生在具二元结构的北东向长江一级阶地与近东西向褶皱构造向斜核部碳酸岩盐地层相交的部位。

Extended Terms:

sodium carbonate 碳酸钠,纯碱
ammonium carbonate 碳酸铵
ammonium carbonate 酸铵(中和剂,缓冲剂)
zinc carbonate 碳酸锌
copper carbonate 碳酸铜

potassium carbonate 碳酸钾
carbonate minerals 碳酸盐矿物

catabolism [kəˈtæbəlizəm] *n.*

Definition: destructive metabolism; the breaking down in living organisms of more complex substances into simpler ones, with the release of energy (opposed to anabolism) 分解代谢作用 fēn jiě dài xiè zuò yòng

Example:

Catabolism metabolic reactions involved in the breakdown of complex molecules to simpler compounds as, for example, in respiration.
这种代谢反应主要是把复杂的分子分解为简单的化合物,例如呼吸作用。

Extended Terms:

protein catabolism 蛋白质分解代谢
antibody catabolism 抗体分解代谢
catabolism degradation 降解
aromatic catabolism 芳香烃代谢
polyamine catabolism 多胺分解代谢

cathodoluminescence [kəˌθəudəˌluːməˈnesəns] *n.*

Definition: light emitted by a substance undergoing bombardment by cathode rays 阴极发光 yīn jí fā guāng

Origin: 1905-1910; cathode +-o-+ luminescence

Examples:

Combined with characteristics of cathodoluminescence images of zircons from this gneiss, it is indicated that metamorphism and magmatism could be happened during period of 293±6Ma~271±5Ma.
结合片麻岩中锆石阴极发光图像特征,说明该地区的变质作用和岩浆作用的时间范围在293±6Ma~271±5Ma之间。

Extended Terms:

cathodoluminescence spectroscopy 阴极发光光谱
cathodoluminescence features 阴极发光特征
cathodoluminescence spectrometer 阴极发光光谱仪
cathodoluminescence microscope 阴极发光显微镜
pulsed cathodoluminescence 脉冲阴极发光

cation ['kætaiən] *n.*

Definition: 1.a positively charged ion that is attracted to the cathode in electrolysis 2.any positively charged atom or group of atoms (opposed to anion) 阳离子 yáng lí zǐ

Origin: 1825-1835; from Greek katión going down (neut. of kation, present partiliple of kateînai), equivalent. to kat-cat-+ -i- go + -on neut. present partiliple suffix

Example:

The butanone glycol ketal was synthesized by the reaction of butanone with glycol in the present of the strong-acid cation-exchange resin.

以强酸性阳离子交换树脂为催化剂,通过丁酮和乙二醇反应合成丁酮乙二醇缩酮。

Extended Terms:

lead cation 铅离子
nutrient cation 阳离子养分
cation selectivity 阳离子选择性
cation modification 阳离子改性
inorganic cation 无机阳离子

---

chalcophile ['kælkəˌfail] *adj.*

Definition: (of a chemical element in the earth) having an affinity for sulfur 亲铜的 qīn tóng de

Origin: chalco-+-phile

Example:

The platinum group elements (Os, Ir, Pt, Ru, Rh, and Pd) are highly siderophile and chalcophile elements, which are compatible elements and show similar geochemical behaviours.

铂族元素(Os,Ir,Pt,Ru,Rh,Pd)具有强亲铁性和强亲铜性,为一组地球化学性质相近的相容元素。

Extended Terms:

chalcophile element 亲铜元素
chalcophile affinity 亲硫性

---

chatoyancy [ʃə'tɔiənsi] *n.*

Definition: reflecting a single streak of light when cut in a cabochon 猫眼 māo yǎn,变彩 biàn cǎi

Origin: 1790-1800; from French, special use of present partiliple of chatoyer to change luster like a cat's eye, equivalent. to chat cat + -oy- v. suffix + -ant

Example:

They display a different pattern of flame structure to an experienced eye, which consists of concentric radiating flames and parallel banded layers displaying a distinct "pseudo-chatoyancy" effect.
对于有经验的人而言,仿制品显示出不同于美乐珠的火焰纹结构,图案呈同心的放射状火焰纹,平行条带层显示出清晰的变彩假象。

Extended Term:

chatoyancy glass 玻璃猫眼

---

chelate ['kiːleit] *n.*

Definition: 1.of or noting a heterocyclic compound having a central metallic ion attached by covalent bonds to two or more nonmetallic atoms in the same molecule
2.of or noting a compound having a cyclic structure resulting from the formation of one or more hydrogen bonds in the same molecule 螯合物 ɑ́o hé wù

Origin: 1820-1830; chel(a) +-ate

Example:

With isopropanol, titanium tetrachloride and phytic acid, etc., as raw materials, type 200 chelate titanate coupling agent is produced through the synthetizing and chelating process.
以异丙醇、四氯化钛、植物油酸等为原料,经合成、螯合等工序制备螯合 200 型钛酸酯偶联剂。

Extended Terms:

chelate complex 螯合物
chelate complex extraction 螯型络合物萃取
chelate extraction 螯合萃取
chelate effect 螯合效应
chelate fibre 螯合纤维

---

chemical ['kemikəl] *adj.*

Definition: a substance produced by or used in a chemical process 化学的 huɑ̀ xué de

Origin: 1570-1580; chemic +-al; r. chimical chemic

Example:

We imagine using an anticatalyst to "freeze out" any chemical reactions in the system.
可想象利用一种反催化剂使体系内的任意化学反应"冻结"。

Extended Terms:

chemical petential 化学位

chemical process 化工过程
chemical therapy 化疗
chemical desulfurization 化学脱硫
chemical reagent 化学试剂

chondrite [ˈkɔndrait] *n.*

Definition: a stony meteorite containing chondrules 球粒陨石 qiú lì yǔn shí

Origin: 1880-1885；chondr-+ -ite

Example:

The chondrite-normalized distribution of the rare earth element (REE) shows light REE enrichment for the majority of samples.
磷灰石中稀土元素含量对于球粒状陨石的标绘表明,绝大多数磷灰石样品对于轻稀土元素有富集现象。

Extended Terms:

carbonaceous 碳质的
carbonaceous chondrite 碳质球粒陨石
ordinary chondrite 普通球粒陨石
chondrite normalized values 球粒陨石标准化值
chemical petrologic of chondrite 球粒陨石化学群

chondrule [ˈkɔndruːl] *n.*

Definition: a small round mass of olivine or pyroxene found in stony meteorites 陨石球粒 yǔn shí qiú lì

Origin: 1885-1890；chondr-+-ule

Example:

The isotopic compositions of lead were determined on the samples of whole rock, chondrule, olivine, pyroxene and troilite from Bo County chondrite (class LL4) of Anhui Province, China.
本文报道了安徽亳县 LL4 型球粒陨石的全岩、球粒、橄榄石、辉石和陨硫铁的铅同位素比值,以及上述部分样品的铀、钍、铅的含量。

Extended Term:

porphyritic chondrule 斑状球粒

chromate [ˈkrəumeit] *n.*

Definition: a salt of chromic acid, as potassium chromate 铬酸盐 gè suān yán

Origin: 1810-1820; chrom(ic acid) + -ate

Example:

In view of the problems in chromate anodizing process of aluminium alloy, this paper analysed the process of degreasing, rinsing, chromate anodizing, sealing and the experiment results.
本文针对铝及铝合金铬酸阳极氧化工艺中出现的问题,论述并分析了除油、水洗、铬酸阳极氧化、封闭等工艺方法和试验结果。

Extended Terms:

sodium chromate 铬酸钠
chromate ettringite 铬酸型钙矾石
chromate plant 铬盐厂
low chromate 低铬
tungstate molybdate and chromate minerals 钨钼铬酸盐矿物

chromophore ['krəuməfɔː] *n.*

Definition: a chemical group capable of selective light absorption resulting in coloration of certain organic compounds 色素离子 sè sù lí zǐ

Origin: 1875-1880; chromo-+ -phore

Example:

In addition, the polymer composed of three different chromophores have been synthesized, the sequential energy and electron transfer along the polymeric chain was also observed.
最后合成出包含三种不同发色团的聚合物,我们可以观察到连续的能量与电子转移现象发生在这个以矽烷基为间隔且交替排列的聚合物系统中。

Extended Terms:

macromolecular chromophores 巨分子发色团; 颜料和一巨分子发色团
Linear Chromophores 一维分子
Two-Dimensional Chromophores 二维分子

class [klaːs, klæs] *n.*

Definition: a number of persons or things regarded as forming a group by reason of common attributes, characteristics, qualities, or traits; kind; sort: *a class of objects used in daily living* 大类 dà lèi(矿物分类单位)

Origin: 1590-1600; earlier classis, pl. classes from Latin: class, division, fleet, army; sing. class back formation from pl.

Example:

Styrene containing unsaturated polyesters is the least expensive chemical class for UV curing.

在 UV 固化涂料中,苯乙烯混合不饱和聚酯是最便宜的种类。

Extended Terms:

subordinate class 下位类
transitive class 可迁类
class identifier 类标识符
information class 信息类
disparate class information 异类信息

---

classification [ˌklæsifiˈkeiʃən] *n.*

Definition: one of the groups or classes into which things may be or have been classified 分类 fēn lèi

Origin: 1780-1790; from Latin classi(s) class + -fication

Example:

Direct to the shortage existing in the classification of surrounding rock stability, provide a new method.
针对目前围岩稳定性分类方法存在的不足,给出一新的分类法。

Extended Terms:

classification algorithms 分类算法
rock classification 岩石分级
classification of minerals 矿物分类
systematic classification 系统分类
classification standard 分类标准

---

clay [klei] *n.*

Definition: earth; mud 黏土 nián tǔ

Example:

Clay is important because it is used around the world to make containers of all kinds.
黏土是很重要的东西,因为全世界都用它来制造各种各样的容器。

Extended Terms:

clay mineral 黏土矿物
saggar clay 火泥箱土
particulate clay 颗粒白土
saturation clay 饱和黏性土
clay polymer 黏土聚合物

---

cleavage [ˈkliːvidʒ] *n.*

Definition: the breaking down of a molecule or compound into simpler structures 解理 jiě lǐ

Origin: 1810-1820；cleave+-age

Example:

In rock mass, there exists rock damage expressing in the form of small opening, air-bubble, fracture, cleavage, etc.

岩石中含有以孔隙、气泡、微观裂隙、解理面等形态表现出来的岩石损伤。

Extended Terms:

eminent cleavage, perfect cleavage 极完全解理
cleavage in trace 极不完全解理
cleavage plane 解理面
good cleavage 完全解理
fair cleavage 中等解理

---

coefficient [ˌkəuiˈfiʃənt] *n.*

Definition: a number or quantity placed (generally) before and multiplying another quantity, as 3 in the expression $3x$ 常数 cháng shù, 系数 xì shù

Origin: 1655-1665；from Neo-Latin coefficient-(s. of coefficiēns). co-, +efficient

Example:

Stress index method and fatigue strength weakened coefficient method are important methods for designing fatigue vessel.

应力指数法和疲劳强度减弱系数法是进行疲劳容器设计的重要方法。

Extended Terms:

weighted coefficient 加权系数
angstrom coefficient 埃氏系数
individual coefficient 个别系数
extinction coefficient 消光系数
estimation coefficient 估值系数

---

colloid [ˈkɔlɔid] *n.*

Definition: a substance made up of a system of particles with linear dimensions in the range of about 10^{-7} to 5×10^{-5} cm dispersed in a continuous gaseous, liquid, or solid medium whose properties depend on the large specific surface area. The particles can be large molecules like proteins, or solid, liquid, or gaseous aggregates and they remain dispersed indefinitely. Compare aerosol, emulsion, gel, sol, suspension 胶粒 jiāo lì,胶体 jiāo tǐ

Origin: 1840-1850; from Greek kóll (a) glue + -oid

Example:

The photographic elements contain a blue-sensitive hydrophilic colloid silver halide emulsion layer.
照像材料含有亲水胶体的感蓝卤化银卤剂层。

Extended Terms:

macromolecular colloid 高分子胶质
colloid mill 胶体磨
colloid rectifier 胶质整流器
homopolar colloid 同极胶体
colloid mill 胶体研磨器

---

colloidal [kə'lɔidəl] *adj.*

Definition: pertaining to or of the nature of a colloid 胶体的 jiāo tǐ de,胶质的 jiāo zhì de

Origin: 1860-1865; colloid + -al

Example:

Colloidal particles are formed in considerable amounts in hard-rock drilling and blasting operation.
在硬岩石钻孔和爆破作业中,大量地生成胶体粒子。
Bubble jamming exhibits universal features that may be compared with those in systems of sand grains, colloidal particles, and glassy molecular liquids.
气泡阻塞展现出某些普适特性,可以和沙粒、胶状颗粒、玻璃状分子液体等系统相比较。

Extended Terms:

colloidal mineral 胶体矿物
colloidal catalyst 胶质催化剂
colloidal structure 胶状结构
colloidal substance 胶体物质
colloidal particles 胶体粒子

---

color ['kʌlə] *n.*

Definition: the quality of an object or substance with respect to light reflected by the object, usually determined visually by measurement of hue, saturation, and brightness of the reflected light; saturation or chroma; hue 色彩 sè cǎi,色泽 sè zé

Origin: 1250-1300; Middle English col(o)ur from Anglo-French (French couleur) from Latin colōr-(s. of color) hue

Example:

The color of the aventurescence depends on the mineral included in the gem.
“砂金石闪光”折射出来的颜色取决于晶体包含的矿物质成分。

Extended Terms:

allochromatic color 他色
color centre 色心
color of mineral 矿物颜色
idiochromatic color 自色
color settings 色彩设置
color restituting 色彩再现

---

columnar [kə'lʌmnə] *adj.*

Definition: shaped like a column 柱状的 zhù zhuàng de

Origin: 1720-1730; from Late Latin columnāris, equivdent to column (a) column + -āris - ar1

Example:

The columnar crystals were different in length at different intervals and problem like fading still existed after PED treatment.
在停止电脉冲处理后停留不同时间的情况下,柱状晶的长度不同,电脉冲处理钢液仍然存在衰退问题。

Extended Terms:

columnar aggregate 柱状集合体
rectangular columnar 方柱形
columnar epithellium 柱状上皮
columnar crystals 柱状晶
columnar grains 柱状晶粒
columnar section 地层柱状

---

combination [ˌkɔmbi'neiʃən] *n.*

Definition: something formed by combining 合并 hé bìng,组合 zǔ hé

Origin: 1350-1400; Middle English combinacyoun (from Middle French) from Late Latin combīnātiōn-(s. of combīnātiō), equivdent to combīnāt (us) combined (See combine, -ate)

+ -iōn -ion

Example:

A honeycomb sandwich panel inside and an aluminum cladding outside for the ideal combination of solidity and lightness, coldness and heat.
内有蜂窝夹心板,外包铝质覆层,乃是坚固性与质地轻盈之完美结合,又使冷热合为一体。

Extended Terms:

combination striations 聚形纹
color combination 配色
horizontal combination 横向合并
vertical combination 纵式联合
combination meter 组合仪表

---

combustibility [kəmˈbʌstəbliti] *n.*

Definition: capable of catching fire and burning; inflammable; flammable 可燃性 kě rán xìng

Origin: 1520-1530; from Late Latin combūstibilis (See combust, -ible)

Example:

Oxidants are substances that support combustion and enhance the combustibility of other materials.
氧化剂是支持燃烧和提高其他材料的可燃性的物质。

Extended Terms:

combustibility test 燃烧性能试验
limited combustibility 有限的可燃性
combustibility classification 燃烧性分级
coal combustibility 煤可燃性
combustibility of wood 木材可燃性

---

compressibility [kəmˌpresəˈbiləti] *n.*

Definition: the quality or state of being compressible 可压缩性 kě yā suō xìng

Origin: 1685-1695; compressible + -ity

Example:

The said cushion is equipped with high bearing capacity, low compressibility, large angle of stress dispersion.
石粉垫层具有承载力高、压缩性低、应力扩散角大、自身消散附加应力能力强的特点。

Extended Terms:

adiabatic compressibility 绝热压缩系数
compressibility effect 压缩性效应
pore compressibility 孔隙压缩性
water compressibility 水的压缩系数
system compressibility 综合压缩系数

conchoidal [kɔŋˈkɔidəl] *adj.*

Definition: noting a shell-like fracture form produced on certain minerals by a blow 贝壳状的 bèi ké zhuàng de

Origin: 1660-1670; conchoid + -al

Example:

Rain-flower agate, is pebble stone has adipose or waxy luster, assumes transparent or the translucent shape, the fracture assumes conchoidal.
雨花玛瑙,有脂肪或蜡状光泽,假定透明或半透明状,断裂处呈贝壳状。

Extended Terms:

conchoidal fracture 贝壳状断口
conchoidal structure 介壳状构造
conchoidal shale 介壳页岩
conchoidal curve 蚌线
conchoidal body 壳状小体

concretion [kənˈkriːʃən] *n.*

Definition: a rounded mass of mineral matter occurring in sandstone, clay, etc., often in concentric layers about a nucleus 结核体 jié hé tǐ

Origin: 1535-1545; from Latin concrētiōn-(s. of concrētiō) (See concrete,-ion)

Example:

The long axes of the ellipsoidal shaped calcareous concretion reflected the direction of movement of mineralizing ground waters.
椭圆形的钙质结核的长轴方向反映了地下矿化水的流动方向。

Extended Terms:

epigenetic concretion 次生结核
prostatic concretion 前列腺凝结体
replacive concretion 置换结核

nasal concretion 鼻石
calcareous concretion 石灰结核

conduction [kən'dʌkʃən] *n.*

Definition: the act of conducting, as of water through a pipe 传导 chuán dǎo

Origin: 1530-1540; from Latin conductiōn-(s. of conductiō) a bringing together, a hiring, equivalent to conduct (us) (See conduct) + -iōn-ion

Example:

The basic modes of heat transfer process relevant to hot dry rock geothermal studies are conduction and convection.
在干热岩石地热能开发中,岩石热传导的基本形式包括热传导和热对流。

Extended Terms:

conduction band 导带
heat conduction 热传导
conduction pump 传导泵
bone conduction 骨传导
decrementless conduction 不递减传导
conduction electrons 传导电子

conductivity [ˌkɔndʌk'tiviti] *n.*

Definition: the property or power of conducting heat, electricity, or sound

Also called specific conductance. Electricity: a measure of the ability of a given substance to conduct electric current, equal to the reciprocal of the resistance of the substance. Symbol: σ 传导性 chuán dǎo xìng

Origin: 1830-1840; conductive + -ity

Example:

Metals are elements that generally have good electrical and thermal conductivity. Many metals have high strength, high stiffness, and have good ductility.
金属就是通常具有良好导电性和导热性的元素。许多金属具有高强度、高硬度以及良好的延展性。

Extended Terms:

heat conductivity 导热系数
acoustic conductivity 声导率
Conductivity Meters 导电率计

conductivity modulation 电导率灯
earth conductivity 大地电导率
eddy conductivity 涡动传导性

constitution [ˌkɔnstiˈtjuːʃən] *n.*

Definition: the composition of something 构成 gòu chéng，构造 gòu zào；成分 chéng fèn

Origin: Middle English (denoting a law, or a body of laws or customs): from Latin constitution, from constituere "establish, appoint"

Example:

Secondly, the conceptual context of static and dynamic collaborative optimization of complicated mechanical system is dissected in this dissertation, and its ontology constitution is studied, afterwards, its field oriented object tree and ontology expression are fulfilled in further.
其次，分析了机械系统静动态协同优化这一专业领域中所包含的概念内容。深入研究了静动态协同优化的本体论构成，构造了静动态协同优化领域对象树及其本体论表达。

Extended Terms:

chemical constitution 化学结构
physical constitution 体格结构
constitution revision 结构修正
constitution factors 结构因素

converse [kənˈvəːs] *adj.*

Definition: opposite or contrary in direction, action, sequence, etc.; turned around 相反的 xiāng fǎn de，逆向的 nì xiàng de

Origin: 1300-1350; Middle English conversen, from Middle French converser from Latin conversārī to associate with

Example:

The results prove these properties of converse-spine nut to be better than those of dowel joint and tenon joint.
结果证明，金属倒刺螺母螺杆连接件接口的接合性能优于直角榫接合和木螺钉接合的性能。

Extended Terms:

converse piezoelectricity 反压电性
converse twist 反向捻
converse theorem 逆定理
converse method 反装法

coordination [kəuˌɔːdiˈneiʃən] *n.*

Definition: harmonious combination or interaction, as of functions or parts
配位 pèi wèi,协同 xié tóng

Origin: 1595-1605; from Late Latin coordinātiōn- (s. of coordinātiō)

Example:

The research of solvent extraction chemistry is actually the coordination chemistry between two phases.
溶剂萃取化学所研究的问题实际上是两相间的配位化学。

Extended Terms:

coordination number(CN)配位数
coordination chemistry 配位化学
coordination number 配位数
coordination compound 配位化合物
coordination agent 配位剂

---

cosmic [ˈkɔzmik] *adj.*

Definition: of or pertaining to the cosmos 宇宙的 yǔ zhòu de

Origin: 1640-1650; from Greek kosmikós worldly, universal, equivalent. to kósm (os) world, arrangement +-ikos -ic

Example:

In addition, the cosmic microwave background has some anomalous features that could potentially be explained by large-scale inhomogeneity.
再加上宇宙微波背景有一些奇怪的特征也可能会用大尺度上的不均匀性来解释。

Extended Terms:

cosmic abundance 宇宙丰度
cosmic mineralogy 宇宙矿物学
cosmic rays 宇宙射线
cosmic string 宇宙弦
cosmic gasdynamics 宇宙气体动力学
cosmic magnetohydrodynamics 宇宙磁铃力学
cosmic rocket 宇宙火箭

---

cosmogonic [kɔzməˈgɔnik, kɔzˈmɔgənik] *adj.*

Definition: a theory or story of the origin and development of the universe, the solar system, or the earth-moon system 天体演化的 tiān tǐ yǎn huà de

Origin: 1860-1865; from Greek kosmogonía creation of the world

Example: Self-cultivation for cosmic balance is a return to a central cosmogonic state of primary and generative chaos (hindus) out of which emerge the myriad things in cycles of Yin and Yang.
自我修养对宇宙平衡来说就是指回到天体演化最初的混沌状态,从中出现了许多事物处于阴阳两极循环中。

Extended Term:

cosmogonic theory 天体演化理论，宇宙起源说

---

cotectic [kəu'tektik] *adj.*

Definition: 共结的 gòng jié de,共熔的 gòng róng de

Origin: co- together + tect + -ic adj.

Examples:

From high-temperature molten, series of silicate melts near cotectic-line were quenched or cooled with different rate.
通过高温熔融,对同结线附近一系列硅酸盐熔体进行淬冷或不同速度冷却。
a Study on the Cotectic Region of $CaGa^{(2+x)}$
关于化合物 $CaGa^{(2+x)}$ 共熔区间的研究

Extended Terms:

cotectic crystallization 共结晶(同结晶)
cotectic line 共析线
cotectic point 共析点
cotectic surface 共析面
cotectic temperature 共析温度

---

cotype ['kəutaip] *n.*

Definition: a syntype 共型 gòng xíng

Origin: 1890-1895; co-+ type = syntype

Example:

In the study, six different cotype varieties of common wheat and F1 derived from 6×6 complete diallel crossing system were selected for the research on the characteristics and heterosis of spike differentiation...

对6个不同生态类型小麦品种及其完全双列杂交F1幼穗分化特点和穗分化杂种优势的研究表明，杂种幼穗分化各主要时期普遍存在杂种优势；不同生态类型组配方式的杂种穗分化优势不同……

Extended Terms:

agroe cotype 农业生态型
cotype specimen 本模标本

---

covalency [kəu'veilənsi] *n.*

Definition: the number of covalent bonds that a particular atom can make with other atoms in forming a molecule 共价 gòng jià

Origin: 1890-1895; co-+ type

Example:

The stabilization of the mixed species is considerably influenced by statistical effect, electrostatic effect, steric effect, cooperative effect and covalency of the coordination bond.
混合络合物的稳定性是由如下因素所决定的，即：统计效应，静电效应，共价键性，立体效应和配体的合作效应。

Extended Terms:

normal covalency 正常共价
normal cross section 标准剖面
coordination covalency 配位共价
bond covalency 键共价
covalency parameter 共价参数

---

creep [kriːp] *n.*

Definition:

1. the gradual movement downhill of loose soil, rock, gravel, etc.; solifluction 渐渐移位 jiàn jiàn yí wèi
2. the slow deformation of solid rock resulting from constant stress applied over long 蠕变 rú biàn

Origin: before 900; Middle English crepen, Old English crēopan; Cognate with Dutch kruipen

Example:

The creep property of nonwoven geotextiles is described well with the model.
该模型可较好地描述非织造土工织物的蠕变性能。

Extended Terms:

creep behaviour 蠕变性
creep strength 蠕变强度
creep strain 蠕变变形
primary creep 初始蠕变
accelerated creep 加速蠕变
creep coefficient 蠕变系数

---

crosshatch [ˈkrɔːshætʃ] *adj.*

Definition: to mark or shade with two or more intersecting series of parallel lines
交叉阴影线 jiāo chā yīn yǐng xiàn，格子状的 gé zi zhuàng de

Origin: 1815-1825；cross-+ hatch

Example:

Behind that grille we can see a crosshatch grille inserts that will probably be chrome on the production model.
后面的进气格栅，我们可以看到插入交叉格栅，将可能是铬的生产模式。

Extended Terms:

crosshatch twinning，grid twin，tartan twin 格子(状)双晶
crosshatch effect 格子状效应
photo crosshatch 照片阴影线
crosshatch boundary 剖面线边界
section crosshatch style 剖面线样式

---

cryptocrystalline [ˌkriptəuˈkristəlain] *adj.*

Definition: having a crystalline structure so fine that no distinct particles are recognizable under the microscope 潜晶的 qián jīng de，隐晶体的 yǐn jīng tǐ de

Origin: 1860-1865；crypto-+ crystalline

Example:

It is infeasible for the volcanic rock with microcrystalline and cryptocrystalline matrix to count its mineral content by conventional slice method.
目前，估算岩石中矿物含量主要是借助薄片显微镜统计和岩石化学成分计算。

Extended Terms:

cryptocrystalline aggregate 隐晶集合体
cryptocrystalline quartz 隐晶石英

cryptocrystalline texture 隐晶质结构
cryptocrystalline material 隐晶质材料
cryptocrystalline magnesite 隐晶质菱镁矿

crystal [ˈkristəl] *n.*

Definition: a solid body having a characteristic internal structure enclosed by symmetrically arranged plane surfaces, intersecting at definite and characteristic angles 晶体 jīng tǐ,水晶 shuǐ jīng

Origin: before 1000; Middle English cristal (le), Old English cristalla from Middle Latin cristallum, Latin crystallum, from Greek krýstallos clear ice, rock crystal, derivation of krystaínein to freeze

Example:

The defining characteristic of a crystal — any crystal — is that it is composed of regularly arranged components.
一种晶体(任何晶体)的定义型特征是,它是由排列规范的成分组成的。

Extended Terms:

liquid crystal 液晶
crystal form 晶形
crystal water 结晶水
hopper crystal 漏斗晶
crystal structure 结晶结构

crystallization [ˌkristəlaiˈzeiʃən] *n.*

Definition: the act or process of crystallizing 结晶 jié jīng

Origin: 1655-1665; crystall- + -ization

Example:

In this paper, it is discussed that sonic field influences the crystallization of the sugar solution, especially on the crystal growth.
本文以蔗糖过饱和溶液为研究对象,研究了声场对糖液结晶,尤其是对晶体生长的影响。

Extended Terms:

cotectic crystallization 共结晶
crystallization water 结晶水
oriented crystallization 定向结晶化
crystallization property 结晶性能

crystallization temperature 结晶温度

crystallo chemical [ˈkristələ ˈkemikəl] *adj.*

Definition: 晶体化学的 jīng tǐ huà xué de

Origin: crystal-+ chemistry

Example:

The crystallo chemical basis and methods of activation of belite and the manufacturing techniques of high belite cement are described.

文章论述了贝利特水泥活化的结晶化学基础和提高贝利特活性、生产高贝利特水泥的主要技术途径。

Extended Terms:

crystallochemical characteristic 晶体化学特征
crystallochemical element 晶体元素
crystallochemical formula 晶体化学式
crystallochemical classification 晶体化学分类

cyclic [ˈsaiklik] *adj.*

Definition:

1.revolving or recurring in cycles; characterized by recurrence in cycles
2.of pertaining to, or constituting a cycle or cycles
3.chemistry of or pertaining to a compound that contains a closed chain or ring of atoms (contrasted with acyclic)
环的 huán de,循环的 xún huán de

Origin: 1785-1795; from Latin cyclicus, from Greek kyklikós circular. See cycle,-ic

Example:

Objective to design and synthesize a series of squamosamide cyclic analogues and to test their antioxidation activity.

目的设计合成一系列番荔枝酰胺环状类似物并考察其抗氧化活性。

Extended Terms:

cyclic twin 环状双晶(轮式双晶)
cyclic accelerator 循环加速器
cyclic code 循环码
cyclic peptide 环肽
cyclic rubber 环化橡胶

cyclical cooling water tower 循环式冷却水塔

---

cyclosilicate [ˌsaikləˈsiləkit] *n.*

Definition: any silicate in which the SiO_4 tetrahedra are linked to form rings 环硅酸盐(又作 ring silicate) huán guī suān yán

Origin: cyclo-+ silicate

Extended Term:

cyclosilicate mineral 环状硅酸盐矿物

---

cylindrical [siˈlindrikəl] *adj.*

Definition: of pertaining to, or having the form of a cylinder 圆柱形的 yuán zhù xíng de

Origin: 1640-1650; from Neo-Latin cylindric (us) (from Greek kylindrikós; See cylinder, -ic) + -al

Example:

Water Ring Vacuum Pump cylindrical shell is installed a pump eccentric impeller-shaped teeth.
水环式真空泵的圆柱形泵壳内安装一个偏心的牙状叶轮。

Extended Terms:

cylindrical bore 圆锥形内孔
cylindrical and helicitical 卷曲结构
cylindrical antenna 圆柱形天线
cylindrical fold 圆柱形褶皱
cylindrical magnetic domain 圆柱形磁畴

---

curve [kəːv] *n.*

Definition: a continuously bending line, without angles 曲线 qū xiàn

Origin: 1565-1575; (from Middle French) from Latin curvus crooked, bent, curved

Example:

The test results showed that the use of the standard working curve can avoid the effect from the matrix.
实验结果表明,在检测过程中,通过使用标准工作曲线,避免了来自基体的影响。

Extended Terms:

solvus exsolution curve 固溶体分解曲线

curve regress 曲线回归
intersecting curve 相贯线,交叉曲线
isothermal curve 等温曲线
offset curve 等距线

dehydration [ˌdiːhaiˈdreiʃən] *n.*

Definition: the act or process of freeing from water; also, the condition of a body from which the water has been removed 脱水作用 tuō shuǐ zuò yòng

Origin: 1854,from de-+hydrate, a chemical term at first, given a broader extension 1880s, related:dehydration (1834)

Example:

Dehydration behaviours of phenol formaldehyde resin-borate hydrogels were studied by the measurement of dehydration proportions.
利用凝胶脱水率的测定,研究了甲阶酚醛树脂与硼酸盐水凝胶的脱水性能。

Extended Terms:

dehydration mechanism 脱水机制
dehydration rate 脱水速率,脱水率
dehydration kinetics 脱水动力学
alcohol dehydration 乙醇脱水
thermal dehydration 热力脱水

---

dendritic [denˈdritik,-kəl] *adj.*

Definition: 树枝状的 shù zhī zhuàng de,树状的 shù zhuàng de

Origin: 1795-1805; dendrite + -ic

Example:

Development of the dendritic pattern is influenced by the type and attitude of rock the erosion encounters.
树枝状水系型式的发展遭受到了侵蚀的岩石类型和层态影响。

Extended Terms:

dendritic aggregate 树枝状集合体

dendritic crystal 枝状晶体
dendritic macromolecule 树枝状高分子
dendritic catalyst 树状催化剂
dendritic tentacles 枝状触角
dendritic spine 树突棘

---

density [ˈdensəti] *n.*

Definition: the state or quality of being dense; compactness; closely set or crowded condition; (Physics): mass per unit volume 密度 mì dù

Origin: 1595-1605; from Latin dēnsitās, equivalent to dēns (us) dense + -itās -ity

Example:

In this area the density value of logging response of volcanic rock correlate well with porosity.
同时发现该区火山岩测井响应的密度值与孔隙度之间具有很好的相关性。

Extended Terms:

relative density 相对密度
low density 低密度
microvessel density 微血管密度
energy density 能量密度
cultivation density 栽培密度

---

determinative [diˈtəːminətiv] *adj.*

Definition: something that determines 鉴定性的 jiàn dìng xìng de, 决定性的 jué dìng xìng de

Origin: 1645-1655; probably from Middle Latin dēterminātīvus fixed, Late Latin: crucial (of a disease), equivalent. to Latin dēterminātּ (us) (See determinate) + -īvus -ive

Example:

In desertification and its opposite process i. e., control of desertification, water plays the determinative role.
荒漠化过程与其逆过程即荒漠化治理中,水在其中起着决定性的作用。

Extended Terms:

determinative mineralogy 鉴定矿物学
determinative bacteriology 鉴定细菌学
determinative factor 决定因子
determinative chart 鉴定图

determinative conclusion 鉴定结论

diagenesis [ˌdaiəˈdʒenisis] *n.*

Definition: Geology:the physical and chemical changes occurring in sediments between the times of deposition and solidification.成岩作用 chéng yán zuò yòng

Origin: 1885-1890; from Neo-Latin; dia-+-genesis

Example:

Formation of sedimentary rocks:weathering, erosion, transportation, deposition and diagenesis.
沉积岩的形成过程:风化作用、侵蚀作用、搬运、沉积和成岩作用。

Extended Terms:

pedogenesis diagenesis 成土成岩作用
diagenesis belt 成岩带
late diagenesis 晚成岩
clastic diagenesis 碎屑岩成岩作用
diagenesis trap 成岩圈闭

diamagnetism [ˌdaiəˈmægnitizəm] *n.*

Definition: The property of being repelled by both poles of a magnet. Most substances commonly considered to be nonmagnetic, such as water, are actually diamagnetic. Though diamagnetism is a very weak effect compared with ferromagnetism and paramagnetism, it can be used to levitate objects. Compare ferromagnetism, paramagnetism. (See also Lenz's law.) 反磁性 fǎn cí xìng, 逆磁性 nì cí xìng,抗磁性 kàng cí xìng

Origin: 1840-1850; dia-+ magnetic

Example:

According to the classic physical theory, the mechanism of magnetic medium diamagnetism was explained.
根据经典物理理论,文章给出了磁介质抗磁性的经典解释方法。

Extended Terms:

molecular diamagnetism 分子抗磁性
diamagnetism current 抗磁性电流
electronic diamagnetism 电子抗磁性
complete diamagnetism 完全抗磁性
atomic diamagnetism 原子抗磁性

diaphaneity [di͵æfəˈniːiti] *n.*

Definition: the quality of being diaphanous; transparency 透明度 tòu míng dù

Origin: 17th century: from Medieval Latin diaphanus, from Greek diaphanēs transparent, from diaphainein to show through, from dia- + phainein to show

Example:

ZND2160 complete crystal shape with full surface, good diaphaneity and low impurity, suitable for circular saws, frame-saw and geological broach.
ZND2160 晶体完整，晶面饱满，透明度高，杂质含量低，适用于圆锯切片、框锯切片、地质钻头。

Extended Terms:

improve diaphaneity 提高透明度
distributive diaphaneity 分析透明度
diaphaneity of urine 尿液透明度
information diaphaneity 信息透明
diaphaneity enhancement of lung field 肺野透明度增强

---

dielectric [͵daiiˈlektrik] *n.*

Definition: a substance or medium that can sustain a static electric field within it 电介质 diàn jiè zhì

Origin: from dia-+ electric

Example:

Research on the dielectric barrier discharge nitrogen nitriding at atmospheric pressure have consumedly progressed.
利用介质阻挡放电进行常压氮气渗氮研究取得突破性进展。

Extended Terms:

dielectric constant 电容率
dielectric constant 介电常数
dielectric property, dielectricity 介电性
dielectric isolation 介质隔离
dielectric effect 介电效应

---

discontinuous [͵diskənˈtinjuəs] *adj.*

Definition: characterized by interruptions or breaks; intermittent 不连续的 bù lián xù de

Example:

Discontinuous single stroke operation, two hands button start. Emergency stop apparatus is set to guarantee safety.
间断式单次行程操作、双手按钮开动。设有紧急停止装置,确保操作安全。

Extended Terms:

discontinuous transformation 不连续相变
discontinuous interference 不连续干扰
discontinuous sterilization 离歇灭菌法
discontinuous distribution 不连续分布
discontinuous gradient 不连续梯度

---

disilicate [dai'silikeit] *n.*

Definition: 双硅酸盐 shuāng guī suān yán

Origin: di- + silicate

Examples:

The results showed that the wastes of zirconium oxychloride can be made into crystalline layered sodium disilicate with good properties.
结果表明,碱性废水和酸性硅渣为主要原料,可制备出性能良好的层状硅酸钠。
The chemical structure, the principle of washing aid function and manufacture method of layered crystal sodium disilicate were described. And its applicationprospect was analysed.
阐述了层状晶态硅酸钠的化学结构、助洗功能原理、基本的生产方法,并对其在洗涤剂方面的应用前景作了分析。

Extended Terms:

lithium disilicate 焦硅酸锂
sodium disilicate 二硅酸钠
tricalcium disilicate 二硅酸三钙
barium disilicate 重硅酸钡
lead disilicate 二硅酸铅

---

dislocation [ˌdisləu'keiʃən] *n.*

Definition: (*Crystallography*) in a crystal lattice a line about which there is a discontinuity in the lattice structure 位错 wèi cuò

Origin: 1350-1400; Middle English dislocacioun (See dislocate, -ion)

Example:

It was found that there exist prismatic dislocations, stacking faults, array of dislocations and dislocation network.

研究发现金刚石中存在层错、棱柱位错、位错列和位错网络等晶体缺陷。

Extended Terms:

dislocation line 位错线
perfect dislocation 全位错
dislocation structure 位错结构
dislocation ring 位错环
dislocation step 位错阶梯

---

dispersed [di'spəːst] *adj.*

Definition: to separate and move apart in different directions without order or regularity; become scattered 分散的 fēn sàn de

Origin: 1350-1400; Middle English dispersen, disparsen (from Middle French disperser) from Latin dispersus (present partiliple of dispergere), equivalent to di- di- + -sper (g)-scatter (s. of-spergere, combining form of spargere to scatter, strew) + -sus suffix

Example:

We make use of rough set to give synthesizing evaluation score to every object through the dispersed data.

我们运用粗糙集方法对离散后的数据进行分析,得出每个待评对象的综合评分。

Extended Terms:

dispersed element 分散元素
dispersed material 分散质
dispersed phase 分散相
dispersed system 分散系统
dispersed flow 分散流

---

displacive [dis'pleisiv] *adj.*

Definition: 位移性的 wèi yí xìng de

Origin: dis-+ plac + -ive

Example:

The errors of lattice constant of compound are able to achieve 10-4, which basically meets the demand for measuring precision in displacive phase transition.

对该化合物晶格常数的测量误差可以达到 10~4,基本上满足了对一般位移性相变测量要求的精度。

Extended Terms:

displacive transformation 切变相变,位移式相变
displacive precipitation 推移沉淀作用
displacive concretion 推移性结核
displacive type ferroelectrics 位移性铁电体

---

distortion [dis'tɔːʃən] *n.*

Definition: an act or instance of distorting 变形 biàn xíng,畸变 jī biàn

Origin: 1575-1585; from Latin distortiōn-(s. of distortiō)

Example:

The analysis of micro zone discovers that there are directly relationships between the distortion structure of eutectic austenitic and lower ductility.
微区分析发现畸变的共晶奥氏体组织与球墨铸铁低韧性之间存在着直接的对应关系。

Extended Terms:

welding distortion 焊接变形
angular distortion 角度变形
linear distortion 线性畸变
attenuation distortion 衰减畸变
quantizing distortion 量化畸变

---

domain [dəu'mein] *n.*

Definition: a connected region with uniform polarization in a twinned ferroelectric crystal 范畴 fàn chóu,区域 qū yù

Origin: 1595-1605; from French domaine, alteration, by assocation with Latin dominium dominium, of Old French demeine from Late Latin dominicum, n. use of neut. of Latin dominicus of a master, equivalent to domin (us) lord + -icus -ic

Examples:

In this section, we show how to update the domain validator class to validate a set of units in a diagram.
在这一部分之中,我们向您展示了怎样更新域确认器类,以确认图表之中的一系列单元。
We wanted to inform you of a policy change to your domain service.
我们想通知你的是一项关于我们域名服务的政策改变。

Extended Terms:

domain structure 畴结构
homeobox domain 同源框
doppler domain 多普勒域
management domain 管理域
functional domain 功能域

---

druse [druːz] *n.*

Definition: an incrustation of small crystals on the surface of a rock or mineral 晶簇 jīng cù

Origin: 1745-1755; from German; compare Middle High German, Old High German druos gland, tumor, German Drüse gland (Middle High German drües, pl. of druos)

Example:

Druse can be the same kind of single crystals of the mineral composition and the crystal can be formed by several different minerals.
晶簇可以由单一的同种矿物的晶体组成,也可以由几种不同矿物的晶体组成。

Extended Terms:

quartz druse 石英晶簇
feldspar druse 长石晶簇
druse crystals 簇晶
autochton druse 同基晶簇

---

ductility [dʌkˈtiləti] *n.*

Definition: capable of being hammered out thin, as certain metals; malleable 延(展)性 yán (zhǎn) xìng

Origin: 1300-1350; Middle English from Latin ductilis, equivalent to duct (us) (present partiliple of dūcere to draw along) + -ilis -ile

Example:

Silicone Bronze has the ductility of copper but much more strength.
硅青铜不但具有铜的可锻性,而且更具有好的强度。

Extended Terms:

impact ductility 冲击韧性
structure ductility 结构延展性
hot ductility 热延性

earthy [ˈəːθi] *adj.*

Definition: (of a mineral) having a dull luster and rough to the touch 土状的 tǔ zhuàng de

Origin: 1350-1400; Middle English erthy

Example:

There is a slight earthy character, which adds to the complexity of this wine.
带有的少许泥土的味道,增加了该酒的复杂性。

Extended Terms:

earthy aggregate 土状集合体
earthy luster 土状光泽
earthy humus 土状腐殖质
earthy phosphate 磷酸碱土类
earthy coal 土状煤

---

ecosystem [ˈiːkəuˌsistəm, ˈekəu-] *n.*

Definition: a system formed by the interaction of a community of organisms with their environment 生态系统 shēng tài xì tǒng

Origin: 1930-1935; eco-+ system

Example:

Ecosystem health is the aim of soil conservation and ecosystem rehabilitation on the Loess Plateau.
生态经济系统健康是黄土高原水土保持与生态建设的目标。

Extended Terms:

terrestrial ecosystem 地表生态系统
forest ecosystem 森林生态系统
prairie ecosystem 草原生态系统
aquatic ecosystem 水生生态系统
agricultural ecosystem 农业生态系统

---

element ['elimənt] *n.*

Definition: one of a class of substances that cannot be separated into simpler substances by chemical means 元素 yuán sù

Origin: 1250-1300; Middle English (from Anglo-French) from Latin elementum one of the four elements, letter of the alphabet, first principle, rudiment

Example:

The atoms of one element differ in structure from those of every other element.
一种元素的原子结构不同于任何一种元素的原子结构。

Extended Terms:

aggregated element 聚集元素
atmophile element 亲气元素
chalcophile element 亲铜元素
dispersed element 分散元素
incompatible elements 不相容元素
rare earth elements 稀土元素
siderophile element 亲铁元素
sulfophile element 亲硫元素
transitional element 过渡型元素
volatile elements 挥发性元素

---

elasticity [ˌelæ'stisəti] *n.*

Definition: the property by virtue of which a material deformed under the load can regain its original dimensions when unloaded 弹性 tán xìng,矿物弹性 kuàng wù tán xìng

Origin: 1660s, from elastic + -ity

Examples:

Most designers faor cobalt-chrome alloy because its higher modulus of elasticity may reduce stresses within the proximal cement mantle.
大部分设计者选择钴铬合金,因其弹性模量较高,可以降低近端骨水泥套中的应力。
Although the polyester possesses great hardness, it still shows good elasticity and good oil resistance, but shows poor water resistance.
尽管弹性体聚酯具有高的硬度,但仍具有弹性,并且耐油性很好,但耐水性差。

Extended Terms:

elasticity fabric 弹性织物
process elasticity 过程弹性

conformational elasticity 构象弹性
elasticity index 弹性指数
pseudo-elasticity 伪弹性
elasticity conjugate 弹性共轭

elbow ['elbəu] *n.*

Definition: the bend or joint of the human arm between upper arm and forearm 肘 zhǒu

Origin: bef.1000; Middle English elbowe, Old English el (n) boga; Cognate with Middle Dutch elle (n) bōghe, Old High German el (l) inbogo (German Ellenbogen); literal "forearm-bend"

Example:

Methods: we proceed stretch experiment on elbow joint ulnar nerve of animal model with reproducing fluorosis animal model.
方法:复制氟中毒动物模型,取动物模型肘关节尺神经进行拉伸实验。

Extended Terms:

elbow twin(knee shaped twin)肘状双晶(膝状双晶)
elbow smash 肘部撞击
elbow punch 肘部猛击
elbow meter 弯管流量计
elbow union 弯头套管

electric [i'lektrik] *adj.*

Definition: pertaining to, derived from, produced by, or involving electricity 导电的 dǎo diàn de,电的 diàn de

Origin: 1640-1650; from Neo-Latin electricus, equivalent to Latin ēlectr (um) amber (See electrum) + -icus -ic

Example:

Finally, we extend the power management and measurement to electric home appliances.
最后,我们延伸电源管理和测量到家庭电器设备。

Extended Terms:

electric conductivity 导电性
electric arc 电弧;弧光放电;电弧炉
electric blender 电动搅拌机
electric filament 电炉丝

electric varnish 绝缘漆

electrical [iˈlektrikəl] *adj.*

Definition: concerned with electricity 有关电的 yǒu guān diàn de, 电气科学的 diàn qì kē xué de

Origin: electric + -al

Example:

Examples of electrical insulators include plastic, glass, air, dry wood, paper, rubber, and helium gas.
电的绝缘体的例子包括塑胶、玻璃、空气、干燥木材、纸、橡胶和氦气。

Extended Terms:

electrical conductivity 导电性
electrical transmission 电传递
Electrical Machinery 电机学
electrical silence 电静息
electrical synapse 电突触

electro [iˈlektrəu] *adj.*

Definition: (electricity; electric; electrically) a combining form representing electric or electricity in compound words 电镀的 diàn dù de, 电的 diàn de

Origin: electr (ic) + -o

Example:

Report is the final written result gained in the electro motor test system.
在电机测试系统中,报表是用户得到的最终书面结果。

Extended Terms:

electro negativity (electronegativity) 电负性
electro otoscopy 电耳镜检查法
electro flocculation 电絮凝
electro precipitator 电除尘器
electro deionization 电去离子

electroluminescence [iˌlektrəuˌljuːmiˈnesəns] *n.*

Definition: luminescence produced by the activation of a dielectric phosphor by an alternating current 场致发光 chǎng zhì fā guāng

Origin: 1900-1905; electro-+ luminescence

Example:

Under these conditions, the plasma becomes the center of a luminous phenomenon (electroluminescence).

在这些条件下,等离子体变成了辉光放电现象(即场致发光现象)的中心。

Extended Terms:

electroluminescence copolymers 共聚体电致发光材料
electroluminescence screen 电致发光屏
electroluminescence memory 电发光存储器
electroluminescence dyes 电致发光染料
electroluminescence sensor 电荧光传感器,电发光传感器

---

electromagnetism [iˌlektrəuˈmæɡnitizəm] *n.*

Definition:

1.the phenomena associated with electric and magnetic fields and their interactions with each other and with electric charges and currents 电磁性 diàn cí xìng
2. also, electromagnetics, the science that deals with these phenomena 电磁学 diàn cí xué

Origin: 1820-1830; electro-+ magnetism

Example:

The waves of electromagnetism, cutting across the secondary winding, cause electrons to flow in the secondary.

当电磁波切断次线圈时,会引起电子在其上流动。

Extended Terms:

electromagnetism compatibility 电磁兼容
marine electromagnetism 海洋电磁学
electromagnetism insulation 电磁隔离
electromagnetism protection 电磁防护贴
electromagnetism stiffness 电磁刚度

---

electron [iˈlektrɔn] *n.*

Definition: a unit of charge equal to the charge on one electron; also called negatron, (*Physics*, *Chemistry*) an elementary particle that is a fundamental constituent of matter, having a negative charge of 1.602×10^{-19} coulombs, a mass of 9.108×10^{-31} kilograms, and spin of

1/2, and existing independently or as the component outside the nucleus of an atom 电子 diàn zǐ

Origin: term first suggested in 1891 by Irish physicist G. J. Stoney (1826-1911); electr(ic) + -on(from the names of charged particles, as ion, cation, anion) with perhaps accidental allusion to Greek elektron amber (See electric)

Example:

This paper describes a new multi ion and electron beam system for preparing optic thin films.
本文介绍一种新型的光学薄膜制备用多离子束电子束系统。

Extended Terms:

electron affinity 电子亲和性
electron configuration 电子构型
electron microprobe 电子微探针
electron transfer 电子转移
electron accelerator 电子加速器

---

electrostriction [iˌlektrəˈstrikʃən] *n.* (=converse piezoelectricity)

Definition: elastic deformation produced by an electric field, independent of the polarity of the field 电致伸缩 diàn zhì shēn suō, 反压电性 fǎn yā diàn xìng

Origin: electro-+ striction

Example:

This phenomenon, which differs from electrostriction, is called Maxwellstress.
这个和电致伸缩不同的现象称为马克士威应力。

Extended Terms:

electrostriction material 电伸缩材料
electrostriction effect 电致伸缩反应
electrostriction vibrator 电致伸缩振动器
Bragg electrostriction 布喇格电致伸缩
electrostriction transducer 电致伸缩换能器

---

eminent [ˈeminənt] *adj.*

Definition: high in station, rank, or repute; prominent; distinguished 显著的 xiǎn zhù de, 明显的 míng xiǎn de

Origin: 1375-1425; late Middle English (from Anglo-French) Latin ēminent-(s. of ēminēns) outstanding (present partiliple of ēminēre to stick out, project), equivalent to ē- e-+ min-(See imminent) + -ent- -ent

Example:

Identifies the source of eminent or potential memory leaks before they occur, preempting performance issues and program crashes.
在出现明显的或潜在的内存泄漏之前,先识别它们的来源,然后再解决性能问题和程序崩溃问题。

Extended Terms:

eminent cleavage, perfect cleavage 极完全解理
eminent domain 征用权
eminent services 功勋卓著
eminent studies 显学
eminent ranks 超品

---

endogenic [ˌendəuˈdʒenik] *adj.*

Definition: arising from or relating to the interior of the earth (opposed to exogenetic) 内生的 nèi shēng de

Origin: endo-+-genetic

Example:

Endogenic Au deposits in the east Hebei Province are varied in genetic types and metallogenic environments.
冀东地区内生金矿床成因类型不同,形成环境各异。

Extended Terms:

endogenic process 内生作用
endogenic factor 内源因素
endogenic texture 内成结构
endogenic deposit 内成矿床
endogenic energy 内能

---

energy [ˈenədʒi] *n.*

Definition: the capacity to do work; the property of a system that diminishes when the system does work on any other system, by an amount equal to the work so done; potential energy; any source of usable power, as fossil fuel, electricity, or solar radiation 能量 néng liàng

Origin: 1575-1585; from Late Latin energīa from Greek enérgeia activity, equivalent to energe-(s. of energeîn to be active) + -ia -y3

Example:

The new theory still assigns definite energy states to an atom.
新的理论仍然规定原子具有确定的能量状态。

Extended Terms:

atomic energy 原子能
Gibbs free energy 吉布斯自由能
potential energy 势能,位能
energy frontogenesis 能量锋生
metallurgical energy 冶金能源

---

enthalpy [en'θælpi] *n.*

Definition: a quantity associated with a thermodynamic system, expressed as the internal energy of a system plus the product of the pressure and volume of the system, having the property that during an isobaric process, the change in the quantity is equal to the heat transferred during the process. Symbol:H (H)焓 hán(热函 rè hán)

Origin: 1925-1930; from Greek enthálp (ein) to warm in (en- + thálpein to warm) + -y3

Example:

Polyethylene glycol(PEG) is a kind of phase change materials with high enthalpy and low thermal hysteresis.
聚乙二醇相变材料是一类相变焓较高、热滞后效应低的储能材料。

Extended Terms:

free enthalpy 自由焓
reaction enthalpy 反应焓
specific enthalpy 比热焓
sensible enthalpy 显焓
dissolution enthalpy 溶解焓

---

entropy ['entrəpi] *n.*

Definition: *Thermodynamics*

1.(on a macroscopic scale) a function of thermodynamic variables, as temperature, pressure, or composition, that is a measure of the energy that is not available for work during a thermodynamic process. A closed system evolves toward a state of maximum entropy.
2.(in statistical mechanics) a measure of the randomness of the microscopic constituents of a thermodynamic system. Symbol:S 熵 shāng

Origin: from German Entropie (1865); See en-2, -tropy

Example:

To question the finality of the principle of entropy is not to dispute the second law of thermodynamics.
对熵原理所导致的结局提出质疑,并不意味着争辩第二热力学定律的真实性。

Extended Terms:

entropy increase 熵增加
entropy diagram 熵图
entropy elasticity 熵弹性
entropy chart 熵图表
entropy of dilution 稀释熵
spectrum entropy 光谱熵

---

environmental [inˌvaiərənˈmentəl] *adj.*

Definition: the air, water, minerals, organisms, and all other external factors surrounding and affecting a given organism at any time 环境的 huán jìng de

Origin: 1595-1605; environ + -ment

Example:

Grassland degradation and desertification have become prominent environmental problems in the Zoigê area.
草地退化、土地沙化等已成为若尔盖地区突出的环境问题。

Extended Terms:

environmental mineralogy 环境矿物学
environmental pollution 环境污染
environmental scanning 环境扫描/分析
environmental strategy 环境战略
environmental security 环境安全

---

epimorphism [ˌepiˈmɔːfism] *n.*

Definition: a homomorphism that maps from one set onto a second set 后继形貌 hòu jì xíng mào

Origin: epi- + -morphism

Example:

This paper defines homology monomorphism, homology epimorphism, homology regular

morphism in the category of topological spaces with point by using homology functor.
利用同调函子,在点标拓扑空间范畴中定义了同调单态、同调满态、同调正则态射等概念。

Extended Terms:

essential epimorphism 本质满射
strict epimorphism 严格满射
strong epimorphism 强满射
weak epimorphism 弱满射
natural epimorphism 标准满射

---

epitaxial [ˌepiˈtæksiəl] *adj.*

Definition: an oriented overgrowth of crystalline material upon the surface of another crystal of different chemical composition but similar structure 外延的 wài yán de

Origin: 1950-1955; from Neo-Latin See epi-, -taxis

Example:

An image sensor including a first epitaxial layer formed over a semiconductor substrate.
该图像传感器包括第一外延层,它形成在半导体衬底上。

Extended Terms:

epitaxial growth 外延生长
epitaxial layer 外延层
epitaxial deposition 外延淀积
epitaxial isolation 外延隔离
epitaxial silicon 外延硅

---

etch [etʃ] *n.*

Definition: to cut (a feature) into the surface of the earth by means of erosion: A deep canyon was etched into the land by the river's rushing waters. 蚀 shí

Origin: 1625-1635; from Dutch etsen, from German ätzen to etch, original cause to eat; coganerte word Old English ettan to graze; akin to eat

Example:

The possible reasons of these results are discussed on the basis of sputtering mechanism. Using enhanced etch can achieve more regular crater.
这些结果的成因在对溅射机制的研究中已讨论过了。使用增强腐蚀剂能形成更规则的弹坑。

Extended Terms:

etch figure 蚀像

etch pit 蚀坑
anisotropic etch 蛤异性腐蚀
etch polishing 腐蚀抛光
sideways etch 侧向腐蚀

eutectic [juːˈtektik] *adj.*

Definition: of greatest fusibility: said of an alloy or mixture whose melting point is lower than that of any other alloy or mixture of the same ingredients 共熔的 gòng róng de

Origin: 1880-1885; from Greek eútēkt (os) easily melted, dissolved (eu- eu-+ tēktós melted) + -ic

Example:

The analysis of micro zone discovers that there are directly relationships between the distortion structure of eutectic austenitic and lower ductility.
微区分析发现畸变的共晶奥氏体组织与球墨铸铁低韧性之间存在着直接的对应关系。

Extended Terms:

binary eutectic 二元共晶
eutectic mixture 共晶混合物
eutectic reaction 共晶反应
eutectic cementite 共晶渗碳体
eutectic composition 共晶成分
aligned eutectic 取向共晶

exchange [iksˈtʃeindʒ] *n.*

Definition: to give and receive reciprocally; interchange 交换 jiāo huàn

Origin: 1250-1300; (v.) Middle English eschaungen from Anglo-French eschaungier from * excambiāre (See ex-, change); (n.) Middle English eschaunge from Anglo-French (Old French eschange), derivative of eschaungier; modern sp. with ex- on the model of dex-1

Example:

In a network, the exchange of messages and responses, with one exchange usually involving a request for information and a response that provides the information.
网络中报文和响应的交换过程,一次交换通常包括对信息的一次请求和对提供信息的一次响应。

Extended Terms:

ion exchange 离子交换
adverse exchange 逆汇,逆汇兑
turbulent exchange 湍流交换
exchange impairment 交换障碍

coordinate exchange 坐标变换

exogenic [ˌeksəuˈdʒenik] *adj.*

Definition: arising from or relating to the surface of the earth (opposed to endogenetic)外生的 wài shēng de

Origin: 1870-1875; exo- + -genetic

Example:

Intense water-rock interaction and coupling of endogenic and exogenic forces are the important factors for the formation and occurrence of geological hazards in the area.
强烈的水岩作用和内外动力耦合作用是该地区地质灾害形成、发生的重要因素。

Extended Terms:

exogenic process 外生作用
exogenic deposits 外生矿床
exogenic leaching 表生淋滤
exogenic modulation 外在调控
exogenic toxicosis 外因性中毒

experimental [ekˌsperiˈmentəl, ekˈs-] *adj.*

Definition: pertaining to, derived from, or founded on experiment 实验的 shí yàn de

Origin: 1400-1450; late Middle English from Middle Latin experīmentālis. See experiment, -al

Example:

The experimental results show that the algorithm is effective in reducing dynamic power consumption.
实验结果显示我们的方法能有效地降低动态功率的消耗。

Extended Terms:

experimental probability 实验概率
experimental mineralogy 实验矿物学
experimental biology 实验生物学
experimental research 实验性研究
experimental process 实验过程

exsolution [ˌeksəˈluːʃən] *n.*

Definition: the process of exsolving 出溶作用 chū róng zuò yòng

Origin: from Latin exsolūtiōn-(s. of exsolūtiō). See ex-, solution

Example:

All metamorphic rocks exhibit an equilibrium texture, but exsolution lamellae of orthopyroxene (pigeonite) occur in all clinopyroxenes in mafic granulites.
这些岩石一般都展示了平衡的矿物共生结构,但在镁铁质麻粒岩的单斜辉石中普遍发育斜方辉石(变辉石)出溶片晶。

Extended Terms:

exsolution lamella 出溶条纹
exsolution structure 固溶体分解结构
exsolution lamellae 出溶纹层
exsolution paragenesis 出溶共生
noncoherent exsolution 不连贯出溶作用

ferrimagnetism [ˌferimæg'netizəm] *n.*

Definition: a phenomenon exhibited by certain substances, such as ferrites, in which the magnetic moments of neighbouring ions are antiparallel and unequal in magnitude. The substances behave like ferromagnetic materials 亚铁磁性 yà tiě cí xìng

Origin: ferri-+ magnet + -ism

Examples:

Ferrimagnetism occurs mainly in magnetic oxides known as ferrites.
亚铁磁性主要存在于磁性氧化物,如铁氧体之中。
There exists a large difference in the line width of uniform resonance estimated for various ferrites by making use of our simplified model of ferrimagnetism.
用我们的简化亚铁磁模型对各种铁氧体算得的一致共振线宽差别很大。

Extended Term:

flasher ferrimagnetism 铁氧体磁性

---

ferromagnetism [ferəu'mægnetizəm] *n.*

Definition: noting or pertaining to a substance, as iron, that below a certain temperature, the Curie point, can possess magnetization in the absence of an external magnetic field; noting or pertaining to a substance in which the magnetic moments of the atoms are aligned 铁磁性 tiě cí xìng

Origin: 1840-1850; ferro-+ magnetic

Example:

Orientation relying on magnetism must have error if there is ferromagnetism around the seismometer.

当地震计周围存在铁磁性物质时,使用传统的磁定向会引起非常大的误差。

Extended Terms:

parasitic ferromagnetism 寄生铁磁性

ferromagnetism semiconductor 铁磁半导体

ferromagnetism crystal 铁磁晶体

ferromagnetism impurity 铁磁杂质

band theory of ferromagnetism 铁磁性能带理论

---

fibrous [ˈfaibrəs] *adj.*

Definition: containing, consisting of, or resembling fibers 纤维(状)的 xiān wéi (zhuàng) de

Origin: 1620-1630; fibr- + -ous

Extended Terms:

fibrous aggregate 纤维状集合体

fibrous fracture 纤维状断口

fibrous plaster 纤维灰泥

fibrous material 纤维材料

fibrous talc 纤维状滑石

---

filmy [ˈfilmi] *adj.*

Definition: thin and light; fine and gauzy 薄膜的 bó mó de;薄的 bó de

Origin: 1595-1605; film + -y

Example:

The price of tall (low) press polyethylene (filmy class), in some areas, have risen nearly 1,000 yuan.

高(低)压聚乙烯(薄膜级),有些地区价格上涨了近 1000 元。

Extended Terms:

filmy aggregate 被膜状集合体

filmy ribbon 膜带
filmy replica 复制膜
filmy urea 包膜尿素
filmy flat shells 膜型扁壳
filmy fern 膜蕨

flexibility [ˌfleksiˈbiliti] *n.*

Definition: capable of being bent, usually without breaking; easily bent 挠性 náo xìng,柔性 róu xìng

Origin: 1375-1425; late Middle English, from Latin flexibilis pliant, easily bent. See flex, -ible

Example:

The effects of casing flexibility on the impact dynamics of a rotor/casing system are studied.
挠性外壳对转子/外壳系统的冲击动态行为影响在本文中被提出。

Extended Terms:

asymmetric flexibility 不对称伸缩性
flexibility monitoring 挠度监测
software flexibility 软件柔性
modal flexibility 模态柔度
flexible manipulator 挠性机械手
flexible gyroscope 挠性陀螺仪

fluorescence [fluəˈresns] *n.*

Definition: the emission of radiation, especially of visible light, by a substance during exposure to external radiation, as light or X-rays. Compare phosphorescence 荧光 yíng guāng

Origin: 1852; fluor(spar) + -escence, on the model of opalescence, in reference to the mineral's newly discovered property

Example:

The dashboard itself is divided into three vacuum fluorescence displays.
仪表板本身化分为三个真空荧光显示器。

Extended Terms:

Inherent fluorescence 固有荧光
delayed fluorescence 延迟荧光
fluorescence detector 荧光检测器
fluorescence quenching 荧光猝灭

fluorescence brightener 荧光增白剂

foliated [ˈfəulieitid] *adj.*

Definition:

1. shaped like a leaf or leaves: foliated ornaments.
2. also, foliate, (Petrology, Mineralogy) consisting of thin and separable laminae 叶(页)片状的 yè piàn zhuàng de

Origin: 1640-1650; foliate + -ed

Example:

I think the foliation is resulted from deformation. The original composition of the foliated rock is granite.
我认为岩石中的这些片理是由于岩石变形造成的,片理化岩石的原岩就是花岗岩。

Extended Terms:

foliated structure 叶片状构造
foliated coal 层状褐煤
foliated manifolds 叶层流形
foliated talc 叶片滑石
foliated copper 薄铜片
foliated hematite 片赤铁矿
foliated fracture 层状断口

formula [ˈfɔːmjulə] *n.*

Definition: a set form of words, as for stating or declaring something definitely or authoritatively, for indicating procedure to be followed, or for prescribed use on some ceremonial occasion 公式 gōng shì

Origin: 1575-1585; from Latin register, form, rule. See form, -ule

Example:

Conclusion: The formula for detection and measurement of oxygen saturation is put forward by using Reflectance Oxygen Saturation Detection System.
结论:利用反射式氧饱和度检测系统提出了测量和计算血氧饱和度的经验公式。

Extended Terms:

mineral formula 矿物化学式
empirical formula 实验公式
distance formula 距离公式

kinematics formula 运动学公式
approximated formula 近似公式

---

forbidden [fə'bidən] *adj.*

Definition: involving a change in quantum numbers that is not permitted by the selection rules: forbidden transition 禁止的 jìn zhǐ de

Origin: bef.1000; Middle English forbeden, Old English forbēodan See for-, bid

Example:

As a new kind of man-made structure function material, photonic crystals could realize thermal infrared camouflage because of its high-reflection photon forbidden band.
光子晶体作为一种新型人工结构功能材料,基于光子禁带的高反射特性可以实现热红外伪装。

Extended Terms:

forbidden band 禁带
forbidden explosive 禁运爆炸物
forbidden transition assignment 禁止变换赋值
forbidden latencies 禁止启动距离
forbidden drugs 违禁药物

---

fracture ['fræktʃə] *n.*

Definition: the characteristic appearance of a broken surface, as of a mineral 断口 duàn kǒu,破裂 pò liè

Origin: 1375-1425; late Middle English, from Middle French Latin frāctūra a breach, cleft, fracture, equivalent to frāct (us) (present partiliple of frangere to break) +-ūra -ure

Example:

The results showed that fracture of the component was mainly due to surface decarburation layer and poor roughness.
检验结果表明,表面脱碳层的存在及表面粗糙是导致构件断裂的主要原因。

Extended Terms:

conchoidal fracture 贝壳状断口
uneven fracture 参差状断口,粗糙断口
fatigue fracture 疲劳断面
piedmont fracture 山前断裂
fracture toughness 断裂韧度

structural fracture 构造裂缝
axle fracture 车轴断裂

framework [ˈfreimwəːk] *n.*

Definition:

1. a skeletal structure designed to support or enclose something
2. a frame or structure composed of parts fitted and joined together 框架 kuàng jià

Origin: 1635-1645; frame + work

Example:

On the basis of them, this paper introduces the concept and status description of SoftMan and designs its framework.
在此基础上我们提出了软件人的概念、状态描述,并设计了软件人的结构模型。

Extended Terms:

framework silicate(=tectosilicate)架状硅酸盐
framework feature 结构特征
framework silicate structure 架状硅酸盐结构
stable framework 坚固框架结构
framework material 骨架材料

fugacity [fjuːˈgæsiti] *n.*

Definition: a property of a gas, related to its partial pressure, that expresses its tendency to escape or expand, given by d = dμ/ RT, where μ is the chemical potential, R is the gas constant, and T is the thermodynamic temperature 逸度 yì dù

Origin: 1625-1635; from Latin fugāci-(s. of fugāx apt to flee, fleet, derivative of fugere to flee + -ous

Example:

The article sums up the constitutional equation of authentic gas and recommends a convenient formula of calculating fugacity.
本文归纳了真实气体的状态方程,并推荐一个计算逸度的简便公式。

Extended Terms:

oxygen fugacity 氧逸度
fugacity coefficient 逸压系数
Lewis-Randall rule of fugacity 路易斯—兰德尔逸度规则
sulphur fugacity 硫逸度
carbondioxide fugacity 二氧化碳逸度

gas [gæs] *n.*

Definition: a substance possessing perfect molecular mobility and the property of indefinite expansion, as opposed to a solid or liquid 气体 qì tǐ

Origin: 1650-1660; coined by J. B. van Helmont (1577-1644), Flemish chemist; suggested by Greek cháos atmosphere

Example:

Air, gas or steam enters the separator at high velocity.
空气、气体或蒸汽以很高的速度进入分离器。

Extended Terms:

inert gas 惰性气体
calibrating gas 校准用气体
gas partition 煤气隔断
gas equilibrium 气体平衡
gas dielectrics 气体电介质

---

gravity ['græviti] *n.*

Definition: the force of attraction by which terrestrial bodies tend to fall toward the center of the earth 重力 zhòng lì,地心引力 dì xīn yǐn lì

Origin: 1500-1510; from Latin gravitāt-(s. of gravitās) heaviness, equivalent to grav (is) heavy, grave + -itāt- -ity

Example:

The effect of gravity at high altitude is random.
在高海拔的地方重力的影响是任意的。

Extended Terms:

specific gravity 比重
negative gravity 负重力
gravity casting 重力铸造
gravity field 重力场

gravity segregation 重力偏析

gem [dʒem] *n.*

Definition: a cut and polished precious stone or pearl fine enough for use in jewelry 宝石 bǎo shí

Origin: 1275-1325; Middle English, gemme from Old French, from Latin gemma bud, jewel; r. Middle English yimme, Old English gim (m)

Example:

Red pyrope is one of the most important gem resources inYunnan.
红色的镁铝榴石是云南最重要的宝石资源之一。

Extended Terms:

gem mineralogy 宝石矿物学
Baroda gem 巴罗达宝石
gem pearl 宝石珍珠
gem stick 宝石棒
artificial gem 人造宝石

gemology [dʒeˈmɔlədʒi] *n.*

Definition: the science dealing with natural and artificial gemstones 宝石学 bǎo shí xué

Origin: 1965-1970; gem + -o- + -logy

Example:

In gemology, it is correct to call any Chrysoberyl that changes color Alexandrite.
在宝石学上,人们都可以把能变色金绿玉称为亚历山大石。

Extended Terms:

gemology specialty 宝石学专业
gemology and material technique 宝石及材料工艺学
characteristic of gemology 宝石学特征
gemology and mineralogy 宝石矿物学
inclusion in gemology 宝石包裹体

generation [ˌdʒenəˈreiʃən] *n.*

Definition: the entire body of individuals born and living at about the same time 世代 shì dài

Origin: 1250-1300; Middle English generacioun from Middle French, from Latin generātiōn-

(s. of generātiō). See generate, -ion

Example:

In 1956, the transistor computer was born and this is the next generation computer.
1956 年,晶体管电子计算机诞生了,这是第二代电子计算机。

Extended Terms:

mineral generation 矿物世代
filial generation 杂交世代
cell generation 细胞世代
backcross generation 回交世代
completed by generations 世代累积

---

genetic [dʒiˈnetik] *adj.*

Definition:

1. (*Biology*) pertaining or according to genetics
2. of, pertaining to, or produced by genes; genic 遗传的 yí chuán de,基因的 jī yīn de

Origin: 1825-1835; gene(sis) + -tic

Example:

Another project underway is the use of genetic engineering techniques to develop a vaccine for some diseases.
正在进行的另一个项目是利用遗传工程技术研制某些疾病的疫苗。

Extended Terms:

genetic mineralogy 成因矿物学
genetic disease 遗传性疾病
genetic marker 遗传;基因标记
genetic effect 遗传效应
genetic diversity 遗传多样性

---

geode [ˈdʒi(ː)əud] *n.*

Definition:

1. a hollow concretionary or nodular stone often lined with crystals
2. the hollow or cavity of this 晶腺 jīng xiàn,晶洞 jīng dòng,异质晶簇 yì zhì jīng cù

Origin: 1670-1680; from French géode from Latin geōdēs from Greek geōdēs earthlike. See geo-, -ode

Example:

The geode natrolite found by authors in Fushan, Dontu county, Anhui Province and Tashan, Liuhe county, Jiangsu Province are characterized by their pure and fine crystals.
本文报道了在安徽当涂县釜山和江苏六合县塔山所发现的晶洞钠沸石,其质地纯净、晶体完好。

Extended Terms:

geode kitty/amethyst geode 紫晶洞
geode natrolite 晶洞钠沸石
geode mordenite 晶洞丝光沸石
geode structure 晶洞状结构
coreless geode type 无核晶洞型
cored geode type 有核晶洞型
miarolitic alkaline granite 晶洞碱性花岗岩

---

geothermometer [ˌdʒiːəuθəˈmɔmitə] *n.*

Definition: a thermometer specially constructed for measuring temperetures at a depth below the surface of the ground 地质温压计 dì zhì wēn yà jì

Example:

The geothermometer is one of the most important methods for reconstructing the thermal history of sedimentary basins.
古温标是恢复沉积盆地热演化历史的重要指标之一。

Extended Terms:

gas geothermometer 气体地热温标
amino-acid geothermometer 氨基酸地质温度计
fluid inclusion geothermometer 液态包体地质温度计
silica geothermometer 二氧化硅地热温标
chemical geothermometer 化学地球温度计

---

glassy [ˈglɑːsi, ˈglæs-] *adj.*

Definition: resembling glass, as in transparency or smoothness; of the nature of glass; vitreous 像玻璃的 xiàng bō li de,透明的 tòu míng de

Origin: 1350-1400; Middle English glasy. See glass,-y

Example:

At first glance, calcite might be confused with quartz where both are clear, colorless, and "glassy".

乍一看来,方解石可能与石英相混淆,因为两者都是透明、无色,且“呈玻璃状”的。

Extended Terms:

glassy luster, vitreous luster 玻璃光泽
glass wool 玻璃棉
glassy state 玻璃态
glassy feldspar 透长石
glassy porcelain 玻璃质瓷器
glassy bond 玻璃黏结

---

granular [ˈgrænjulə] *adj.*

Definition: of the nature of granules; grainy 粒状 lì zhuàng

Origin: 1785-1795; granule + -ar

Example:

The results show that polygonal ferrite, granular bainite and a large amount of stabilized retained austenite can be obtained after hot rolling.
结果表明,热轧后能够获得多边形铁素体、粒状贝氏体和大量稳定的残留奥氏体组织。

Extended Terms:

granular endoplasmic reticulum 颗粒内质网
granular layer 颗粒层
granular structure 粒状结构
granular variation 粒度变异
granular pearlite 粒状珠光体
granular snow 粒状雪

---

graphic [ˈgræfik] *adj.*

Definition: pertaining to the use of diagrams, graphs, mathematical curves 图形(表)的 tú xíng (biǎo) de

Origin: 1630-1640; from Latin graphicus of painting or drawing from Greek graphikós able to draw or paint, equivalent to gráph (ein) to draw, write + -ikos -ic; c. carve

Example:

Test the local network traffic, in memory drawing, and then to graphic display.
测试本机的网络流量,在内存中绘图,然后以图形显示。

Extended Terms:

graphic demonstration 图形显示

graphic texture 文象结构
graphic sign 图形记号
graphic character 图形字符
graphic data 图形数据

greasy [ˈgriːzi] *adj.*

Definition: greaselike in appearance or to the touch 油脂(状)的 yóu zhī (zhuàng) de

Origin: greas + -y

Example:

This consists of conversion of greasy wool into clean wool.
这由含脂的羊毛转换成干净的羊毛组成。

Extended Terms:

greasy luster, oily luster 油脂光泽
greasy combing wool 精梳原毛
greasy wool 含脂原毛
greasy wool weight 含脂羊毛重
greasy stain 油垢

group [gruːp] *n.*

Definition: a division of stratified rocks comprising two or more formations 族 zú (矿物分类单位)

Origin: late 17th century: from French groupe, from Italian groupe, of Germanic origin; related to crop

Example:

It comes under this group. 它属于这一类。

Extended Terms:

amphibole group 角闪石族
leukotriene group 白三烯族

growth [grəuθ] *n.*

Definition: the act or process, or a manner of growing; development 生长 shēng zhǎng

Origin: 1550-1560; See grow, -th; probably cognate with ON grōthr

Example:

In the Cr_7C_3, the {013} growth twin was discovered by the electron diffraction patterns and the secondary twin might be possibly found.
从电子衍射花样上发现了 Cr_7C_3 中存在{013}生长孪晶,并可能存在着二次孪晶。

Extended Terms:

growth hillocks 生长丘
growth steps 生长台阶
growth twin 生长双晶
deposit growth 积灰生长
sustained growth 持续增长

hackly [ˈhækli] *adj.*

Definition: Rough or jagged, some minerals break with a hackly fracture.
参差不齐的 cēn cī bù qí de,粗糙的 cū cāo de

Origin: 1790-1800; hackle + -y

Example:

Described for the first time in this paper the hackly microcline, a typomorphicmineral in potash metasomatic uranium deposits.
本文首次系统总结了钾交代型铀矿床的标型矿物——犬牙状微斜长石的五项标型特征。

Extended Terms:

hackly fracture 锯齿状断口
hackly surface 粗糙不平表面

---

halide [ˈhælaid] *n.*

Definition: a chemical compound in which one of the elements is a halogen 卤化物 lǔ huà wù

Origin: 1875-1880; hal(ogen) + -ide

Example:

Gallium and arsenic can be removed as halide by volatilization.
大量的砷和镓可以卤化物形式挥发除去。

Extended Terms:

halide minerals 卤化物矿物
cyan halide 卤化氰
methyl halide 甲基卤
halide crystal 卤化物晶体
acid halide 酸性卤化物

---

hardness [ˈhɑːdnis] *n.*

Definition: the state or quality of being hard 硬度 yìng dù

Origin: before 900; Middle English hardnes, Old English heardnes. See hard,-ness

Example:

Uses in each kind of medium degree of hardness solid material, in cement industry, limestone in broken bits, paste, chamotte, mix material and so on.
主要用于各种中等硬度固体物料的细碎,水泥工业中,细碎石灰石、石膏、熟料、混合材料中。

Extended Terms:

Mohs' hardness 莫氏硬度(莫氏硬度)
mineral hardness scale 矿物硬度计
Mohs' hardness scale(MHS)莫氏硬度计
indentation hardness 压痕硬度
scratch hardness 抗刮硬度

---

heat [hiːt] *n.*

Definition: the state of a body perceived as having or generating a relatively high degree of warmth 热度 rè dù,热量 rè liàng

Origin: before 900; Middle English hete, Old English hitu; akin to German Hitze; See hot

Example:

Sound carries well over water; The airwaves carry the sound; Many metals conduct heat.
声音在水中传导很快;广播传导声音;许多金属传导热量。

Extended Terms:

heat capacity 热容
heat loss 热损耗
heat value 热值
heat sink 散热器
heat radiation 热辐射

heat reservoir 热库,储热器

hexagonal [hek'sægənəl] *adj.*

Definition: noting or pertaining to a system of crystallization in which three equal axes intersect at angles of 60° on one plane, and the fourth axis, of a different length, intersects them perpendicularly 六方形的(晶体)liù fāng xíng de (jīng tǐ)

Origin: 1565-1575; hexagon + -al

Example:

Mimetite group minerals are a group of isomorphic series minerals with hexagonal system, which include mimetite, vanadinite and pyromorphite.
砷铅矿族矿物是一族六方晶系的类质同象系列矿物,包括砷铅矿、钒铅矿和磷氯铅矿。

Extended Terms:

hexagonal close-packed 六方最紧密堆积
hexagonal lattice 六方晶格
hexagonal column 六角柱体
hexagonal platelet 六角板体
hexagonal system 六方晶系
hexagonal lenticulation 六角形透镜光栅
hexagonal mesh 六方格

hopper ['hɔpə] *n.*

Definition: A funnel-shaped chamber or bin in which loose material, as grain or coal, is stored temporarily, being filled through the top and dispensed through the bottom.漏斗 lòu dǒu

Origin: 1200-1250; Middle English See hop, -er

Example:

This machine is suitable for autotransporting the granules, and be coordinatly used with dryer or extruder hopper.
本机适用于颗粒料物流的自动输送,可与干燥机或挤出机料斗连接使用。

Extended Terms:

hopper crystal 漏斗晶
feed hopper 装料斗
hopper weigher 料斗秤
chip hopper 装锅漏漏斗

card hopper 卡片传送斗

hourglass [ˈauəglɑːs, -glæs] *n.*

Definition: an instrument for measuring time, consisting of two bulbs of glass joined by a narrow passage through which a quantity of sand or mercury runs in just an hour 砂钟 shā zhōng

Origin: 1505-1515; hour + glass

Example:

Another asymmetry is that the stalled shock front can deform, causing the explosion to develop an hourglass shape.
另一种不对称原因是失速的震波波前变形,这会导致爆炸发展成沙漏的形状。

Extended Terms:

hourglass texture 砂钟构造
hourglass reflector 沙漏型反射器
hourglass valley 沙漏形河谷
hourglass cursor 沙漏光标
hourglass energy 沙漏能
hourglass nebula 沙漏星云

hue [hjuː] *n.*

Definition: a gradation or variety of a color; tint 主色调 zhǔ sè diào,色度 sè dù

Origin: before 900; from Middle English hewe, from Old English hīw form, appearance, color; Gothic hiwi form, appearance; akin to Old English hār gray (See hoar)

Example:

Video equalizer, allows you to adjust the brightness, contrast, hue, saturation and gamma of the video image.
"视频均衡器":允许您调整亮度、对比度、色调、饱和度和伽玛的视频影像。

Extended Terms:

extraspectral hue 谱外色
manual hue 手控色调
Munsell hue 芒赛尔色调
hue offset 色调偏移
shade hue 暗色调

hydrogen [ˈhaidrədʒən] *n.*

Definition: a colorless, odorless, flammable gas that combines chemically with oxygen to form water; the lightest of the known elements. Symbol; H; atomic weight; 1.00797; atomic number; 1; density; 0.0899 g/l at 0℃ and 760 mm pressure 氢 qīng

Origin: 1785-1795; from French hydrogène. See hydro-, -gen

Example:

The isotope fractionation effect is primary factor affecting accuracy of hydrogen isotope analysis.
同位素分馏效应是影响氢同位素准确分析的主要因素。

Extended Terms:

hydrogen bond 氢键
hydrogen maser 氢微波激射器氢脉泽
hydrogen bonding 氢键
diffusible hydrogen 扩散氢
atomic hydrogen 原子氢
liquid hydrogen 液氢

---

hydrolysed [ˈhaidrəlaiz] *adj.*

Definition: to subject or be subjected to hydrolysis 水解的 shuǐ jiě de

Origin: 1875-1880; hydro(lysis) + -lyze

Example:

The results show that hydrolysed keratin after proper modification obtains excellent decolourizing effects on the majority of dyestuffs.
结果表明,水解角蛋白经适当改性,对大多数种类的染料具有良好的脱色效果。

Extended Terms:

hydrolysed whey protein 乳清蛋白水解物
hydrolysed wheat gluten 可溶性小麦蛋白
hydrolysed protein 水解蛋白

---

hydroscopic [ˌhaidrəsˈkɔpik] *adj.*

Definition: an optical device for viewing objects below the surface of water 吸附的 xī fù de

Origin: 1670-1680; hydro- + -scope

Example:

Sepiolite hydroscopic agent trial—produced by addition of some agents is tested.

添加一两种化学药剂,试验研制出海泡石吸湿剂。

Extended Terms:

hydroscopic water 吸附水;湿存水
hydroscopic property 吸湿性
hydroscopic swelling 吸湿膨胀

---

hydrothermalism [ˌhaidrəuˈθəməlizm] *n.*

Definition: 热液作用 rè yè zuò yòng

Origin: hydro-+ thermal + -ism

Example:

Intensive tectonic activity, frequent magmatic hydrothermalism and abundant gold-bearing rock created advantageous geological conditions for remobilizationand enrichment of gold.
区内构造活动剧烈,岩浆热液活动频繁,岩石多不同程度变质,金源岩丰富,为金的活化与富集成矿提供了优越的地质条件。

Extended Terms:

magma hydrothermalism 岩浆热液作用
magmatic hydrothermalism 岩浆热液作用

---

hydroxide [haiˈdrɔksaid] *n.*

Definition: a chemical compound containing the hydroxyl group 氢氧化物 qīng yǎng huà wù

Origin: 1820-1830; hydr- + oxide

Example:

The graphite is mixed with clay from Mississippi in which ammonium hydroxide is used in the refining process.
石墨要与产自密西西比河床的黏土混合,在精炼过程中,还要用到氢氧化铵。

Extended Terms:

cadmium hydroxide 氢氧化镉
copper hydroxide 氢氧化铜
ferric hydroxide 氢氧化铁
potasium hydroxide 氢氧化钾
aluminium hydroxide 氢氧化铵
oxides and hydroxides minerals 氧化物氢氧化物矿物

---

hypergene [ˈhaipəˌdʒiːn] *adj.* =supergene

Definition: 表生的 biǎo shēng de，浅成的 qiǎn chéng de

Origin: hyper-+ gene

Examples:

The reef underwent four diagenetic period, such as syngenetic, eogenetic, telogenetic, and hypergene diagenetic period.
生物礁经历了同生成岩、早成岩、晚成岩和表生成岩四个阶段。
In shallow hypergene zone, the shift of elements is of long-term and complicated feature and is influenced by a lot of factors, which plays an important role in destabilizaton of slope.
在浅表生带中，元素迁移在斜坡失稳孕育过程中起显著的伴生作用，直观、细致地反映出斜坡失稳演化过程的长期、复杂、受多种因素控制的特征。

Extended Terms:

hypergene mineral 表生矿物
secondary stratfication 次生层理
hypergene mobility 浅成活动性
hypergene structure 表生构造
hypergene zone 浅成带

inclusion [inˈkluːʒən] *n.*

Definition: a solid body or a body of gas or liquid enclosed within the mass of a mineral 包裹体 bāo guǒ tǐ，包体 bāo tǐ

Origin: 1590-1600; 1945-1950 from Latin inclūsiōn-(s. of inclūsiō) a shutting in, equivalent to inclūs (us) (See incluse) + -iōn- -ion

Example:

Results: Most of the product didn't form inclusion body in cell.
结果：绝大部分产物在细胞内没有形成包含体。

Extended Terms:

cognate inclusion 同源包裹体
gaseous inclusion 气体包裹体

inclusion granule 内含颗粒
coaly inclusion 煤包体
inclusion conjunctivitis 包含体结膜炎
pseudo-secondary inclusion 假次生包体
primary inclusion 原生包体

idiochromatic [ˌidiəukrəuˈmætik] *adj.*

Definition: (of a mineral) deriving a characteristic color from its capacity to absorb certain light rays 自色的 zì sè de,本质的 běn zhì de

Origin: idio-+ chromatic

Extended Terms:

idiochromatic color 自色
idiochromatic crystal 本质色晶体
idioblastic mineral 自色矿物

incommensurate [ˈinkəˌmenʃərət] *adj.*

Definition: not commensurate; disproportionate; inadequate 不相称的 bù xiāng chèn de

Origin: 1640-1650; in- + commensurate

Example:

The incommensurate modulation structures, domain structures and their evolution of CBN28 crystals were studied by transmission electron microscopy.
用透射电子显微镜研究了该单晶中的无公度调制结构、畴结构及其演变。

Extended Terms:

incommensurate structure 不相称结构
incommensurate crystal 无公度晶体
incommensurate modulation 非相称性调变
commensurate-incommensurate transition 公度非公度转变

incompatible [ˌinkəmˈpætəbl] *adj.*

Definition: not compatible; unable to exist together in harmony 不相容的 bù xiāng róng de

Origin: 1555-1565; from Medieval Latin incompatibilis.See in-, compatible

Example:

The components of solid propellants are of necessity highly reactive and essentially incompatible.

固体推进剂成分必须是很活泼的,基本上是不相容的。

Extended Terms:

incompatible elements 不相容元素
incompatible development 不协调的发展
incompatible land use 不协调的土地用途
incompatible method 不兼容方法
incompatible procedure 不兼容程序
incompatible minerals 不相容矿物

---

index [ˈindeks] *n.*

Definition: something that directs attention to some fact, condition, etc.; a guiding principle 指标 zhǐ biāo;索引 suǒ yǐn

Origin: 1350-1400; from Middle English, from Latin informer, pointer, equivalent to in- in- + -dec-(combining form of dic-, show, declare, indicate; akin to teach) + -s noun sing. ending

Example:

The luster depends on the refractive index of all kind of mineral which consists of the rock, also depends on the micro-structure of the stone surface.
光泽度的大小一方面取决于组成岩石的各种矿物的折射率的大小,另一方面与石材表面的微观结构密切相关。

Extended Terms:

index mineral 指示矿物
index gears 分度齿轮
index head 分度头
refractive index 折射指数
stock index 股市指数

---

inert [iˈnəːt] *adj.*

Definition: having little or no ability to react, as nitrogen that occurs uncombined in the atmosphere 惰性的 duò xìng de

Origin: 1640-1650; from Latin inert- (s. of iners) unskillful, equivalent to in- in- + -ert-, combining form of art- (s. of ars) skill. See art1

Example:

The inert deposits include silicon, calcium, Cl and K and their influence on production is obvious.

这些惰性物质包括：硅、钙、氯和钾等元素，它们对生产的影响是显而易见的。

Extended Terms:

inert gas (atmophile element) 惰性气体(亲气元素)
inert gelatin 惰胶
inert plasticizer 惰性增塑剂
inert dust 惰性尘末
inert element 惰性元素

---

inosilicate [inəˈsilikit] *n.* (=chain silicate)

Definition: any silicate having a structure consisting of paired parallel chains of tetrahedral silicate groups, every other of which shares an oxygen atom with a group of the other chain, the ratio of silicon to oxygen being 4 to 11 链状硅酸盐 liàn zhuàng guī suān yán

Origin: from Greek īno-(combining form of ī's fiber, sinew) + silicate

Extended Terms:

inosilicate mineral 链状硅酸盐矿物
alum inosilicate-based fiber 铝硅酸盐玻璃纤维

---

insulator [ˈinsjuleitə, ˈinsə-] *n.*

Definition: a material of such low conductivity that the flow of current through it is negligible 绝缘体 jué yuán tǐ

Origin: 1795-1805; insulate + -or

Example:

An insulator is a poor conductor because it has a high resistance to such flow.
绝缘体是不良导体，因为它是对电流的通过呈高电阻的物质。

Extended Terms:

heat insulator 热绝缘体
suspension insulator 悬式绝缘子
electrical insulator 电绝缘体
laminated insulator 层状绝缘物
glass insulator 玻璃绝缘子

---

intercalation [inˌtəːkəˈleiʃən] *n.*

Definition:

1. the act of intercalating; insertion or interpolation, as in a series 插入 chā rù
2. something that is intercalated; interpolation 夹层 jiā céng

Origin: 1570-1580; from Latin intercalātiōn- (s. of intercalātiō). See intercalate, -ion

Extended Terms:

rock intercalation 岩石夹层
hemipelagic intercalation 半深海沉积夹层
polymer intercalation 聚合物插层
intercalation compound 夹层复合物
muddy intercalation 泥质夹层

---

intergrowth [ˈintəgrəuθ] *n.*

Definition: growth or growing together, as of one thing with or into another 互生 hù shēng (交生 jiāo shēng)

Origin: 1835-1845; inter-+ growth

Example:

The result indicated that granularity magnetite was uneven. Most of Ascharite particles is fine and intergrowth with magnetite and Zermattite.
结果表明磁铁矿的粒度粗细不均。多数硼镁石的粒度细,常以微晶粒状集合体与磁铁矿和蛇纹石共生。

Extended Terms:

dactylotype intergrowth 指状交生
lamellar intergrowth 片晶连生
graphic intergrowth 文像共生
Bggild intergrowth 博吉尔德连生(斜长石出溶形成)
Huttenlocher intergrowth 休顿洛契连生(斜长石出溶形成)

---

interlayer [ˈintəleiə] *n.*

Definition: bed; stratum 夹层 jiā céng,隔层 gé céng

Origin: inter-+ layer

Example:

The cross section of the oxidized alloys consists of oxide scale, interlayer and substrate.
氧化后的合金由表面氧化层、中间层和基体组成。

Extended Terms:

interlayer water 层间水
interlayer dielectric 层间绝缘
interlayer connection 层间连接

interlayer crossflow 层间对流
cathode interlayer 阴极介层

intermetallic [ˌintə(ː)miˈtælik] *adj.*

Definition: 金属间(化合)的 jīn shǔ jiān (huà hé) de

Origin: 1560-1570; from Latin metallicus from Greek metallikós of, for mines. See metal,-ic

Example:

Composite alloy coatings of intermetallic compound dispersed and distributed in amorphous alloy possess excellent corrosion resistance and abrasion performance.
非晶基上弥散分布着金属间化合物的复相合金镀层,具有优异的耐蚀性及良好的耐磨性。

Extended Terms:

intermetallic compound(alloy)金属互化物
intermetallic semiconductor 金属间半导体
intermetallic interaction 金属间相互作用
intermetallic compound superconductor 金属间化合物超导体
intermetallic compound phase of sueralloy 高温合金材料的金属间化合物相

interstice [inˈtəːstis] *n.*

Definition: a small or narrow space or interval between things or parts, especially when one of a series of alternating uniform spaces and parts 空隙 kòng xì

Origin: 1595-1605; from Latin interstitium, equivalent to interstit-, var. s. of intersistere to stand or put between + -ium -ium

Example:

The interstice of foreland slope was mainly dissolved pore and some primary pore.
前缘斜坡带储层类型为孔隙型,主要以残余原生孔及次生孔为主。

Extended Terms:

discontinuous interstice 不连续间隙
primary interstice 初生孔隙
tetrahedral interstice 四面体空隙
secondary interstice 次生孔隙
isolated interstice 隔离空隙

interstitial [ˌintəˈstiʃəl] *adj.*

Definition: an imperfection in a crystal caused by the presence of an extra atom in an otherwise complete lattice. Compare vacancy 空隙的 kòng xì de

Origin: 1640-1650; from Latin interstiti (um) interstice + -al

Example:

In petroleum reservoirs, however, the rocks are usually saturated with two or more fluids, such as interstitial water, oil, and gas.
然而在储油层中,岩石通常被两种或更多的流体饱和,如间隙水、油和气。

Extended Terms:

interstitial pneumonia 间质性肺炎
interstitial cell 间质细胞
interstitial diffusion 填隙式扩散
interstitial ion 填隙离子
interstitial sedimentary water 隙间沉积水

---

ion [ˈaiən] *n.*

Definition: an electrically charged atom or group of atoms formed by the loss or gain of one or more electrons, as a cation (positive ion), which is created by electron loss and is attracted to the cathode in electrolysis, or as an anion (negative ion), which is created by an electron gain and is attracted to the anode. The valence of an ion is equal to the number of electrons lost or gained and is indicated by a plus sign for cations and a minus sign for anions, thus: Na^{+}, Cl^{-}, Ca^{2+}, S^{2-}离子 lí zǐ

Origin: the term introduced by Michael Faraday in 1834

Example:

The experimental results show that surface mortar layer can significantly reduce the chloride ion diffusion coefficient of concrete.
实验结果表明,致密的表面砂浆层能显著地降低混凝土本体的氯离子扩散系数。

Extended Terms:

ion exchange 离子交换
positive ion 阳离子
ion plating 离子电镀
zwitter ion 两性离子
ion laser 离子激光器

---

ionicity [ˌaiəˈnisəti] *n.*

Definition: 电离度 diàn lí dù,离子性 lí zǐ xìng

Origin: ion + -ici + -ty

Examples: A method to calculate new valence electronegativity and the ionicity of chemical bond of the transition metal elements is also presented here.
提出了计算过渡元素价态共价半径的公式,以及计算过渡元素新价态电负性和化学键的离子性的方法。
These mechanisms are the internal displacement of the ionic charge, the internal displacement of the electronic charge and the charge in ionicity due to strain.
这些机制是离子电荷的内部位移、电子电荷的内部位移和由于应变引起的离子性的变化。

Extended Terms:

ionicity parameter 离子性参数
ion exchange 离子交换(作用)
ion exclusion 离子排斥
ion meter 电离压力表
ion plating film 电离镀膜

---

ionisation [ˌaiənaiˈzeiʃən, -niˈz-] *n.*

Definition: the formation of ions as a result of a chemical reaction, high temperature, electrical discharge, particle collisions, or radiation 电离 diàn lí

Origin: ion + -isa + -tion

Example:

The cutting beam is a short ranged but devastating ionisation beam usually employed for gouging out recalcitrant moonlets.
切割光束是一种短程但极具破坏力的离子光束,通常用于切割坚硬的小卫星。

Extended Terms:

ionisation potential 电离电势(电离电位)
air ionisation 空气离子化
ionisation unit 电离单元
successive ionisation energy 逐级电离能
ionisation chamber 电离室

---

iridescence [ˌiriˈdesəns, ˈaiəri-] *n.*

Definition: iridescent quality; a play of lustrous, changing colors 晕彩 yùn cǎi (晕色 yùn sè)

Origin: 1795-1805; irid-+ -escence

Example:

Type of feldspar mineral in the plagioclase series that is often valued as a gemstone and as ornamental material for its red, blue, or green iridescence.
斜长石系列的长石类矿物,由于有红、蓝或绿色等闪光,常被当作宝石或装饰材料。

Extended Term:

iridescence/rainbow 彩虹

---

isomorphism [ˌaisəuˈmɔːfizəm] *n.* (allomerism)

Definition: the state or property of being isomorphous or isomorphic 类质同像 lèi zhì tóng xiàng

Origin: 1820-1830; isomorph(ous) + -ism

Example:

The gallium and germanium present in the form of isomorphism in tetrahedron coordination with sulphur in sphalerite.
镓、锗与硫呈四面体配位,形成闪锌矿型晶体结构,以类质同象形式伴生于闪锌矿中。

Extended Terms:

isovalent isomorphism 等价类质同像
linear isomorphism 线性同构
weak isomorphism 弱同构
perfect isomorphism 完全类质同象
equivalent isomorphism 等价类同象

---

isotherm [ˈaisəuθəːm] *n.*

Definition: a curve on which every point represents the same temperature 等热 děng rè; 等温线 děng wēn xiàn

Origin: 1855-1860; back formation from isothermal

Example:

An isotherm is a line on a map that joins locations having the same mean temperatures.
等温线是在地图上把具有相同平均温度的地方连接起来的线。

Extended Terms:

saturation isotherm 饱和等温线
solubility isotherm 溶度等温线
zero isotherm 零度等温线

partition isotherm 分配等温线
distribution isotherm 分配等温线

isotope [ˈaisəutəup] *n.*

Definition: any of two or more forms of a chemical element, having the same number of protons in the nucleus, or the same atomic number, but having different numbers of neutrons in the nucleus, or different atomic weights. There are 275 isotopes of the 81 stable elements, in addition to over 800 radioactive isotopes, and every element has known isotopic forms. Isotopes of a single element possess almost identical properties. 同位素 tóng wèi sù

Origin: 1910-1915; iso- + -tope from Greek tópos place

Example:

And the study on application of lead isotope tracing for city environment pollution sources can get a good result.
铅同位素示踪理论应用于城市环境污染源的研究,能够取得比较理想的结果。

Extended Terms:

isotope mineralogy 同位素矿物学
isotope enrichment 同位素浓缩
carbon isotope 碳同位素
isotope carrier 同位素载体
isotope chart 同位素表

isotype [ˈaisəutaip] *n.*

Definition: a drawing, diagram, or other symbol that represents a specific quantity of or other fact about the thing depicted 等型 děng xíng

Origin: 1880-1885; iso-+ type

Example:

The oxygen isotype method was used to test the hydrothermal chimney and sulfide samples collected from the Mariana Island Arc, the Mariana Trough, the Okinawa Trough and the Galapagos Rift.
使用氧同位素方法,测试了西太平洋马里亚纳岛弧、马里亚纳海槽、冲绳海槽和东太平洋加拉帕戈斯裂谷的海底热液烟囱和硫化物全岩样品。

Extended Terms:

isotype method 同型法
isotype switch 同种型转换

isotype control antibodies 同型对照抗体
isotype control 同型对照
isotype exclusion 同型排斥

jamesonite [ˈdʒeimsənait] *n.*

Definition: a metallic, dark-gray mineral, lead and iron antimony sulfide, formerly mined for lead 羽毛矿 yǔ máo kuàng

Origin: 1815-1825; named after Robert Jameson(1774-1854), Scottish scientist, See -ite

Example:

The medium for jamesonite treatment with Slurry Electrolysis Process is selected by systematic comparison and experiments.
通过比较和系统的试验,确定矿浆电解法处理脆硫锑铅矿的介质体系。

Extended Term:

jamesonite concentrate 脆硫锑铅矿精矿

---

joint [jɔint] *n.*

Definition: a fracture plane in rocks, generally at right angles to the bedding of sedimentary rocks and variously oriented in igneous and metamorphic rocks, commonly arranged in two or more sets of parallel int ersecting systems 节理 jié lǐ;关节 guān jié

Origin: 1250-1300; 1900-1905 for def.6; Middle English, from Old French joint, jointe from Latin junctum, juncta

Example:

Evolution of rock joints into transfixion is described by the time series from strain-stress curves, and the state of joint system in sandstone is chaotic before its ultimate state.
形成岩石全应力应变曲线的时间序列反映了岩石内部节理裂隙趋向贯通的演化进程,砂岩节理裂隙系统在峰前阶段处于不同程度的混沌状态。

Extended Terms:

joint cavity 关节腔
joint plane 节理面

joint set 节理组
joint surface 节理面
joint system 节理系
joint wall 节理面

kame [keim] *n.*

Definition: an irregular mound or ridge of gravel, sand, etc., deposited by water derived from melting glaciers 冰砾阜 bīng lì fù

Origin: 1860-1865 for this sense; special use of Scots, N dial. kame comb (Middle English (dial.) camb, kambe, Old English camb, comb); See comb

Extended Terms:

kame moraine 冰砾碛
kame ridge 冰砾阜脊
kame plateau 冰砾阜高地
residual kame 蚀余冰砾阜
kame plain 冰砾阜平原

---

karat ['kærət] *n.*

Definition: A measure of the fineness (i.e., purity) of gold. It is spelled carat outside the United States but should not be confused with the unit used to measure the weight of gems, also called carat. A gold karat is 124 part, or 4.1667 percent, of the whole, and the purity of a gold alloy is expressed as the number of these parts of gold it contains. Thus, an object that contains 16 parts gold and 8 parts alloying metal is 16-karat gold, and pure gold is 24-karat gold. 克拉 kè lā

Origin: from Old French, from Medieval Latin carratus, from Arabic qīrāt from Greek keration (a little horn, from keras horn)

Example:

The dessert is made with a blend of 28 rare and exotic cocoas from around the world, whipped cream, black truffle shavings, and 23 karat edible gold.
这款甜点混合了世界上 28 种珍贵且具有异国情调的可可粉,生奶油,并且用黑松露装饰,里面还有 23 克拉可食用黄金。

Extended Terms:

high karat gold 高开黄金
karat needles(needles)试金棒

kinetics [ki'nɛtiks] *n.*

Definition: the branch of chemistry or biochemistry concerned with measuring and studying the rates of reactions 动力学 dòng lì xué

Origin: 1864, from Greek kinetikos "moving, putting in motion," from kinetos "moved," v. kinein "to move"

Example:

study on nonisothermal crystallization kinetics of PTT
PTT 的非等温结晶动力学研究

Extended Terms:

attachment kinetics 附着能(吸附能)
consistent subcritical kinetics 相容次临界动力学
copolymerization kinetics 共聚反应动力学
corrosion kinetics 腐蚀动力学
densification kinetics 致密化动力学

kink [kiŋk] *n.*

Definition: a sharp twist or curve in something that is otherwise straight 扭折 niǔ zhé

Origin: late 17th century, from Middle Low German kinke, probably from Dutch kinken "to kink"

Example:

The results show a lot of kink bands formed at about 90° with respect to draw axis and(110), (200)reflection planes slipped in the kink bands.
结果表明,在与拉伸轴约成 90°形成许多变形带,变形带中(110)和(200)衍射面发生滑移。

Extended Terms:

kink band 膝折带
kink mode 扭曲模
kink site 扭折位
kink fold 膝折褶皱
intragranular kink 晶内扭折

knee [niː] *n.*

Definition: the joint between the thigh and the lower leg in humans（人的）膝关节（rén de）xī guān jié，膝盖 xī gài

Origin: Old English cnēow，cnēo，of Germanic origin；related to Dutch knie and German knie，from an Indo-European root shared by Latin genu and Greek gonu

Example:

His left knee was hurt in a traffic accident.
他的左膝在一次交通事故中受伤了。

Extended Terms:

knee shaped twin（elbow twin）膝状双晶（肘状双晶）
knee-brace 膝形拉条
knee-girder 肘状梁
sliding knee 滑动膝杆
diagonal knee 对角接铁

labradorescence [ˌlabrədɔːˈrɛs(ə)ns] *n.*

Definition:（*Mineralogy*）the brilliant iridescence exhibited by some specimens of labradorite and other feldspars［矿］拉长晕彩（kuàng）lā cháng yùn cǎi

Origin: labrador + escence

Example:

relation between labradorescence and internal structure of labradorite
拉长石晕彩与内部结构的关系

---

lamella [ləˈmɛlə] *n.*

Definition: a thin layer，membrane，scale，or plate-like tissue or part，especially in bone tissue（尤指骨组织）薄片，薄膜，瓣（yóu zhǐ gǔ zǔ zhī）báo piàn，báo mó，bàn

Origin: late 17th century，from Latin，diminutive of lamina "thin plate"

Example:

having the form of a thin plate or lamella
有薄片或鳃瓣状形体的

Extended Terms:

annulate lamella 环形片层
boehm lamellae 勃姆薄层
concentric lamella 同心性骨板
deformation lamella 变形壳层
exsolution lamellae 出溶层

Landau [lɑːn'dau] *n.*

Definition: Lev (Davidovich) (1908-1968), Soviet theoretical physicist, born in Russia. Active in many fields, Landau was awarded the Nobel Prize for Physics in 1962 for his work on the superfluidity and thermal conductivity of liquid helium
朗道·莱夫(1908—1968,苏联理论物理学家,生于俄罗斯,活跃于多个领域,1962年因液氦超流态和导热性研究获诺贝尔物理学奖)lǎng dào lái fū

Example:

The vinyl covering proved popular, and some form of vinyl trim, embellished with opera windows and Landau bars, would be seen on Thunderbird roofs for the next two decades.
乙烯基车顶果然大行其道,而且,一些配有后侧壁板小窗和双排座格栅的简化版乙烯基车顶在此后二十年间的雷鸟车上依然可见。

Extended Terms:

Landau carriage 朗道运输
Ginzburg-Landau vortices 有限量估计

---

lath [lɑːθ, læθ] *n.*

Definition: a thin flat strip of wood, especially one of a series forming a foundation for the plaster of a wall or the tiles of a roof or made into a trellis or fence (灰泥、瓦片等基部结构)木条,板条 mù tiáo, bǎn tiáo

Origin: Old English lætt, of Germanic origin; related to Dutch lat and German Latte, also to lattice

Example:

to change the shape of(a structure, such as a wall) by applying lath and plaster or boarding
用板条、灰泥或大木板来改变(建筑物,如墙)的形状

Extended Terms:

lath shaped 长板
bottom lath 船底板条
double lath 双面灰板条
gypsum lath 石膏板条
martensite lath 马氏体板条

---

leaching [ˈliːtʃiŋ] *n.*

Definition: to dissolve out soluble constituents from (ashes, soil, etc.) by percolation 浸出 jìn chū,淋滤 lín lǜ

Origin: 1425-1475; late Middle English leche leachate, infusion, probably Old English læc (e), lec (e), akin to leccan to wet, moisten, causative of leak

Example:

The treatment of metal or the separation of metal from ores and ore concentrates by liquid processes, such as leaching, extraction, and precipitation.
水冶金术通过液体过程,像过滤、提炼、沉淀等方法处理金属,从矿石中或矿石浓缩物中分离出金属。

Extended Terms:

atmospheric pressure leaching 常压浸出
bacterial leaching 细菌浸矿
batch leaching 分批浸出;间歇浸出
calcine leaching 焙砂浸出
carbonate leaching 碳酸盐浸取

---

lichen [ˈlaik(ə)n, ˈlitʃ(ə)n] *n.*

Definition: a simple slow-growing plant which typically forms a low crust-like, leaf-like, or branching growth on rocks, walls, and trees(count noun) 地衣 dì yī

Origin: early 17th century: via Latin from Greek leikhēn

Example:

Lichen is the dominant life-form that covers the surfaces of stones and trees.
地衣是生长在石头和树木表面的主要生物。

Extended Terms:

crustose lichen 壳状地衣
fruticose lichen 灌木状地衣

heteromerous lichenes 异层地衣
homoeomerous lichenes 同层地衣

lightness [ˈlaitnis] *n.*

Definition: the natural agent that stimulates sight and makes things visible 光 guāng

Example:

An achromatic color refers to any lightness between the extremes of black and white.
一种非彩色的颜色，指各种各样介于黑色和白色两个极端之间的浅淡的颜色。

Extended Terms:

lightness contrast 明度对比
lightness index 明度指数
lightness constancy 明度守恒
cored for lightness 为减轻重量而钻空
lightness scales 亮度标度

liquidus [ˈlikwidəs] *n.*

Definition: a curve in a graph of the temperature and composition of a mixture, above which the substance is entirely liquid 液相线 yè xiāng xiàn

Origin: Latin, literally "liquid"

Example:

The p-t path of the ascent of the Hannuoba basaltic magma is consistent with the experimental basalt liquidus, which indicates that the magma ascending from the upper mantle up to the ground surface had in the main maintained the liquidstate.
岩浆上升的 p-t 路线与玄武岩的液相线一致，说明原生岩浆由上地幔源区主要呈液体状态快速上升达到地表。

Extended Terms:

liquidus line 液线
liquidus temperature 液线温度
liquidus sintering 液相线烧结
liquidus curve 液相线
liquidus surface 液相面

lithophile [ˈliθəufail] *adj.*

Definition: (of a chemical element) concentrated in the earth's crust, rather than in the core or mantle 亲石的 qīn shí de

Origin: 1920-1925; litho- + -phile

Example:

Uranium is a lithophile element.
铀是亲石元素。

Extended Terms:

lithophile element 亲石元素
lithophile ore 亲石性矿化元素

---

luminescence [luːmiˈnɛs(ə)ns] *n.*

Definition: the emission of light by a substance that has not been heated, as in fluorescence and phosphorescence 发冷光(如荧光、磷光) fā lěng guāng

Origin: late 19th century: from Latin lumen, lumin-"light" + -escence (denoting a state)

Example:

luminescence produced by physiological processes (as in the firefly)
生物体的生理作用发出的光,如萤火虫发出的光。

Extended Terms:

X-ray luminescence X-射线发光
cathode-ray luminescence 阴极射线发光
crystal luminescence 晶体发光
electro-photo luminescence 电控光致发光
electrochemical luminescence 电化学发光

---

luster [ˈlʌstə] *n.*

Definition: a gentle sheen or soft glow, especially that of a partly reflective surface 光泽 guāng zé

Origin: early 16th century: from French lustre, from Italian lustro, from the verb lustrare, from Latin lustrare "illuminate".

Example:

any of various volcanic glasses distinguished by their dull pitchlike luster
(松脂石)某种火山玻璃,以其暗似沥青的光泽而著名

Extended Terms:

adamantine luster 金刚光泽

earthy luster 土状光泽
glassy luster, vitreous luster 玻璃光泽
metallic luster 金属光泽
non-metallic luster 非金属光泽
oily luster, greasy luster 油脂光泽
pearly luster 珍珠光泽
pitchy luster 沥青光泽
resinous luster 树脂光泽(松脂光泽)
semi-metallic luster 半金属光泽
submetallic luster 半金属光泽
waxy luster 蜡状光泽

magmatism [ˈmægmətizm] *n.*

Definition: (*Geology*) the motion or activity of magma [地质]岩浆活动 yán jiāng huó dòng ,岩浆作用 yán jiāng zuò yòng

Origin: 1400-1450; late Middle English from Latin: dregs, leavings from Greek mágma kneaded mass, salve, equivalent to mag-(base of mássein to knead, press; see mass) + -ma n. suffix of result

Example:

Early Paleozoic Collisional Orogeny and Magmatism on Northern Margin of the Qaidam Basin
柴达木盆地北缘早古生代碰撞造山及岩浆作用

Extended Terms:

aggressive magma 侵进岩浆
anomalous magma 异常岩浆
bench magma 台阶岩浆
emulsive magma 乳浊岩浆
magnesium magma 氢氧化镁浮悬液

---

magnetic [mægˈnetik] *adj.*

Definition: having the properties of a magnet; exhibiting magnetism 磁的 cí de;有磁性的

yǒu cí xìng de

Origin: early 17th century: from late Latin magneticus, from Latin magneta

Example:

Quantitative Technology and Application Research on Magnetic Flux Leakage Inspection of Pipeline Defects
管道缺陷漏磁检测量化技术及其应用研究

Extended Terms:

magnetic-energy-storage 磁能存储
magnetic-field 磁场
magnetic-film 磁(性薄)膜
magnetic-flux 磁通(量), 磁性焊剂
magnetic susceptibility 磁化率

---

magnetism [ˈmægniˌtizəm] *n.*

Definition: a physical phenomenon produced by the motion of electric charge, which results in attractive and repulsive forces between objects 磁力现象 cí lì xiàn xiàng

Origin: early 17th century: from modern Latin magnetismus, from Latin magneta

Example:

the magnetism produced by the battery attracts the metal
电池产生的磁力能够吸引金属

Extended Terms:

apparent magnetism 视在磁性
blue magnetism 蓝磁性(南极磁性)
earth's magnetism 地磁
free magnetism 自由磁性
galvano magnetism 电磁; 电磁学

---

malleability [ˌmæliəˈbiliti] *n.*

Definition: ability of being hammered or pressed permanently out of shape without breaking or cracking 展性 zhǎn xìng

Origin: late Middle English (in the sense "able to be hammered"): via Old French from medieval Latin malleabilis, from Latin malleus "a hammer"

Example:

One of the most promising properties of glassy metals is their high strength combined with high

malleability.
玻璃金属的一个最有前途的特性,是它的高强度与高延展性相结合。

Extended Terms:

malleability test 锻造性试验
malleability wrought iron 展性锻铁

massive [ˈmæsiv] *adj.*

Definition: (of rocks or beds) having no discernible form or structure (岩石或地层)均匀构造的 jūn yún gòu zào de,大块的 dà kuài de

Origin: late Middle English: from French massif,-ive, from Old French massis, based on Latin massa

Example:

massive reef-building coral having a convoluted and furrowed surface
大块的造礁珊瑚,有盘旋的和沟回的表面

Extended Terms:

massive martensite 大块马氏体
massive transformation 块状转变
massive case 厚重表壳
massive texture 块状组织;整体(块状)结构
massive aggregate 块状集合体

mechanical [miˈkænikəl] *adj.*

Definition: working or produced by machines or machinery 机械的 jī xiè de,机器的 jī qì de

Origin: late Middle English (describing an art or occupation concerned with the design or construction of machines): via Latin from Greek mēkhanikos (See mechanic) + -al

Example:

Here we're assembling all the key parts of the mechanical, electrical and electronic systems of the videotape recorder.
我们在这里装配录像机的机械、电机和电子系统的主要零件。

Extended Terms:

mechanical automation 机械自动化
mechanical engineering 机械工程
mechanical materialism 机械唯物论
mechanical property 力学性质

mechanical twin 滑移双晶(机械双晶)

metabolism [məˈtæbəlizəm] *n.*

Definition: the chemical processes that occur within a living organism in order to maintain life 新陈代谢 xīn chén dài xiè

Origin: late 19th century: from Greek metabolē "change" (from metaballein "to change") + -ism

Example:

A crystalline amino acid occurring in proteins is important in protein metabolism
蛋白质中的一种晶状氨基酸,对蛋白质的新陈代谢有重要作用。

Extended Terms:

aromatic hydrocarbon metabolism 芳烃代谢
autotrophic metabolism 自养代谢(作用)
bacterial metabolism 细菌代谢
basal metabolism 基础代谢
biological metabolism 生物代谢(作用)

metallic [miˈtælik] *adj.*

Definition: of, relating to, or resembling metal or metals 金属的 jīn shǔ de, 金属般的 jīn shǔ bān de

Origin: late Middle English: via Latin from Greek metallikos, and metallon

Example:

a silvery ductile metallic element found primarily in bauxite
一种有银色光泽和延展性的金属元素,主要见于矾土中

Extended Terms:

metallic cohesion 金属结合力
metallic crystal 金属晶体
metallic substance 金属物质
metallic luster 金属光泽
metallic mineral 金属矿物

metamorphism [ˌmetəˈmɔːfizəm] *n.*

Definition: (*Geology*) alteration of the composition or structure of a rock by heat, pressure, or other natural agency (地质岩石)变质作用 biàn zhì zuò yòng

Origin: meta-+ morph + -ism

Example:

Quartzite: a rock formed from the metamorphism of quartz sandstone
石英岩:一种由石英沙岩的变质作用而形成的岩石。
The process of the changing from any previously existing rocks into metamorphic rocksis is called metamorphism.
将原来岩石改变为变质岩的这种作用称为变质作用。

Extended Terms:

allochromatic metamorphism 他化变质, 增减变质
cataclastic metamorphism 碎裂变质作用
contact metamorphism 接触变质作用
diagenetic metamorphism 成岩变质
regional metamorphism 区域变质作用

---

metasomatism [ˌmɛtəˈsəʊmətiz(ə)m] *n.*

Definition: (*Geology*) change in the composition of a rock as a result of the introduction or removal of chemical constituents[地质]交代作用 jiāo dài zuò yòng, 交代变质 jiāo dài biàn zhì

Origin: late 19th century: from meta-(expressing change) + Greek sōma, somat-"body" + -ism

Example:

Discussion on the lithospheric thinning of the North China craton: delamination or thermal erosion and chemical metasomatism?
关于华北克拉通燕山期岩石圈减薄的机制与过程的讨论:是拆沉,还是热侵蚀和化学交代?

Extended Terms:

fluorine metasomatism 氟素交代作用
diffusive metasomatism 扩散交代作用
sulfur metasomatism 硫交代
infiltration metasomatism 渗滤交代作用
injection metasomatism 贯入交代作用

---

metastable [ˌmɛtəˈsteib(ə)l] *adj.*

Definition: (*Physics*) (of a state of equilibrium) stable provided it is subjected to no more than small disturbances[物理](平衡状态)亚稳的 yà wěn de, 准稳的 zhǔn wěn de

Origin: 1895-1900; meta-+ stable

Example:

Study on the Microstructure of Laser RS Co-base Metastable Alloy
钴基合金激光快速熔凝亚稳组织的结构研究

Extended Terms:

metastable zone 亚稳区,亚稳带
metastable peak 亚稳态峰
metastable equation 亚稳态方程
metastable structure 亚稳结构
metastable circuit 准稳态电路

---

meteorite [ˈmiːtiərait] *n.*

Definition: a piece of rock or metal that has fallen to the earth's surface from outer space as a meteor. Over 90 per cent of meteorites are of rock while the remainder consist wholly or partly of iron and nickel 陨石 yǔn shí

Origin: 1815-1825; meteor + -ite

Example:

a scar on the earth's surface left from the impact of a meteorite
陨石对地表冲击所造成的痕迹

Extended Terms:

carbonaceous meteorite 碳质陨石
iron meteorite 铁陨星
iron-stony meteorite 铁石陨石
stony meteorite 石陨石
meteorite crater 陨石坑

---

metric [ˈmetrik] *adj.*

Definition: of or based on the metre as a unit of length; relating to the metric system 米的 mǐ de;米制的 mǐ zhì de,公制的 gōng zhì de

Origin: mid-19th century (as an adjective relating to length): from French métrique, from mètre (See metre)

Example:

In the metric system, measurements are made in meters and liters.

在公制中,用米和升作计量单位。

Extended Terms:

metric subspace 度量子空间
metric product 度量积
metric ton 公吨
Metric Convention 米制公约
metric compactum 度量紧统

micrographic [ˌmaikrəuˈgræfiks] *adj.*

Definition: examination or study with the microscope 显微文像结构的 xiǎn wēi wén xiàng jié gòu de

Origin: 1650-1660; micro- + -graphy

Example:

The sorption behaviors of actinides and long-lived fission products on rock and minerals are studied by a micrographic method based on autoradiograpy techneque.
应用自射线照相和岩相照相技术联合图像法研究了锕系元素和长寿命裂变产物在岩石和矿物上的吸附行为(直接观察到不同矿物和化学成分对核素的吸附)。

Extended Terms:

micrographic texture 显微文像结构
micrographic intergrowth 共生
micrographic test 显微镜试验

microhardness [ˌmaikrəuˈhɑːdnis]

Definition: 微观硬度 wēi guān yìng dù

Origin: micro-+ hard + -ness

Example:

Relationship between microhardness of artificial crystals and structures.
人工晶体的显微硬度与结构之间的关系。

Extended Terms:

microhardness testing 微观硬度测量
microhardness head 显微硬度锥头
microhardness value 微观硬度值
Vicker's microhardness 维克斯显微硬度(维氏显微硬度)
microhardness instrument 显微硬度计

milky [ˈmilkiː] *adj.*

Definition: containing or mixed with a large amount of milk 含乳的 hán rǔ de;掺奶的 chān nǎi de

Origin: before 900; Middle English; Old English meol(o)c, (Anglian) milc; c. Greek Milch, akin to Latin mulgēre, Greek amélgein to milk

Example:

herbs or shrubs having milky and often colored juices and capsular fruits
草本或灌木具有乳状和经常带有色彩的汁液和蒴状的果实

Extended Terms:

milky glass 乳白玻璃
milky surface 乳白镀层
milky glaze 乳白釉
milky opalescence 乳蛋白晕彩
milky quartz 乳水晶,油脂状石英,乳石英

---

mineral [ˈminərəl] *n.*

Definition: a solid inorganic substance of natural occurrence 矿 kuàng,矿物 kuàng wù

Origin: late Middle English: from medieval Latin minerale, neuter (used as a noun) of mineralis, from minera "ore"

Example:

Many African countries have rich mineral resources.
非洲许多国家蕴藏着大量矿物资源。

Extended Terms:

antistress minerals 反应力矿物
borate minerals 硼酸盐矿物
classification of minerals 矿物分类
carbonate minerals 碳酸盐矿物
clay minerals 黏土矿物
colloidal minerals 胶体矿物
halide minerals 卤化物矿物
heavy minerals 重矿物
hypergene mineral, supergene minerals 表生矿物
index minerals 指示矿物
inosilicate minerals 链状硅酸盐矿物

morphology of minerals 矿物形态
metacolloidal minerals 变胶体矿物
mineral formula 矿物化学式
mineral generation 矿物世代
mineral hardness scale 矿物硬度计
mineral physics 矿物物理
native single element minerals 自然元素矿物
nesosilicate minerals 岛状硅酸盐矿物
new minerals 新矿物
nitrate minerals 硝酸盐矿物
non-metallic minerals 非金属矿物
organic minerals 有机矿物
oxides and hydroxides minerals 氧化物氢氧化物矿物
paragenetic association of minerals 矿物共生组合
paragenetic minerals 共生矿物
phosphate minerals 磷酸盐矿物
phyllosilicate minerals 层状硅酸盐矿物
placer minerals 重砂矿物
sorosilicate minerals 群状硅酸盐矿物
rock-forming minerals 造岩矿物
silicate minerals 硅酸盐矿物
stress minerals 应力矿物
sulfate minerals 硫酸盐矿物
synthetic minerals(artificial minerals) 合成矿物(人工矿物)
tectosilicate minerals 架状硅酸盐矿物
transparent minerals 透明矿物
tungstate molybdate and chromate minerals 钨钼铬酸盐矿物
typomorphic minerals 标型矿物

---

mineralogy [ˌminəˈrælədʒi] *n.*

Definition: the scientific study of minerals 矿物学 kuàng wù xué

Origin: 1375-1425; late Middle English, from Middle French, Old French mineral, from Middle Latin minerāle (*n.*), minerālis (*adj.*), equivalent to miner(a) mine, ore

Example:

Study on Mineralogy of Synthetic Jadeite Jade Stone
合成硬玉玉石的矿物学研究

Extended Terms:

applied mineralogy 应用矿物学

cosmic mineralogy 宇宙矿物学
determinative mineralogy 鉴定矿物学
environmental mineralogy 环境矿物学
experimental mineralogy 实验矿物学
gem mineralogy 宝石矿物学
genetic mineralogy 成因矿物学
isotope mineralogy 同位素矿物学

molybdate [məˈlibdeit] *n.*

Definition: (*Chemistry*) a salt in which the anion contains both molybdenum and oxygen, especially one of the anion MoO_4^{2-} [化]钼酸盐 mù suān yán

Origin: late 18th century: from molybdic (acid), a parent acid of molybdates, + -ate

Example:

Kinetic Study of Propylene Oxidation over Europium Molybdate and Erbium Molybdate Catalysts
丙烯在钼酸铕和钼酸铒催化剂上氧化动力学研究

Extended Terms:

americium molybdate 钼酸镅
ammonium molybdate 钼酸铵
barium molybdate 钼酸钡
calcium molybdate 钼酸钙
tungstate molybdate and chromate minerals 钨钼铬酸盐矿物

modulated [ˈmɔdjuleitid] *adj.*

Definition: to regulate by or adjust to a certain measure or proportion; soften; tone down 调制的 tiáo zhì de

Origin: 1550-1560; < L modulātus (ptp. of modulārī to regulate (sounds), set to music, play an instrument).

Example:

The process of retrieving intelligence (data) from a modulated carrier wave is the reverse of modulation.
从调制过的载波信号中检出信息(数据)的过程,是调制的逆过程。

Extended Terms:

modulated spectrum 调制光谱

frequency modulated 调频的
noise modulated 噪声调制的
pulse modulated 脉冲调制的
modulated structure 调制结构

molecular [mə'lɛkjulə] *adj.*

Definition: of, relating to, or consisting of molecules 分子的 fēn zǐ de

Origin: 1785-1795; earlier molecula equivalent to Latin mōlē (s) mass + -cula -cule

Example:

At any given instant the distribution of molecular speeds is always constant under the same conditions.
在相同的条件下,分子速度分布在任何时候都是恒定的。

Extended Terms:

molecular formula 分子式
molecular weight 分子量
molecular orbital theory 分子轨道理论
molecular sieve 分子筛
molecular theory 分子论

monohydrate [ˌmɔnəu'haidreit] *n.*

Definition: (*Chemistry*) a hydrate containing one mole of water per mole of the compound [化]一水合物 yī shuǐ hé wù,一水化物 yī shuǐ huà wù

Origin: 1850-1855; mono-+ hydrate

Example:

Synthesis of L-(2R,3R)-(-)-Dibenzoyl Tartaric Acid Monohydrate
一水合-L-(2R,3R)-(-)-二苯甲酰酒石酸的合成

Extended Terms:

alloxan monohydrate 水合阿脲;阿脲-水合物
glucose monohydrate 水合葡萄糖
sulfuric monohydrate 一水硫(化)合物
sulfuric acid monohydrate 一水(合)硫酸
doxycycline monohydrate 一水合强力霉素

monomer [ˈmɒnəmə] *n.*

Definition: (*chemistry*) a molecule that can be bonded to other identical molecules to form a polymer [化]单体 (huà) dān tǐ

Origin: 1910-1915; mono- + -mer

Example:

The liquid crystal/monomer mixtures are prepared using a ferroelectric liquid crystal and a diacrylate monomer.
液晶/单体混合物由铁电液晶和双丙烯酸单体制成。

Extended Terms:

copolymerizable monomer 可共聚的单体
fiber-grade monomer 化纤单体
styrene monomer 乙烯基苯，苯乙烯
unconverted monomer 未聚合的单体
vinyl monomer 乙烯(型)单体

---

morphology [mɔːˈfɒlədʒi] *n.*

Definition: the study of the forms of things, in particular (尤指)形态学 xíng tài xué

Origin: mid-19th century: from Greek morphē "form" + -logy

Example:

Inflectional morphology is used to indicate number, case, tense and person, etc.
屈折形态学惯于说明数、格、时态和人称等。

Extended Terms:

crystal morphology 晶体形态学
dendrite morphology 枝状结构(形态)
derivational morphology 派生形态学
morphology of minerals 矿物形态
skeletal morphology 骸晶形貌

---

multiple [ˈmʌltipl] *adj.*

Definition: having or involving several parts, elements, or members 多个的 duō gè de; 多部分的 duō bù fèn de; 多成分的 duō chéng fèn de

Origin: mid-17th century: from French, from late Latin multiplus, alteration of Latin multiplex (See multiplex)

Example:

The number 8 is a multiple of 4.
八是四的倍数。

Extended Terms:

multiple twinning 聚片双晶
multiple-arch 多(连)拱
multiple-barrel 火箭发动机组
multiple-beam 复(多)光束
multiple-connector 多路插头，多路接头
multiple-contact 多触头(接点)的

naphthene ['næfθiːn] *n.*

Definition: (*Chemistry*) any of a group of cyclic aliphatic hydrocarbons (e.g. cyclohexane) obtained from petroleum [化]环烷属烃 huán wán shǔ tīng

Origin: 1840-1850; naphth- + -ene

Example:

The results also show Liaohe oil aloes turn heavier from low sulphur middle-paraphin base to low sulphur naphthene middle base.
分析结果同时也证明，辽河原油确实在变重，由低硫中间—石蜡基向低硫环烷—中间基方向发展。

Extended Terms:

alkylated naphthene 烷化环烷
five cardon ring naphthene 五碳环烷
polycyclic naphthene 多环环烷
naphthene base crude oil 环烷基油
naphthene index 环烷烃指数

---

native ['neitiv] *adj.*

Definition: (of a metal or other mineral) found in a pure or uncombined state (金属或其他

矿物）天然的 tiān rán de；呈天然纯态的 chéng tiān rán chún tài de

Origin: late Middle English; from Latin nativus, from nat-"born", from the verb nasci

Example:

There are more than 200 species of native birds, mammals and reptiles in the largest collection of Australian wildlife. They are displayed in a beautiful bushland setting.
最大的澳洲野生动物园有 200 多种当地鸟类、哺乳类和爬行类两栖动物，它们寄居在一个美丽的灌木地带。

Extended Terms:

native silver 天然银
native copper 自然铜
native graphite 天然石墨
native gold 自然金
native single element minerals 自然元素矿物

navicular [nə'vikjulə] *n.*

Definition: a boat-shaped bone in the ankle or wrist, especially in the ankle, between the talus and the cuneiform bones 舟（状）骨 zhōu（zhuàng）gǔ

Origin: late Middle English; from French naviculaire or late Latin navicularis, from Latin navicula "little ship", diminutive of navis

Example:

It permits excision of the entire talus, and the only tarsal joints that it cannot reach are those between the navicular and the second and first cuneiforms.
通过该切口可将整个距骨切除，该切口唯一不能显露的关节是舟状骨与第一、二楔骨之间的关节。

Extended Terms:

navicular bone 舟骨
navicular fossa 舟状窝
navicular abdomen 舟形腹

nematoblastic [ˌnemətə'blæstik] *n.*

Definition: 纤状变晶 xiān zhuàng biàn jīng

Origin: = fibroblastic

Examples:

The structures often observed in Dushan jade include granular texture, cataclastic texture, porphyroid

texture, metasomatic texture, nematoblastic texture, and mylonitization structure occasionally.
独山玉常具有粒状结构、碎裂结构、残斑结构、熔蚀交代结构,似斑状结构,偶见针状变晶结构和糜棱结构。

Serpentinite is nematoblastic texture, massive structure, made of chrysotile.
蛇纹岩具纤状结构、块状构造,由石棉组成。

Extended Term:

nematoblastic texture 纤状变晶结构

---

neotype [ˈniːoutaip] *n.*

Definition: a specimen selected to replace a holotype that has been lost or destroyed 新型 xīn xíng

Origin: 1850-1855; neo- + -type

Example:

The 4DA1 engine is a neotype engine which will be put into operation recently by JAC automobile company.
4DA1 型发动机是 JAC 公司拟投入生产的新型发动机。

Extended Terms:

neotype specimen 新定模式标本
neotype industrialization 新型工业化
neotype immunosuppressant 新型免疫抑制剂

---

nesosilicate [ˌniːsəuˈsiləkeit] *n.*

Definition: any silicate, as olivine, in which the SiO_4 tetrahedra are not interlinked 岛状硅酸盐 dǎo zhuàng guī suān yán

Origin: from Greek nêso (s) island + silicate

Example:

The synthetic compound of $Ca_{10}Si_6O_{21}Cl_2$ is a colorless, transparent, prismatic crystal, forming C_3S_2 and β-C_2S when reacting with water vapour at about 1100℃. It is a kind of nesosilicate minerals with isolated group of [Si_2O_7] as its basic skeleton.
化合物 $Ca_{10}Si_6O_{21}Cl_2$ 是一种五色透明柱状晶体,在 1100℃左右与水汽作用时生成 C_3S_2 和 β-C_2S,它是一种以[Si_2O_7]为基团的岛状硅酸盐矿物。

Extended Term:

nesosilicate mineral 岛状硅酸盐矿物

---

neutralization [ˌnjuːtrəlaiˈzeiʃən] *n.*

Definition: render (something) ineffective or harmless by applying an opposite force or effect 中和反应 zhōng hé fǎn yìng，中和作用 zhōng hé zuò yòng

Origin: mid-17th century: from French neutraliser, from medieval Latin neutralizare, from Latin neutralis (See neutral)

Example:

Neutralization degree of acrylic acid is 60%
丙烯酸的中和度为 60%

Extended Terms:

anode neutralization 阳(极电)路中和法
capacity neutralization 电容中和
plate neutralization 屏极(电路)中和
radar neutralization 雷达失效
shunt neutralization 并联中和

---

nitrate [ˈnaiˌtreit, -trit] *n.*

Definition: a salt or ester of nitric acid, containing the anion NO_3^- or the group-NO_3 硝酸盐 xiāo suān yán，硝酸酯 xiāo suān zhǐ

Origin: late 18th century: from French (See nitre, -ate)

Example:

Silver nitrate is a salt.
硝酸银是一种盐类化合物。

Extended Terms:

bed nitrate 硝石矿床
benzenediazonium nitrate 硝酸重氮苯
calcium carbonate-ammonium nitrate 碳酸钙合硝酸铵
cellulose nitrate 硝酸纤维素
nitrate minerals 硝酸盐矿物

---

non-crystallizing [ˌnɔn-ˈkristəlaiziŋ] *adj.*

Definition: not become crystalline in form 非晶化 fēi jīng huà

Origin: 1590-1600; crystall- + -ize

Example:

The results show that all compositions of the powder are uniformally mixed at atom level and have a high non-crystallizing degree.
研究结果表明:喷雾干燥法制得的复合氧化物粉末各种组元是在原子水平上的均匀混合,且非晶化程度高。

Extended Term:

non-crystallizing degree 非晶化程度

---

non-metallic [ˌnɔnmiˈtælik] *adj.*

Definition: an element or substance that is not a metal 非金属的 fēi jīn shǔ de

Origin: 1560-1570; from Latin metallicus from Greek metallikós of, for mines. See metal, -ic

Example:

an oxygen sensor assembled with non-metallic electrode
一种非金属工作电极的氧传感器

Extended Terms:

non-metallic products 非金属产品
non-metallic reduction agent 非金属还原剂
non-metallic sheath 非金属包皮
non-metallic resistor 非金属电阻器
non-metallic mineral 非金属矿物

---

nonpolar [ˈnɔnˈpəulə] *adj.*

Definition: containing no permanently dipolar molecules; lacking a dipole 非极性的 fēi jí xìng de

Origin: 1890-1895; non-+ polar

Example:

Oils have a high carbon and hydrogen content and are nonpolar substances.
油的碳、氢元素含量很高,为非极性物质。

Extended Terms:

nonpolar solvent 非极性溶剂
nonpolar liquid 非极性液体

---

nonstoichiometry [ˌnɔnstɔikiˈɔmitri] *n.*

Definition: 非化学计量性 fēi huà xué jì liàng xìng

Origin: non-+ stoichiometry

Example:

the influence of nonstoichiometry Bi_2O_3 on the growth of $Bi_4Ti_3O_{12}$ powder
非化学计量 Bi_2O_3 对制备 $Bi_4Ti_3O_{12}$ 粉体增长的影响

Extended Terms:

nonstoichiometry ratio 非化学计量比
oxygen nonstoichiometry 氧非化学计量
bi-nonstoichiometry bi 的非化学计量
nonstoichiometry metal oxide 非化学计量比氧化物
b-site nonstoichiometry b 位非化学计量比

---

nontoxic [nɔnˈtɔksik] *adj.*

Definition: not poisonous or toxic 无毒的 wú dú de

Origin: 1946, from non-+ toxic.

Example:

The results proved that Dame Bath Milk is harmless, nontoxic and no allergic reaction.
结果表明:Damie 沐浴乳液无毒性、不致敏,长期使用无任何副作用。

Extended Terms:

nontoxic plasticizer 无毒增塑剂
nontoxic salt 无毒盐类
nontoxic nodule 非中毒性结节
nontoxic goiter 非毒性甲状腺肿
nontoxic chelate 无毒络合物

---

normal [ˈnɔːməl] *adj.*

Definition: (*Geology*) denoting a fault or faulting in which a relative downward movement occurred in the strata situated on the upper side of the fault plane [地质]正(断层)的 (dì zhì) zhèng (duàn céng) de

Origin: mid-17th century (in the sense "right-angled"): from Latin normalis, from norma "carpenter's square" (See norm). Current senses date from the early 19th century

Example:

The Active Features of the Hangingwall on the Southeastern Edge Listric Normal Fault in the Weihe Basin.
渭河盆地东南缘铲形正断层上盘活动特征。

Extended Terms:

normal curve 标准曲线
normal depth 正常深度
normal device 电位电极系
normal dispersion 正常色散
normal distribution 正态分布
normal electric field 正常电场
normal epoch 正常时期
normal fault 正断层
normal field 正常场
normal fold 正褶皱
normal geomagnetic porality epoch 正磁极期
normal gravity 正常重力
normal magnetic field 正常磁场
normal metamorphism of coal 煤正常变质
normal order 正常层序
normal projection 正轴投影
normal solution 规定溶液
normal state 正常态
normal stress 法向应力
normal succession 正常层序
normal temperature 正常温度
normal velocity 法向速度
normal vibration 简正振动

oil [ɔil] *n.*

Definition: any of various thick, viscous, typically flammable liquids which are insoluble in water but soluble in organic solvents and are obtained from animals or plants 油类 yóu lèi

Origin: Middle English: from Old Northern French olie, Old French oile, from Latin oleum (olive oil); compare with olea "olive"

Example:

Revenues from oil are the biggest single component part in the country's income.
来自石油的收益是这个国家收入中最高的一项。

Extended Terms:

oil accumulating area 聚油面积
oil accumulation 油池
oil basin 油田
oil bearing 含油的
oil bearing rock 油母岩
oil bearing structure 储油构造
oil damping 油减震
oil deposit 石油矿床
oil field 油田
oil field water 油田水
oil gas 油气
oil geology 石油地质学
oil horizon 含油层
oil indication 石油显示
oil layer 含油层
oil measure 油层;油贮
oil pool 油池
oil production 采油
oil sand 油砂
oil saturation 油饱和率
oil seepage 油苗
oil shale 油母页岩
oil stone 油石
oil trap 油捕
oil well 油井

---

ooze [uːz] *n.*

Definition: (*Geology*) a deposit of white or grey calcareous matter largely composed of foraminiferan remains, covering extensive areas of the ocean floor [地质]软泥(dì zhì) ruǎn ní

Origin: Old English wōs "juice or sap"; the verb dates from late Middle English

Example:

Drip or ooze systems are common for pot watering.

滴灌和渗灌系统一般也用于盆栽灌水。

Extended Terms:

heavy-metal ooze 重金属软泥
diatom ooze 硅藻软泥

oolitic [ˌəuəuˈlitik] *n.*

Definition: (*Geology*) limestone consisting of a mass of rounded grains (ooliths) made up of concentric layers [地质]鲕粒岩 (dì zhì) ér lì yán

Origin: early 19th century: from French oölithe, modern Latin oolites (See oo-, -lite)

Example:

The petrologic study of oolitic limestone of zhangxia series of middle Cambrian in Tangshan, Hebei.
河北唐山地区中寒武纪张夏组鲕粒灰岩的岩石学研究

Extended Terms:

oolitic structure 鲕状构造
oolitic limestone 鲕状灰岩;鲕状石炭岩
oolitic facies 鲕粒岩相; 鲕石相
oolitic iron ore deposit 鲕状铁矿床
oolitic aggregate 鲕状集合体

opalescence [ˌəupəˈlesənt] *n.*

Definition: showing many small points of shifting colour against a pale or dark ground 蛋白光(乳光)dàn bái guāng (rǔ guāng)

Origin: early 19th century, from opalescent

Example:

Study on the Correlation Function of Rayleigh Scattering and Critical Opalescence Scattering
瑞利散射与临界乳光散射关联函数的研究。

Extended Terms:

milky opalescence 乳蛋白晕彩
critical opalescence 临界乳光
critical opalescence scattering 临界乳光散射

opaque [əuˈpeik] *adj.*

Definition: not able to be seen through; not transparent 不透明的 bù tòu míng de;不透光的

bù tòu guāng de

Origin: late Middle English opake, from Latin opacus "darkened". The current spelling (rare before the 19th century) has been influenced by the French form

Example:

a form of opaque or dark-colored diamond used for drills
一种不透明的或深色的钻石,用作钻(头)

Extended Terms:

opaque illuminator 不透明照明器
opaque crystal 墨晶
opaque glass 不透明玻璃
opaque enamel 乳浊搪瓷
opaque substrate 不透明底基

---

ore [ɔː] *n.*

Definition: a naturally occurring solid material from which a metal or valuable mineral can be extracted profitably 矿 kuàng,矿石 kuàng shí,矿砂 kuàng shā

Origin: Old English ōra "unwrought metal", of West Germanic origin; influenced in form by Old English ār "bronze" (related to Latin aes "crude metal, bronze")

Examples:

Workers abstract metal from ore.工人从矿砂中提炼金属。
So if you get an ore of the right type of radioactive element, and you wait millions and millions (or even billions) of years, you make a giant underground store of helium!
所以要是你获得了放射性元素的正确类型矿石后,等上数十亿年(甚至数万亿年)就能获得地下隐藏着的大量的氦!

Extended Terms:

ore bearing 含矿的
ore bed 矿层
ore bin 矿仓
ore block 矿段
ore bringer 运矿岩
ore bunch 矿袋
ore chimney 管状矿体
ore column 矿柱
ore complex 矿石复合
ore core 矿心

ore course 富矿体
ore dampness 矿石湿度
ore deposit 矿床
ore dilution 矿石贫化
ore dressing 选矿
ore fines 粉矿
ore formation 矿石建造
ore grade 矿石品级
ore hardness 矿硬
ore leaching 矿石浸出
ore losses 矿石损失
ore mineral 金属矿物
ore nest 矿巢
ore pillar 矿柱
ore pipe 管状矿体
ore pocket 矿袋
ore reserve 矿储藏量
ore rock 含矿岩
ore shoot 富矿体
ore sort 矿石工业品级
ore stock 矿储藏量

---

orientation [ˌɔːrienˈteiʃən] *n.*

Definition: the determination of the relative position of something or someone (especially oneself)(尤指自身)定向(yóu zhǐ zì shēn) dìng xiàng,定位 dìng wèi;确定方向 què dìng fāng xiàng

Origin: mid-19th century:apparently from orient

Example:

relationship among morphology, ATC value of alloy coat and crystal orientation of base plate of tinplate.
镀锡板基板结晶取向与合金层形貌及 ATC 值的关系

Extended Terms:

orientation of crystal 晶体定位
oriented thin section 定向薄片
basal orientation [晶]基向
absolute orientation 绝对定向, 对地定向

deformation orientation 变形取向

organic [ɔːˈɡænik] *adj.*

Definition: of, relating to, or derived from living matter (与)生物体有关的(yǔ) shēng wù tǐ yǒu guān de;(与)有机体有关的 (yǔ) yǒu jī tǐ yǒu guān de

Origin: late Middle English: via Latin from Greek organikos "relating to an organ or instrument"

Example:

Organic compounds form the basis of life.
有机化合物构成生命的基础。

Extended Terms:

organic synthesis 有机合成
organic petrology 有机岩石学
organic chemistry 有机化学
organic reagents 有机试剂
organic mineral 有机矿物

orthosilicate [ˌɔːθəˈsiləkeit] *n.*

Definition: 正硅酸盐 zhèng guī suān yán

Origin: ortho-+ silicate

Example:

In this paper, the effect of the tetraethyl orthosilicate[$Si(OC_2H_5)_4$] adding to TiO_2-V_2O_5-M_2O (M=K, Na) series, humid-sensing ceramic materials have been studied.
研究了添加正硅酸酯[$Si(OC_2H_5)_4$]对 TiO_2-V_2O_5-K_2O(Na_2O)湿敏陶瓷材料性能的影响。
On such basis, a new porous structured heteropoly acid catalyst is obtained by salt-gel reaction, using tetraethyl orthosilicate as the silicon source.
在此基础上,以正硅酸乙酯为硅源,利用"盐—凝胶"法得到一种新型介孔结构的杂多酸催化剂。

Extended Terms:

barium orthosilicate 原硅酸钡
cadmium orthosilicate 原硅酸镉
calcium orthosilicate 原硅酸钙
copper orthosilicate 硅酸铜矿; 透视石
ethyl orthosilicate 正硅酸乙酯

oxide [ˈɒksaid] *n.*

Definition: (*Chemistry*) a binary compound of oxygen with another element or group[化]氧化物 (huà) yǎng huà wù

Origin: late 18th century: from French, from oxygène "oxygen" + -ide (as in acide "acid")

Example:

an oxide containing five atoms of oxygen in the molecule
分子中含有五个氧原子的一种氧化物

Extended Terms:

oxides and hydroxides minerals 氧化物氢氧化物矿物
acicular type zinc oxide 针状氧化锌
acidic oxide 酸性氧化物
activating oxide 活性氧化物
active aluminum oxide 活性氧化铝

packet [ˈpækit] *n.*

Definition: a paper or cardboard container, typically one in which goods are packed to be sold 纸盒 zhǐ hé;硬纸盒 yìng zhǐ hé

Origin: mid 16th century: diminutive of pack, perhaps from Anglo-Norman French; compare with Anglo-Latin paccettum

Example:

The route filter packet based on the source and destination of the packet
路由器针对包的来源和目的过滤包

Extended Terms:

data packet 数据包
debugging packet 调试程序包
experiment packet 实验仪器组
filtering packet (过)滤袋
packet aggregate 束状集合体

parameter [pə'ræmitə] *n.*

Definition: a numerical or other measurable factor forming one of a set that defines a system or sets the conditions of its operation 参(变)数 cān (biàn) shù;参(变)量 cān (biàn) liàng

Origin: mid 17th century: modern Latin, from Greek para-"beside" + metron "measure"

Example:

a fixed or invariable value, parameter or data item
一种固定的或不可变的值、参数或数据项

Extended Terms:

acceleration parameter 加速参数
action parameter 行动参数
activation parameter 活化参量[参数]
actual parameter 实在参数
order parameter 有序参数

---

particle ['pɑːtikl] *n.*

Definition: (*Physics*) any of numerous subatomic constituents of the physical world that interact with each other, including electrons, neutrinos, photons, and alpha particles [物理]粒子 (wù lǐ) lì zǐ

Origin: late Middle English: from Latin particula "little part", diminutive of pars, part-

Example:

relativistic quantum-mechanical equations of particle and multiparticle system
单粒子与多粒子体系的相对论量子力学方程

Extended Terms:

accelerated particle (被)加速粒子
active particles 放射性[活性]粒子
albedo particle[地物]反照粒子
anisodimensional particle 不对称形粒子
suspended particle 悬浮粒

---

paragenetic [ˌpærə'dʒenitik] *n.*

Definition: (*Geology*) a set of minerals which were formed together, especially in a rock, or with a specified mineral [地质](尤指一组矿物在岩石内或与某特定矿物)共生的 gòng shēng de

Origin: 1850-1855; para-, genesis

Example:

The minerals of Jinchuan deposit have complex paragenetic relationship and lots of accompanying elements.
金川矿床矿物共生关系复杂,伴生元素多。

Extended Terms:

paragenetic mineral 共生矿物
paragenetic association of minerals 矿物共生组合
paragenetic rock 共生围岩
paragenetic elements 共生元素
paragenetic relationship 共生关系

---

phanerocrystalline [ˌfænərəˈkristəlain] *adj.*

Definition: (of a rock) having the principal constituents in the form of crystals visible to the naked eye 显晶质的 xiǎn jīng zhì de

Origin: 1860-1865; Greek phaneró (s) visible, manifest + crystalline

Example:

Extended Terms:

phanerocrystalline aggregate 显晶集合体
phanerocrystalline texture 显晶质结构
phanerocrystalline variety 显晶质类
phanerocrystalline bomb 显晶火山弹
phanerocrystalline-adiagnostic texture 微晶质结构
phanerocrystalline-adiagnostic 隐晶质;微晶质

---

phosphate [ˈfɒsfeit] *n.*

Definition: (*Chemistry*) a salt or ester of phosphoric acid, containing PO_4^{3-} or a related anion or a group such as &b1; OPO(OH)2[化]磷酸盐 lín suān yán

Origin: late 18th century: from French, from phosphore "phosphorus"

Example:

a common complex mineral consisting of calcium fluoride phosphate or calcium chloride phosphate; a source of phosphorus
一种常见的复合矿物,由钙氟化物磷酸盐或钙氯化物磷酸盐组成,可以从中提取磷

Extended Terms:

acid phosphate 酸式磷酸盐
acid calcium phosphate 酸性磷酸钙
acid manganous phosphate 酸式磷酸锰
acidic sodium aluminum phosphate 酸式磷酸钠铝
phosphate minerals 磷酸盐矿物

---

potential [pəˈtenʃəl] *n.*

Definition: (*Physics*) the quantity determining the energy of mass in a gravitational field or of charge in an electric field[物理]势 shì,位 wèi;电势 diàn shì,电位 diàn wèi

Origin: late Middle English: from late Latin potentialis, from potentia "power", from potent-"being able" (See potent). The noun dates from the early 19th century

Example:

electromotive force or potential difference, usually expressed in volts
电压电位或电位差,常以伏特为单位。

Extended Terms:

surface potential 表面电势
potential energy 势能
evoked potential 诱发电位
ionisation potential 电离电势(电离电位)

---

phyllosilicate [ˌfiləˈsilikeit] *n.*

Definition: any silicate mineral having the tetrahedral silicate groups linked in sheets, each group containing four oxygen atoms, three of which are shared with other groups so that the ratio of silicon atoms to oxygen atoms is two to five 层状硅酸盐 céng zhuàng guī suān yán

Origin: 1945-1950; phyllo-+ silicate

Example:

Cu, Zn and Cd grew together and consisted in phyllosilicate in matrix soils, whereas they existed in Mn oxides in cutans.
Cu、Zn 和 Cd 在基质土壤中是共生的,且主要赋存在粘粒层状硅酸盐中。

Extended Terms:

phyllosilicate mineral 层状硅酸盐矿物
phyllosilicate minerals 叶硅酸盐矿
phyllosilicate, layer silicate 层状硅酸盐

---

polysynthetic [ˌpɔlisinˈθetik] *adj.*

Definition: denoting or relating to a language characterized by complex words consisting of several morphemes, in which a single word may function as a whole sentence 多式综合的 duō shì zōng hé de

Origin: 1795-1805; from Greek polysýnthet (os) much compounded + -ic.

Example:

ALCOA tabular alumina and tabular corundum prepared by fusion method are α-Al_2O_3 polycrystal aggregations with table-shaped feature-polysynthetic twin with parting plane.
ALCOA 板状氧化铝和电熔法板状刚玉系具有板片状晶体形貌的 α-Al_2O_3 多晶集合体——具有裂理面的聚片双晶。

Extended Terms:

polysynthetic composition 多式综合组合
polysynthetic language 多式综合语
polysynthetic twin 多合孪晶

---

piezoelectricity [paiˌiːzəuiˌlekˈtrisiti] *n.*

Definition: electric polarization in a substance (especially certain crystals) resulting from the application of mechanical stress 压电(现象) yā diàn

Origin: late 19th century: from Greek piezein "press, squeeze" + electricity

Example:

Piezoelectricity of Polyvinylidene Fluoride (PVDF) with Crystalline Changed from γ-Phase into β-Phase
γ 晶型转变为 β 晶型的聚偏氟乙烯(PVDF)的压电性

Extended Terms:

converse piezoelectricity 反压电性
piezoelectricity element 压电元件
piezoelectricity effect 压电效应
Theory of Piezoelectricity 压电理论
piezoelectricity ceramic power supply 压电陶瓷驱动电源

---

pisolitic [ˈpizəlait, ˈpaisə-] *n.*

Definition: (Geology) a sedimentary rock, especially limestone, made up of small pea-shaped pieces[地质](尤指豌豆状小石灰石构成的) 豆岩 dòu yán

Origin: early 19th century: from modern Latin pisolithus (See pisolith) + -lite

Example:

Their main mineral is kaolinite, usually in pisolitic, oolitic or clastic structure.
主要矿物是高岭石,常见豆状、鲕状和碎屑状构造。

Extended Terms:

pisolitic tuff 豆状凝灰岩
pisolitic limestone 豆状灰岩
pisolitic structure 豆状构造
pisolitic iron ore 豆状铁矿
pisolitic aggregate 豆状集合体

---

plasticity [plæ'stisəti] *n.*

Definition: the quality of being easily shaped or moulded 可塑性 kě sù xìng

Origin: 1782, from plastic + -ity

Example:

It is their plasticity at certain temperatures that gives plastics their main advantage over many other materials.
正是在一定温度下塑料所具有的塑性使它比其他许多材料更优越。

Extended Terms:

surface plasticity 表面塑性
thermal plasticity 热(可)塑性
inverted plasticity 搅胀性; 膨胀性
inverted plasticity 反塑性
morphological plasticity 形态可塑性

---

platy ['pleiti] *adj.*

Definition: (of an igneous rock) split into thin, flat sheets, often resembling strata, as a result of uneven cooling. (岩石)裂成平坦薄片的 liè chéng píng tǎn báo piàn de;板状的 bǎn zhuàng de,扁平状的 biǎn píng zhuàng de

Origin: early 20th century: colloquial abbreviation of modern Latin Platypoecilus (former genus name), from Greek platus "broad" + poikilos "variegated"

Example:

The Development and Prospect of Platy New Wall Material in Heibei Province

河北省板状新墙材的发展与前景

Extended Terms:

platy shaped particle 片状颗粒
platy structure 片状结构
platy crystal 片状晶体
double-layer platy 双层成形术
platy rock-mass 板裂岩体

---

pluton [ˈpluːtɔn] *n.*

Definition: (*Geology*) a body of intrusive igneous rock[地质]深成岩体 shēn chéng yán tǐ, 火成岩体 huǒ chéng yán tǐ

Origin: 1930s: back-formation from plutonic

Example:

wall rock of the Hongge pluton and its age
红格岩体的围岩及其时代

Extended Terms:

pluton of mesozone 中带深成岩体
granitoid pluton 花岗岩体
pluton age 岩体时代
ore forming pluton 成矿岩体
synchronous pluton 同期深成岩体

---

poikilitic [ˌpɔikiˈlitik] *adj.*

Definition: (*Geology*) relating to or denoting the texture of an igneous rock in which small crystals of one mineral occur within crystals of another
[地质]嵌晶结构的 qiàn jīng jié gòu de, 嵌晶状的 qiàn jīng zhuàng de

Origin: mid 19th century: from Greek poikilos "variegated" + -ite + -ic

Example:

GRV 99027 is a new member of the martian lherzolitic shergottites (L-S) consisting of poikilitic, non-poikilitic and melt pocket textures.
GRV 99027 陨石是二辉橄榄岩质辉玻无球粒陨石(L-S)的火星陨石新成员，具有嵌晶、非嵌晶和冲击熔融袋结构。

Extended Terms:

poikilitic texture 嵌晶状结构

poikilitic structure 嵌晶构造
seriate poikilitic texture 不等粒嵌晶结构

polarization [ˌpəʊləriˈzeiʃən] *n.*

Definition: (*Physics*) restrict the vibrations of (a transverse wave, especially light) wholly or partially to one direction[物理]使产生偏振 shǐ chǎn shēng piān zhèn

Origin: 1812, from polarize + -ation

Example:

The polarization mechanism of ferroelectric materials includes displacement polarization and turning-direction polarization.
铁电材料的极化机制包括:位移极化,转向极化。

Extended Terms:

polarization vector 偏振矢量
polarization grille 偏振栅
polarization photometer 偏振光度计
electric polarization 电极化
vacuum polarization 真空极化

powdery [ˈpaʊdəri] *adj.*

Definition: consisting of or resembling powder 粉的 fěn de,粉状的 fěn zhuàng de

Origin: Middle English: from Old French poudre, from Latin pulvis, pulver-"dust"

Example:

As a whole stainless steel horizontal tank type mixer, this machine is widely used for mixing of powdery or paste material in pharmaceutical, chemical and foodstuff industries.
本机为全不锈钢卧式槽形混合机,在制药、化工、食品、电子、饲料、颜料、染料等工业中用于混合粉状或糊状的物料。

Extended Terms:

powdery aggregate 粉末状集合体
powdery diffractometry 粉末衍射
powdery analysis 粉末分析
powdery structure 粉末状结构
powdery material 粉状物料

primary [ˈpraiməri] *adj.*

Definition: earliest in time or order of development 最初的 zuì chū de，原始的 yuán shǐ de

Origin: late Middle English (in the sense original, not derivative): from Latin primarius, from primus "first". The noun uses date from the 18th century

Example:

current through primary coil induces current in secondary coil
通过原线圈在次级线圈中产生电流

Extended Terms:

primary inclusion 原生包体
primary mineral 原生矿物
primary twin 原生双晶

---

property [ˈprɔpəti] *n.*

Definition: an attribute, quality, or characteristic of something 特性 tè xìng，性质 xìng zhì，性能 xìng néng

Origin: Middle English: from an Anglo-Norman French variant of Old French propriete, from Latin proprietas, from proprius "one's own, particular"

Example:

the property of heat to expand metal at uniform rates
热以均匀速度使金属膨胀的性能

Extended Terms:

acidic property 酸性
additional properties 附加性能，补充性能
additivity property 相加性，加成性，加和性
adhesive property 胶黏性，黏结性，黏着性
anomalous flow property 反常流动性

---

prospecting [prɔsˈpektiŋ] *n.*

Definition: to search or explore a region for gold or the like 探矿 tàn kuàng

Origin: 1400-1450; late Middle English prospecte from Latin prōspectus outlook, view

Example:

The drillship for marine geological prospecting has successfully drilled a well for deep water oil

prospecting in the Yellow Sea.
这艘海洋地质钻探船已成功地在黄海打了一口深水石油勘探井。

Extended Terms:

biogeochemical prospecting[地化]生物地球化学找矿
deep prospecting 深层钻探
electrical prospecting 电法勘探
geobotanical prospecting[地质][采矿]地面植物勘探
prospecting mineralogy 找矿矿物学

---

pseudomorphism [ˈsjuːdəumɔːfinizəm] *n.*

Definition: a crystal consisting of one mineral but having the form of another 假晶 jiǎ jīng, 假同晶 jiǎ tóng jīng

Origin: mid 19th century: from pseudo-"false" + Greek morphē "form"

Example:

The processed seismic data had clear reflection character, high resolution and eliminated the pseudomorphism caused by non zero phase wavelet.
处理后的地震体基本上消除了地震子波非零相位化引起的假象,地质反射特征清晰,分辨率明显提高。

Extended Terms:

pseudomorphic replacement 假象交代
transition pseudomorphism 转变假象
alteration-pseudomorphism 蚀变假象

---

pyroclastic [ˌpairouˈklæstik] *adj.*

Definition: relating to, consisting of, or denoting fragments of rock erupted by a volcano 火山碎屑的 huǒ shān suì xiè de

Origin: 1885-1890; pyro-+ clastic

Example:

A Study of Sedimentary Section of the Subaerial Pyroclastic Flows
陆上火山碎屑流沉积剖面的研究

Extended Terms:

pyroclastic rock 火山碎屑岩
pyroclastic deposits 火山喷发碎屑堆积物
pyroclastic lava 火山碎屑熔岩

pyroclastic tephra 火山碎屑物
pyroclastic plateau 火山碎屑岩台地

pyroelectricity [ˌpaiərəuiˌlekˈtrisiti] *n.*

Definition: having the property of becoming electrically charged when heated 热电性 rè diàn xìng

Origin: 1825-1835; pyro-+ electricity

Example:

relating to or exhibiting pyroelectricity
热电的与热电有关的,或显示热电现象

Extended Terms:

false pyroelectricity 假热电性
tertiary pyroelectricity 第三热电性
pyroelectricity of polymers 高聚物热电性

quadrate [ˈkwɔdrit] *n.*

Definition: a squarish bone with which the jaw articulates, thought to be homologous with the incus of the middle ear in mammals 方骨 fāng gǔ

Origin: late Middle English (as an adjective): from Latin quadrat-“made square”, from the verb quadrare, from quattuor “four”

Example:

This thesis briefly introduced the design of concretefilled quadrate tubular steel column in the structural system of Xiamen Hilton Hotel.
本文简要介绍了厦门希尔顿酒店结构体系中方形钢管混凝土柱的设计。

Extended Terms:

quadrate bone 方骨
quadrate algebra 方代数
quadrate cartilage 小翼软骨

quadratic [kwə'drætik] *adj.*

Definition: involving the second and no higher power of an unknown quantity or variable 二次的 èr cì de,平方的 píng fāng de

Origin: mid 17th century: from French quadratique or modern Latin quadraticus, from quadratus "made square", past participle of quadrare

Example:

The discrete-time direct adaptive optimal control problem of jump linear quadratic (JLQ) model is investigated.

研究离散时间跳变线性二次(JLQ)模型的直接自适应最优控制问题。

Extended Terms:

piecewise quadratic 分段平方的
pure quadratic 纯二次的
quadratic loop 平方环
quadratic equation 二次方程
quadratic mean 均方值

---

quadruped ['kwɔdruped] *n.*

Definition: an animal, especially a mammal, having four feet 四肢动物 sì zhī dòng wù

Origin: 1640-1650; quadruped-(s. of quadrupēs)

Example:

relating to the caudal end of the body in quadrupeds or the dorsal side in human beings and other primates

(动物学)四足动物身上处于或者靠近后部的,或者灵长动物身上朝向脊骨的。

Extended Term:

卵生四足类 oviparous quadrupeds

---

quaternary system [kwə'tən(ə)ri, 'sistəm] *n.*

Definition: the Quaternary period or the system 四元体系 sì yuán tǐ xì

Origin: late Middle English (as a noun denoting a set of four): from Latin quaternarius, from quaterni "four at once", from quater "four times", from quattuor "four"

Example:

A Study on the Phase Equilibrium of the Quaternary System K_2CO_3-Na_2CO_3-Li_2CO_3-H_2O at

288 K

K_2CO_3-Na_2CO_3-Li_2CO_3-H_2O 四元体系 288K 的相平衡研究

Extended Terms:

quaternary stromatolite 第四纪叠层石
quaternary covering 第四系覆盖
quaternary pore 第四系孔隙
quaternary carbonates 第四纪碳酸盐岩
quaternary stratigraphical 第四纪地层

---

quartering [ˈkwɔːtəriŋ] *n.*

Definition: the action of dividing something into four parts(把某物)四等分 sì děng fēn

Origin: 1585-1595; quarter + -ing

Example:

Since a large particle size can easily cause division deviation, material over 10mm is not suitable for coning and quartering method. In this instance a riffle divider should be used.
由于大颗粒样品堆成圆锥时容易产生偏析,造成缩分偏差,样品粒度大于 10mm 不宜使用圆锥四分法,最好使用二分器。

Extended Terms:

quartering attachment 车削曲柄轴装置
quartering machine 曲柄轴钻孔机
quartering process 四分法工艺过程
quartering system 一刻钟制度

radiating [ˈreidieitiŋ] *adj.*

Definition: to extend, spread, or move like rays or radii from a center 放射 fàng shè

Origin: 1610-1620; radiātus (present participle of radiāre to radiate light, shine)

Example:

The best reaction conditions are: the mole ratio of pentyl alcohol and cinnamic acid is 2:1 (mol/mol), the amount of catalyst is 0.25g, the power of miorwave is 595W, the radiating

time of microwave is 6 min, the transforming percentage may reach 94.7%.
最佳反应条件为:戊醇和肉桂酸的摩尔比为2:1(mol/mol),催化剂0.25g,微波功率595W,微波辐射时间6min,转化率94.7%。

Extended Terms:

radiating aggregate 放射状集合体
radiating structure 放射状组织
radiating collar 散热管
radiating spokes 放射幅
radiating capacity 辐射本领

---

radius [ˈreidiəs] *n.*

Definition: a straight line from the centre to the circumference of a circle or sphere 半径 bàn jìng

Origin: late 16th century (in sense 2): from Latin, literally staff, spoke, ray

Example:

We are measuring the radius of the circle.
我们正在测量圆的半径。

Extended Terms:

angular radius 角半径
atomic radius 原子半径
average radius 平均半径
base radius 基圆半径
bending radius 弯曲半径

---

recrystallization [ˌriːkristəlai ˈzeition] *n.*

Definition: to become crystallized again 重结晶作用 chóng jié jīng zuò yòng

Origin: 1790-1800; re-+ crystallize

Example:

recrystallization of polysilicon films by Ar^+ laser irradiation
多晶硅薄膜的 Ar^+ 激光再结晶

Extended Terms:

natural recrystallization 自然重结晶
coarse recrystallization annulus 粗晶环
metamorphic recrystallization 变质重结晶

recrystallization mechanism 结晶机理
dynamic recrystallization 动态再结晶

reflectance [riˈflektəns] *n.*

Definition: (*Physics*) the measure of the proportion of light or other radiation striking a surface which is reflected off it [物理]反射比 fǎn shè bǐ,反向射系数 fǎn xiàng shè xì shù

Origin: 1925-1930; reflect + -ance

Example:

reflectance of soil clay minerals and its application in pedology
土壤黏土矿物反射特性及其在土壤学上的应用

Extended Terms:

boundary reflectance 界面反射比
coupled reflectance 耦合的反射系数(能力)
diffuse reflectance 漫反射系数
direct reflectance 单向反射比
directional reflectance 定向反射比

refractive [riˈfræktiv] *adj.*

Definition: of or involving refraction 折射的 zhé shè de;与折射有关的 yǔ zhé shè yǒu guān de

Origin: 1925-1930; reflect + -ance

Example:

Glass containing lead oxide has a high refractive index.
含铅氧化物的玻璃折射率高。

Extended Terms:

refractive medium 折射媒质
refractive index 折射率
refractive modulus 折射模数
fully refractive 全折射的
singly refractive 单折射的

refractory [riˈfræktəri] *adj.*

Definition: (of a substance) resistant to heat; hard to melt or fuse (物质)耐高温的(wù zhì)nài gāo wēn de;难熔炼的 nán róng liàn de

Origin: early 17th century:alteration of obsolete refractary, from Latin refractarius "stubborn"

Example:

Application of Nano Technology in Refractory Material
纳米技术在耐火材料中的应用

Extended Terms:

chrome-base refractory 铬基耐火材料
chrome-magnesite refractory 铬镁耐火材料
electro-cast refractory 电炉熔铸耐火材料
extreme temperature refractory 超高温耐火材料
fireclay refractory 黏土耐火材料

---

reniform [ˈrenifɔːm] *adj.*

Definition: (chiefly Mineralogy & Botany) kidney-shaped (主矿物学与植物学)肾形的 shèn xíng de

Origin: mid 18th century:from Latin ren "kidney" + -iform

Example:

"Xuanlong Style" ironstone formation is an Early middle proterozoic reniform ironstone and oolitic ironstone formation which is one of the earliest ironstone discovered on the earth.
"宣龙式"铁岩建造是中元古代早期形成的肾状、鲕状铁岩建造,它属于地球上较早的铁岩建造。

Extended Terms:

reniform aggregate 肾状集合体
reniform structure 肾状构造
reniform spot 肾形点
reniform ironstone 肾状铁岩

---

reverse [riˈvəːs] *n.*

Definition: a complete change of direction or action 逆向 nì xiàng;逆转 nì zhuǎn

Origin: Middle English:from Old French revers, reverse (nouns), reverser (verb), from Latin reversus "turned back", past participle of revertere, from re-"back" + vertere "to turn"

Example:

The Study of Surface Reconstruction Technologies for Reverse Engineering and Web-based

Application System Development.
逆向工程中的曲面重构技术研究及基于 Web 的应用系统开发。

Extended Terms:

reverse fault 逆断层
bevel gear reverse 锥齿轮换向法
continuous reverse 连续换向
power reverse 动力反向
reverse zoning 反向环带

rheology [riˈɒlədʒi] *n.*

Definition: the branch of physics that deals with the deformation and flow of matter, especially the non-Newtonian flow of liquids and the plastic flow of solids 流变学 liú biàn xué

Origin: 1920s: from Greek rheos "stream" + -logy

Example:

study on Rheology of Concrete Face Rockfill Dam
混凝土面板堆石坝流变研究

Extended Terms:

fluid rheology 流体流变学
industrial rheology 工业流变学
molecular rheology 分子流变学
phenomenological rheology 唯象流变学
polymer rheology 聚合物流变学

saturation [ˌsætʃəˈreiʃən] *n.*

Definition: the state or process that occurs when no more of something can be absorbed, combined with, or added 饱和度 bǎo hé dù, 色的纯度 sè de chún dù

Origin: mid 16th century: from late Latin saturatio(n-), from Latin saturare "fill, glut"

Example:

In recording, the state of magnetic materials subjected to a magnetic field. With a sufficiently

large field the material saturates and further increases in magnetic field produce no further changes in magnetic flux.
在磁记录技术中,磁性材料的磁化状态,当磁场足够强时,材料饱和,此时,进一步增加磁场强度不会改变其磁通量。

Extended Terms:

base saturation 碱饱和;盐基饱和
colour saturation 色饱和度
fluid saturation[地质]流体饱和度
interstitial water saturation[水]饱和间隙水量
magnetic saturation 磁(性)饱和

---

scaly [ˈskeili] *adj.*

Definition: covered in scales 鳞片状的 lín piàn zhuàng de

Origin: 1520-1530; scale + -y

Example:

Bionics Design for Scaly Nonsmooth Surfaces
鳞片形非光滑表面的仿生设计

Extended Terms:

scaly bark 鳞状树皮
scaly structure 鳞片状构造
scaly leaf 鳞叶
scaly graphite 鳞片石墨
scaly talc 鳞片状滑石

---

schistic [ˈʃistik] *adj.*

Definition: 片状的 piàn zhuàng de

Origin: schist + -ic

Examples:

The formation I is characterized hornblende—epidote—pyroxene assemblage, lower contents of schistic mineral and authigenous pyrite, and relatively higher abundance in Cd and affected by V, Cd, Cr, Sc which are active and inactive elements.
层 I 为角闪石—绿帘石—辉石组合,以低含片状矿物和自生黄铁矿为特征,受活动性强和不活动性元素 V,Cd,Cr,Sc 的共同影响,高含 Cd 等。

Oil shales are of mostly schistic grey-black and mainly composed of hard clayminerals with a

little quartz, a few feldspar scraps and bituminous flecks.
油页岩矿石多呈灰黑色片状,主要成分为硬质黏土类矿物,含少量石英、长石细屑及沥青质点。

Extended Term:

schistic aggregate 片状集合体

S

secondary [ˈsekəndəri] *adj.*

Definition: coming after, less important than, or resulting from someone or something else that is primary 第二的 dì èr de;第二位的 dì èr wèi de;副的 fù de

Origin: 1350-1400; secundārius

Example:

The Influence of Cr to Secondary Hardening Effect of Mo-Steel
铬对钼钢的次生硬化的影响

Extended Terms:

secondary derivatives 次生衍生物
secondary collapse 次生崩塌
secondary inclusion 次生包体
secondary mineral 次生矿物
secondary twin 次生双晶

secretion [siˈkriːʃən] *n.*

Definition: a substance discharged in such a way 分泌体 fēn mì tǐ

Origin: mid 17th century: from French sécrétion or Latin secretio(*n.*) "separation", from secret-"moved apart", from the verb secernere

Example:

Study on Gastric Acid Secretion and Related Mechanism of Immunoregulation
胃酸分泌及其免疫调控机制的相关研究

Extended Terms:

secretion responsiveness 分泌反应性
external secretion 外分泌
defensive secretion 防御性分泌物
secretion clearance 分泌物清除

section ['sekʃən] *n.*

Definition: the shape resulting from cutting a solid along a plane 截面 jié miàn,断面 duàn miàn

Origin: late Middle English (as a noun): from French section or Latin sectio (*n.*), from secare "to cut". The verb dates from the early 19th century

Example:

variations of the surface velocity of the kuroshio on section G in the east China Sea from 1956-1975

东海 G 断面上二十年(1956—1975)来黑潮表层流速的变动

Extended Terms:

asymmetrical tunnel section 不对称隧道断面
broken-out section 破断面, 切面
effective cross section 有效横截面
equatorial section 中纬切面
vertical section 纵截面
polished section 光片

---

sedimentation [ˌsedimen'teiʃən] *n.*

Definition: the deposition or accumulation of sediment 沉积作用 chén jī zuò yòng

Origin: mid 16th century: from French sédiment or Latin sedimentum "settling", from sedere "sit"

Example:

Geologic-geochemical features of the silicalite indicate that it is a product of submarine exhalative sedimentation which was responsible for the formation of the deposit.

硅质岩的地质地球化学特征表明,它是海底喷气沉积作用的产物,拉尔玛金—铜—铀建造矿床也具有海底喷流的沉积特征。

Extended Terms:

sedimentation velocity 沉降速度
chemical sedimentation 化学沉积(作用)
coagulating sedimentation 凝结沉降(作用); 混凝沉淀
preliminary sedimentation 预沉(作用)
rhythmic sedimentation[地质]韵律沉积

---

seismological [ˌsaizməˈlɔdʒikəl] *adj.*

Definition: the branch of science concerned with earthquakes and related phenomena 地震学上的 dì zhèn xué shàng de

Origin: 1855-1860; seismo- + -logy

Example:

an outline of seismological detection techniques for magma chambers
岩浆囊的地震学探测技术概述

Extended Terms:

seismological evidence 地震实迹
seismological observation 地震观测

---

sequence [ˈsiːkwəns] *n.*

Definition: a particular order in which related events, movements, or things follow each other 顺序 shùn xù

Origin: late Middle English (in sense 4): from late Latin sequentia, from Latin sequent- "following", from the verb sequi "follow"

Example:

The content of the programme should follow a logical sequence.
项目的内容应该按照逻辑顺序排列。

Extended Terms:

additive sequence 可加序列
admissible sequence 容许序列
adjusted homology sequence 调节同调序列
approximating sequence 近似序列
arbitrary sequence 任意序列

---

silicate [ˈsilikit] *n.*

Definition: any of the many minerals consisting of silica combined with metal oxides, forming a major component of the rocks of the earth's crust 硅酸盐矿物 guī suān yán kuàng wù

Origin: 1805-1815; silic(a) + -ate

Example:

These silicate tetrahedrons join into chains.

这些硅酸盐四面体缠结成链状。

Extended Terms:

alkaline silicate 碱性硅酸盐
alumina silicate 水合硅酸铝
phyllosilicate(sheet silicate)层状硅酸盐
silicate minerals 硅酸盐矿物
single chain silicate 单链状硅酸盐

---

silicic [siˈlisik] *adj.*

Definition: (*Geology*)(of rocks) rich in silica[地质](岩石)富含硅石的(dì zhì)(yán shí) fù hán guī shí de

Origin: 1810-1820; silic(a) + -ic

Example:

a salt of boric and silicic acids
一种硼酸和硅酸形成的盐

Extended Terms:

silicic acid 硅酸
silicic anhydride 硅酸酐
silicic soil 硅质土壤
ortho-silicic acid 原硅酸
silicic acid anhydride 硅酐

---

siliciclastic [siˌlisəˈklæstik] *adj.*

Definition: (*Geology*) relating to or denoting clastic rocks consisting largely of silica or silicates. [地质]硅质碎屑的 guī zhì suì xiè de

Origin: silicic + clasti + -ic

Examples:

Examples of common clastic sedimentary rocks include siliciclastic rocks such as conglomerate, sandstone, siltstone and shale.
常见的碎屑沉积岩的实例有硅质碎屑岩,如砾岩、砂岩、粉砂岩和页岩

Occurrence of both carbonate and siliciclastic sources, active hydrodynamicregimes, alternate dry and wet climates, and the fluctuation of sea level are main factors for forming mixed sediments.
形成混合沉积的条件是具备碳酸盐和矽质碎屑两类物源、活跃的水动力和干湿交替的气候,

此外,海平面的相对波动也会造成积极的影响。

Extended Terms:

siliciclastic rock 硅质碎屑岩
siliciclastic source 硅质碎屑物源

---

silicon ['silikən] *n.*

Definition: the chemical element of atomic number 14, a non-metal with semiconducting properties, used in making electronic circuits. Pure silicon exists in a shiny dark grey crystalline form and as an amorphous powder(Symbol:Si)(化学元素)硅(符号:Si) guī

Origin: early 19th century:alteration of earlier silicium, from Latin silex, silic-"flint", on the pattern of carbon and boron

Example:

The silicon with halogenideization composes easily the new chemical compound.
硅易于和卤化物化合成新的化合物。

Extended Terms:

amorphous silicon 非晶硅
aluminium silicon 硅铝合金
calcium silicon 硅钙合金
controlled silicon 可控硅
silicon oxygen tetrahedron 硅氧四面体

---

silky ['silki] *adj.*

Definition: of or resembling silk, especially in being soft, fine, and lustrous 丝绸一样的 sī chóu yí yàng de;柔软的 róu ruǎn de;光滑的 guāng huá de

Origin: 1605-1615; silk + -y

Example:

A. Silky fiber obtained from the fruit of the silk-cotton tree and used for insulation and as padding in pillows, mattresses, and life preservers.
木棉是一种绢状纤维,从木棉树的果实中提取出来,用作隔音材料,也可用作枕头,褥垫和救生用具中的垫料。

Extended Terms:

silky leaved 丝状叶的
silky fracture 丝状断口
silky cloudiness 丝状混浊(性)

silky luster 丝绢光泽

skeletal [ˈskelitl] *adj.*

Definition: of, relating to, or functioning as a skeleton (与)骨骼(有关)的(yǔ) gǔ gé (yǒu guān) de

Origin: 1850-1855; skelet(on) + -al

Example:

The conversion of 1-hexene attained 82.4% and the selectivity of 2-hexene and 3-hexene were 21.2% and 61.2% respectively under optimum conditions without skeletal isomerization.
在最佳条件下1-己烯转化率达到82.4%,2-己烯选择性21.2%,3-己烯61.2%,没有发现骨架异构化。

Extended Terms:

skeletal substance 骨架物质
skeletal limestone 骨骸石灰岩
skeletal texture 骸晶结构
skeletal chain complex 简单链复形
skeletal morphology 骸晶形貌

solid [ˈsɔlid] *adj.*

Definition: firm and stable in shape; not liquid or fluid 固体的 gù tǐ de

Origin: late Middle English: from Latin solidus; related to salvus "safe" and sollus "entire"

Example:

a gaseous suspension of fine solid or liquid particles
浮质细小的固体或液体微粒的气态悬浮物

Extended Terms:

active solid 活性固体(可用作吸附剂)
aerated solids 气溶胶;充气固体
amorphous solid 非晶质固体
anisotropic solid 各向异性固体
solid state chemistry 固态化学

solvus [ˈsɔlvəs] *n.*

Definition: 溶线,固溶相线

Examples:

As shown in Figure 1, precipitates microstructure confirms that near solvus temperature precipitation can limit grain boundary precipitation.

如图1所示,沉淀物微观结构证明:在固溶相线温度附近的沉淀可以限制晶界沉淀。

However, at high temperature, when the matrix is in the state of lows aturation (near solvus), precipitations in grain and on grain boundary arecontrolled by driving force (thermodynamics).

然而,高温时,当基体出于低饱和状态时(接近固溶相线),颗粒中和晶界的沉淀由热力学控制。

Extended Terms:

solvus tempreture: 溶线温度

solvus exsolution curve 固溶体分解曲线

calcite-dolomite solvus geothermometer 方解石—白云石溶线地温计

---

sorosilicate [ˌsɔrəˈsilikit] *n.*

Definition: any of the silicates in which each silicate tetrahedron shares one of its four oxygen atoms with a neighboring tetrahedron, the ratio of silicon to oxygen being two to seven 双岛状硅酸盐 shuāng dǎo zhuàng guī suān yán

Origin: sōró (s) heap + silicate

Example:

It possessed rich mineral deposits, including gold.

它拥有丰富的矿藏,包括黄金。

Extended Terms:

sorosilicate mineral 群状硅酸盐矿物

---

species [ˈspiːʃiːz] *n.*

Definition: a group subordinate to a genus and containing individuals agreeing in some common attributes and called by a common name 种(矿物分类基本单位) zhǒng

Origin: late Middle English: from Latin, literally appearance, form, beauty, from specere "to look"

Example:

Many species are now threatened with extinction.

许多物种面临灭绝之虞。

Extended Terms:

accessory species 次要种; 辅助树种

accompanying species 伴生种
adsorbing species 吸附品种
adventitious species 侵入种，外来种
atomic species 原子种类

sphere [sfiə] *n.*

Definition: a round solid figure, or its surface, with every point on its surface equidistant from its center 圆形实心体 yuán xíng shí xīn tǐ,球体 qiú tǐ;球面 qiú miàn

Origin: Middle English: from Old French espere, from late Latin sphera, earlier sphaera, from Greek sphaira "ball"

Example:

The Earth is not a perfect sphere.
地球并不是一个完全的球体。

Extended Terms:

combinatorial sphere 组合球面
complex sphere 复球(面)
director sphere 准球面
earth's sphere 地球
equivalent sphere 当量球体，等效球体

spherulitic [ˈsferiˌlait, ˈsfiə-] *n.*

Definition: (*Chiefly Geology*) a small spheroidal mass of crystals (especially of a mineral) grouped radially around a point(主地质)球粒 (zhǔ dì zhì) qiú lì

Origin: early 19th century: from spherule + -ite

Example:

No clear spherulitic morphology was observed in polarizing optical micrograph of PHB-diol.
在 PHB-diol 的偏光显微镜照片中观察不到明显的球晶形态。

Extended Terms:

spherulitic texture 球粒结构
spherulitic crystal structure 球晶结构
spherulitic graphite 球状石墨
spherulitic iron 球状石墨铸铁
spherulitic growth 球晶生长

spiral [ˈspaiərəl] *adj.*

Definition: winding in a continuous curve of constant diameter about a central axis, as though along a cylinder; helical 圆柱螺旋形的 yuán zhù luó xuán xíng de

Origin: mid 16th century (as an adjective): from medieval Latin spiralis, from Latin spira "coil"

Example:

A coiled spring forms a spiral.
一条卷起来的弹簧形成一个螺旋形。

Extended Terms:

cooling spiral 冷却螺管
Cornu's spiral (=clothoid) 卡牛螺线，回旋曲线
corrugation spiral (=flute spiral) 轧辊螺槽
combustion spiral 燃烧旋管
spiral steps 螺旋生长台阶

---

stalagmitic [ˌstæləgˈmitik] *n.*

Definition: a mound or tapering column rising from the floor of a cave, formed of calcium salts deposited by dripping water and often uniting with a stalactite 石笋 shí sǔn

Origin: late 17th century: from modern Latin stalagmites, from Greek stalagma "a drop", based on stalassein

Example:

In many of the karstic caves once occupied by Middle or Late Pleistocene hominids, there exist travertine layers and/or stalagmitic carbonates of other forms, which are suitable for U-series dating.
中、晚更新世洞穴古人类遗址中，多有与遗物有明确层位关系的钙板或其他形式的钟乳石类碳酸岩，是测定铀系年龄的理想材料。

Extended Term:

stalagmitic calcite 石笋状方解石

---

stoichiometry [ˌstɔikiˈɔmitri] *n.*

Definition: (*Chemistry*) the relationship between the relative quantities of substances taking part in a reaction or forming a compound, typically a ratio of whole integers [化] 元素量法 yuán sù liàng fǎ, 化学计量 huà xué jì liàng

Origin: early 19th century: from Greek stoikheion "element" + -metry

Example:

Investigation of the Stoichiometry of Bismuth Tellurite Bi_2TeO_5 Single Crystals.
亚碲酸铋 Bi_2TeO_5 单晶的化学计量比的研究。

Extended Terms:

stoichiometric chemistry 计量化学
industrial stoichiometry 化工计算
melt-stoichiometry 熔体化学计量学
non-stoichiometry 非化学计量性
nonstoichiometric crystal 非化学计量晶体

---

streak [striːk] *n.*

Definition: a long, thin line or mark of a different substance or colour from its surroundings 条纹 tiáo wén,条痕 tiáo hén,色带 sè dài

Origin: Old English strica, of Germanic origin; related to Dutch streek and German Strich, also to strike. The sense "run naked" was originally US slang.

Examples:

Scientists guess the mysterious crimson streak is either evidence of a recent collision or a gas leak coming from Haumea's hot interior.
科学家猜测这些神秘的深红色条纹可能是守护石星的炙热内核向外喷射气体时造成的冲击所导致的。
The galaxy is running out of raw gas and dust, with the dark streak representing the last material available for star formation.
该星系正在消耗完自己的原料气体和尘埃,中间的暗条显示了形成这颗恒星的最后一些可用物质。

Extended Terms:

mill streaks 轧制条痕
nonradial streak 非径向条痕
pay streak 富矿脉
pitch streak 树脂条纹
radial streak 径向条痕

---

stress [stres] *n.*

Definition: pressure or tension exerted on a material object 压力 yā lì,重压 zhòng yā

Origin: Middle English (denoting hardship or force exerted on a person for the purpose of compulsion): shortening of distress, or partly from Old French estresse "narrowness, oppression", based on Latin strictus "drawn tight"

Example:

The distribution of stress is uniform across the bar.
杆上压力分布是均衡的。

Extended Terms:

stress-deviation 应力偏差
stress-induced 应力诱发的，应力感生的
stress-intensity 应力度
additional normal stress 附加正应力
stress mineral 应力矿物

---

striation [strai'eiʃən] *adj.*

Definition: marked with striae 有条纹的 yǒu tiáo wén de;有突起的 yǒu tū qǐ de;有凹槽的 yǒu āo cáo de

Origin: 1840-1850; striate + -ion

Example:

The chief character of fracture is fatigue striation and dimple.
断口分析表明,合金低周疲劳断口的主要特征是平行的条纹及韧窝。

Extended Terms:

combination striations 聚形纹
twinning striation 双晶条纹
striated pebble 擦痕卵石

---

stumpy ['stʌmpi] *adj.*

Definition: short and thick; squat 粗而短的 cū ér duǎn de;矮胖的 ǎi pàng de

Origin: 1590-1600; stump + -y

Example:

Development of the Cutting Reducer of Stumpy Cantilever Tunneler for Semi-Coal rock
矮型悬臂式半煤岩掘进机切割减速器研制

Extended Terms:

stumpy horn 短粗角
stumpy carbon cell 竖碳极电池

---

sublimation [ˌsʌbliˈmeiʃən] *n.*

Definition: (*Chemistry*) a solid deposit of a substance which has sublimed [化]升华物 shēng huá wù

Origin: late Middle English (in the sense "raise to a higher status"): from Latin sublimat- "raised up", from the verb sublimare

Example:

a fine powder produced by condensation or sublimation of a compound
细粉末由化合物浓缩或升华而得到的精细粉末

Extended Terms:

fractional sublimation 分级升华
vacuum sublimation 真空升华
sublimation from the frozen state 冷凝态升华(低温干燥)
sublimation of carbon 碳的升华
sublimed sulfur 升华硫黄

---

subspecies [ˈsʌbˌspiːʃiːz] *n.*

Definition: a taxonomic category that ranks below species, usually a fairly permanent geographically isolated race 亚种(矿物分类单位) yà zhǒng

Origin: 1690-1700; sub-+ species

Example:

A subspecies can synthesize more than one type of ICP.
一个亚种可合成一种以上的 ICP。

Extended Terms:

geographic subspecies 地理亚种
subspecies differentiation 亚种分化
nominate subspecies 指名亚种
soil subspecies 土壤变种
polytopic subspecies 异地同型亚种

---

sulfate [ˈsʌlfeit] *n.* =salphate

Definition: (*Chemistry*) a salt or ester of sulphuric acid, containing the anion SO_4^{2-} or the divalent group &b1; O_5O_2O &b1 [化]硫酸盐(或酯) liú suān yán

Origin: late 18th century: French, from Latin sulfur

Example:

a white crystalline double sulfate of aluminum: the ammonium double sulfate of aluminum
一种铝的双硫酸盐:铝铵双硫酸盐,白色晶状

Extended Terms:

sulfate-free 无硫酸盐的
sulfate minerals 硫酸盐矿物
sulfate sulfur 硫酸盐型硫
sulfate scale 硫酸盐垢
sulfate-sodium type 硫酸钠水型

---

sulfide [ˈsʌlfaid] *n.* =salphide

Definition: a binary compound of sulfur with another element or group [化]硫化物 liú huà wù

Origin: sulf + -ide

Example:

application of PDS technology for sulfide removal
PDS 脱硫技术的应用

Extended Terms:

mercuric sulfide 硫化汞
sulfide catalyst 硫化物催化剂
hydrogen sulfide 硫化氢
red mercuric sulfide 红色硫化汞
alkali sulfide 硫化碱

---

superlattice [ˌsjuːpəˈlætis] *n.*

Definition: an ordered arrangement of certain atoms in a solid solution which is superimposed on the solvent crystal lattice 超晶格 chāo jīng gé

Origin: super-+ lattice

Example:

Characteristics of Semiconductor/Superlattice Distributed Bragg Reflector.
半导体/超晶格分布布拉格反射镜的特性研究。

Extended Terms:

superlattice reflection 超点阵反射
superlattice structure 超点阵结构

superlattice type magnet 规则点阵型磁铁
superlattice dislocation 超点阵位错
superlattice symmetry 超点阵对称性

supersaturation [ˈsjuːpəˌsætʃəˈreiʃən] *n.*

Definition: (*Chemistry*) increase the concentration of (a solution) beyond saturation point [化]使过饱和 shǐ guò bǎo hé

Origin: 1750-1760; super-+ saturate

Example:

This equation has the feedback property of Supersaturation→nucleation→depletion cycle.
该速率表达式具有过饱和→成核→耗尽循环的反馈特征。

Extended Terms:

supersaturation state 过饱和状态
supersaturation rate 过饱和速率
metastable supersaturation 亚稳过饱和
supersaturation environment 过饱和环境

susceptibility [səˌseptəˈbilitiː] *n.*

Definition: (*Physics*) the ratio of magnetization to a magnetizing force [物理]磁化率 cí huà lǜ;电极化率 diàn jí huà lǜ

Origin: 1635-1645; susceptibilitās, equivalence to susceptibil (is) susceptible + -itās- -ity

Example:

the calculation of apparent susceptibility and its application
视磁化率的计算及其应用

Extended Terms:

antiferromagnetic susceptibility 反铁磁磁化率
antiknock susceptibility 抗爆感应性
atomic susceptibility 原子磁化率
magnetic susceptibility 磁化率
diamagnetic susceptibility 抗磁磁化率

suspended [səˌseˈpendid] *adj.*

Definition: to keep from falling, sinking, forming a deposit, etc. 悬浮的 xuán fú de

Origin: Middle English: from Old French suspendre or Latin suspendere, from sub-“from below” + pendere “hang”

Example:

After the explosion dust was suspended in the air.
爆炸后灰尘飘浮于空中。

Extended Terms:

suspended joint 悬式接头
suspended colloid 悬(浮)胶(体)
suspended particle 悬浮粒
suspended structure 悬挂结构
suspended particulate 悬浮颗粒

---

synthetic [sin'θetik] *adj.*

Definition: noting a gem mineral manufactured so as to be physically, chemically, and optically identical with the mineral as found in nature 综合的 zōng hé de,合成的 hé chéng de

Origin: Greek synthetikos of composition, component, from syntithenai to put together

Example:

Some synthetic oils are green oil. It is inevitable that green synthetic lubricants will replace mineral oils, which can be helpful to solve the problem of petroleum crisis.
一些合成油是绿色润滑油,因此,以绿色合成润滑油取代矿物基润滑油将是必然的发展趋势,也是石油能源危机的一种补充研究方向。

Extended Terms:

synthetic mineral 合成矿物
synthetic aperture radar 合成隙孔雷达
synthetic protection 综合保护
synthetic mechanism 合成机理
synthetic parameters 综合参数

tabular ['tæbjələ] *adj.*

Definition: (of a crystal) relatively broad and thin, with two well-developed parallel faces

(结晶)薄片状的 báo piàn zhuàng de

Origin: mid 17th century (in sense 2): from Latin tabularis, from tabula

Example:

the crystalline feature of ALCOA tabular alumina and tabular corundum by fusion method
板状氧化铝和电熔法板状刚玉的晶体形貌

Extended Terms:

tabular structure 板状构造
tabular crystal 片状晶体
a tabular rock 一块扁平的岩石
tabular alumina 板状氧化铝
tabular grain 扁平颗粒

---

tapered [ˈteipəd] *adj.*

Definition: to become smaller or thinner toward one end 尖面的 jiān miè de,锥形的 zhuī xíng de

Origin: before 900, tapur

Example:

a tapered, serrated tool used to shape or enlarge a hole
用来钻孔,或扩大孔的尖细带齿的工具

Extended Terms:

amplitude taper 振幅锥度
back taper 倒锥
Brown and Sharpe taper 布朗沙普锥度
linear taper 线性电阻分布特性
metric taper 公制锥度, 公制退拔

---

tarnish [ˈtɑːniʃ] *n.*

Definition: a film or stain formed on an exposed surface of a mineral or metal(矿物,金属上的)失泽膜 shī zé mó

Origin: late Middle English (as a verb): from French terniss-, lengthened stem of ternir, from terne "dark, dull"

Example:

Brass tarnishes quickly in wet weather.

天气潮湿时黄铜很快会失去光泽。

Extended Terms:

tarnish-proof 不失去光泽的，不变色的
tarnish-resistant 不生锈的，不变色的
tarnish resistance 抗蚀力，耐蚀性
tarnish inhibitor 晦暗抑制剂
tarnish resisting alloy 抗变色合金

---

tartan [ˈtɑːtn] *n.*

Definition: a woollen cloth woven in one of several patterns of colored checks and intersecting lines, especially of a design associated with a particular Scottish clan 格子(尤指苏格兰格子呢)gé zi

Origin: late 15th century (originally Scots): perhaps from Old French tertaine, denoting a kind of cloth; compare with tartarin, a rich fabric formerly imported from the east through Tartary

Example:

He also produced the first color photograph, a picture of a tartan ribbon.
他还制造出了人类第一张彩色照片，拍的是一条格子呢缎带。

Extended Terms:

tartan structure 格子结构
tartan twinning 格子双晶

---

tectosilicate [ˌtektəuˈsilikit] *n.*

Definition: any silicate in which each tetrahedral group shares all its oxygen atoms with neighboring groups, the ratio of silicon to oxygen being 1 to 2. 架状硅酸盐 jià zhuàng guī suān yán

Origin: tēktó (s) molten + silicate

Example:

The microstructure of $CaOAl_2O_3SiO_2$ (CAS) glass system is studied by molecular dynamics method. When Ca/Al equals to 1/2, which is called tectosilicate glass system, CAS is not a totally full network as conventional theories, but with some Non Bridging Oxygen (NBO). The theoretical result proves Stebins's experimental conclusion.
采用分子动力学模拟的方法研究了 $CaOAl_2O_3SiO_2$ 系玻璃的微观结构，发现 Ca/Al = 1/2 时 $CaOAl_2O_3SiO_2$ 系玻璃(网硅酸盐体系)并不像传统理论认为的那样是一个完整的三维网络，而是存在一定量的非桥氧，从而从理论上进一步证实了 Stebins 等人的实验结果。

Extended Term:

tectosilicate mineral 架状硅酸盐矿物

temperature [ˈtempəritʃə] *n.*

Definition: the degree or intensity of heat present in a substance or object, especially as expressed according to a comparative scale and shown by a thermometer or perceived by touch 温度 wēn dù, 气温 qì wēn

Origin: late Middle English: from French température or Latin temperatura, from temperare "restrain". The word originally denoted the state of being tempered or mixed, later becoming synonymous with temperament. The modern sense dates from the late 17th century.

Example:

A thermometer gauges the temperature.
温度计可测量温度。

Extended Terms:

Celsius temperature 摄氏温度
temperature effect 温度效应
air temperature 气温
Rankine temperature 兰氏温度
Curie temperature 居里温度

tenacity [təˈnæsiti] *n.*

Definition: not readily letting go of, giving up, or separated from an object that one holds, a position, or a principle 黏滞性, 韧性 zhān zhì xìng, rèn xìng

Origin: early 17th century: from Latin tenax, tenac-(from tenere "to hold") + -ious

Example:

Study on Optimizing Tenacity and Elongation at Break of Bright Sewing PET Staple Fiber.
有光缝纫线型涤纶短纤维强伸优化的研究

Extended Terms:

breaking tenacity 断裂强度, 扯断力
dry tenacity 干燥强度
specific tenacity 比抗张力, 比强度
wet tenacity 湿强度
tenacity of the lubricating film 润滑油膜的韧度

terrestrial [tə'restriəl] *adj.*

Definition: of, on, or relating to the earth (与)地球(有关)的 dì qiú de;地球上的 dì qiú shàng de

Origin: late Middle English (in the sense "temporal, worldly, mundane"): from Latin terrestris (from terra "earth") + -al

Example:

Terrestrial longitude is measured in degrees east or west of the Greenwich meridian.
地球的经度是用格林尼治子午线向东或向西的度数来表示的。

Extended Terms:

terrestrial plant 陆生植物
terrestrial isotope 地球上的同位素
terrestrial evolution 陆地进化
terrestrial ecosystem 地表生态系统
terrestrial photogrammetry 地面摄影测量

---

ternary ['təːnəri] *adj.*

Definition: composed of three parts 三个组成的 sān gè zǔ chéng de;三重的 sān chóng de

Origin: late Middle English: from Latin ternarius, from terni "three at once"

Example:

Study on Synthesis and Properties of Ternary Laryerd Carbide Ti_3SiC_2
三元层状碳化物 Ti_3SiC_2 的制备及性能研究

Extended Terms:

ternary cycle 三次循环
ternary alloy(s) 三元合金
ternary electrolyte 三元电解质
ternary eutectic 三元共晶
ternary system 三元[体]系

---

tetrahedron [ˌtetrə'hedrən] *n.*

Definition: a solid having four plane triangular faces; a triangular pyramid 四面体 sì miàn tǐ

Origin: late 16th century: from late Greek tetraedron, neuter (used as a noun) of tetraedros "four-sided"

Example:

The internal structure of silica has a pyramid(tetrahedron) unit.
硅的内部结构是一个四面体。

Extended Terms:

silicon oxygen tetrahedron 硅氧四面体
coordinate tetrahedron 坐标四面形(体)
desmic tetrahedrons 连锁四面体
elementary tetrahedron 元四面体
local tetrahedron 局部四面体

texture [ˈtekstʃə] *n.*

Definition: the feel, appearance, or consistency of a surface or a substance(物质、表面的)结构 jié gòu,构造 gòu zào;纹理 wén lǐ;肌理 jī lǐ

Origin: late Middle English (denoting a woven fabric or something resembling this): from Latin textura "weaving", from text-"woven", from the verb texere

Example:

The texture of the rock's grains was described as being like sugar.
这种岩石颗粒的质地被说成像糖一样。

Extended Terms:

atoll texture 环形结构
core texture 核心结构
crystalloblastic texture 变晶结构
flaky texture 薄片状结构
fracture texture 断口结构
gneissose texture 片麻状结构
granulitic texture 变粒结构
hyalopilitic texture 玻晶交织结构
intergranular texture 粒间结构

thallus [ˈθæləs] *n.*

Definition: a plant body that is not differentiated into stem and leaves and lacks true roots and a vascular system. Thalli are typical of algae, fungi, lichens, and some liverworts 原植体 yuán zhí tǐ,叶状体 yè zhuàng tǐ,菌体 jūn tǐ

Origin: early 19th century: from Greek thallos "green shoot", from thallein "to bloom"

Example:

An algal cell filled with chlorophyll, formed in the thallus of a lichen.
共生藻细胞是在苔藓的叶状体里发现的带有叶绿素的藻类细胞。

Extended Terms:

endogenous thallus 内生叶状体
exogenous thallus 外生叶状体
heterotrichous thallus 异丝体的叶状体
nematoparenchymatous thallus 线形薄壁组织叶状体
stratified thallus 叠生叶状体

---

thermal ['θəːməl] *adj.*

Definition: of or relating to heat 热的 rè de;热量的 rè liàng de;由热造成的 yóu rè zào chéng de

Origin: mid 18th century (in the sense "relating to hot springs"): from French, from Greek thermē "heat"

Examples:

thermal measurement and control of furnaces in metallurgical industry
冶金炉热工测量和控制
Lit from below by Saturn's internal thermal glow, clearings in the planet's cloud system appear as white pearls.
由土星内部的热发光从下面照亮,星球云系中的空地显得好像是白色的珍珠。

Extended Terms:

thermal cracking 热裂化
thermal expansion 热膨胀
thermal stress 热应力
thermal conductivity 热传导率
thermal expansion 热膨胀

---

thermodynamics [ˌθəːməudai'næmiks] *n.*

Definition: the branch of physical science that deals with the relations between heat and other forms of energy (such as mechanical, electrical, or chemical energy), and, by extension, of the relationships and interconvertibility of all forms of energy 热(动)力学 rè lì xué

Origin: 1850-1855; thermo-+ dynamics

Example:

James Prescott(1818-1889),British physicist who established the mechanical theory of heat and discovered the first law of thermodynamics.

詹姆斯・普雷斯科特(1818−1889)是英国物理学家,为热量机械理论奠定了基础,并发现了热力学第一定律。

Extended Terms:

chemical thermodynamics 化学热力学
engineering thermodynamics 工程热力学
irreversible thermodynamics 不可逆过程热力学
marine thermodynamics 海洋热力学
metallurgical thermodynamics 冶金热力学

---

thermoluminescence [ˌθəːməuˌluːmiˈnesns] *n.*

Definition: the property of some materials which have accumulated energy over a long period of becoming luminescent when pretreated and subjected to high temperatures, used as a means of dating ancient ceramics and other artifacts 热发光 rè fā guāng

Origin: 1895-1900; thermo-+ luminescence

Examples:

The thermoluminescence strengths and peak temperatures of beryls can be regarded as the typomorphic characteristics.

绿柱石的热释光强度和峰温可作为其标型特征之一。

the thermoluminescence research systems for deep volcanic rockssection, this will possibly demarcate eruptible gyration and evolvement character

对巨厚的火山岩剖面进行系统的热释光研究,可能标定其喷发旋回和演化特征

Extended Terms:

thermoluminescence dosimetry 热致发光剂量测定法
thermoluminescence dosimeter 热致发光剂量计
thermoluminescence method 热发光法
thermoluminescence dating 热释光判断年代法
thermoluminescence curve 热致曲线

---

theory [ˈθiəri] *n.*

Definition: a set of principles on which the practice of an activity is based 理论 lǐ lùn;学理 xué lǐ,原理 yuán lǐ

Origin: late 16th century (denoting a mental scheme of something to be done): via late Latin from Greek theōria "contemplation, speculation", from theōros "spectator"

Examples:

Combine theory with practice.
把理论和实际结合起来。
So far it is only concepts and theory, but we will shortly look into how this all works out in practice.
到目前为止还仅仅是概念和理论,但我们很快会看到这些能在实践中应用。

Extended Terms:

the theory of relativity 相对论(学说)
theory of games 对策论; 博弈论
Born-Mayer theory 玻恩—梅尔理论
cosmogonic theory 天体演化理论
molecular orbital theory 分子轨道理论

---

thin section [θin ˈsekʃən] *n.*

Definition: a thin, flat piece of material prepared for examination with a microscope, in particular a piece of rock about 0.03 mm thick, or, for electron microscopy, a piece of tissue about 30 nm thick 薄片 báo piàn

Examples:

The classic theory may be divided into Vorousov open section and Wumanski closed section thin-walled bar theory.
经典理论主要可分为符拉索夫开口薄壁理论与乌曼斯基闭口薄壁杆件约束扭转理论。
Comparative observations were made on the morphology and anatomy of secretory cavities in leaves of 22 genera, 40 species and two varieties by the whole mounting, paraffin and thin section method.
利用整体透明、石蜡和薄切片方法对芸香科 22 属,40 种和 2 变种植物叶分泌囊的形态结构和分布进行了比较研究。

Extended Terms:

thin-section casting 薄壁铸件
extra thin section 超薄切片
polished thin section 磨光薄片
thin-section ball bearing 薄壁套圈的球轴承
thin section analysis 薄片分析

---

topochemical [ˌtɔpəuˈkemikəl] *adj.*

Definition: 局部化学的 jú bù huà xué de

Origin: topo-+ chemical

Examples:

It is shown that expansible ettringite forms in the topochemical reaction, and crystallizes in-situ on the surface of unhydrated aluminate minerals and orients radially.
发现能够产生膨胀的钙矾石是由局部化学反应所生成的;它们在未水化的铝酸盐矿物表面原地结晶,呈放射状生长。
The expansion mechanism of hydroxide expanding phase for oil well cement slurry are discussed on the basis of principle of chemical thermodynamics and topochemical reaction model.
要基于膨胀剂原地反应模型和化学热力学原理,讨论了晶体生长压的源动力及水泥浆体体积膨胀与结晶压的关系。

Extended Terms:

topochemical polymerization 局部化学聚合
topochemical concept 拓扑化学概念
topochemical reaction 拓扑化学反应
topochemical control 局部化学控制

---

topology [təu'pɒlədʒi] *n.*

Definition: the study of geometrical properties and spatial relations unaffected by the continuous change of shape or size of figures 地志学 dì zhì xué,拓扑学 tuò pū xué

Origin: late 19th century: via German from Greek topos "place" + -logy

Example:

This topology is a star or extended star logical topology.
这种拓扑是星状或延伸式星状的逻辑拓扑。

Extended Terms:

Homogeneous topology 一致拓扑,均匀拓扑
hull-kernel topology 包—核拓扑
identification topology 同化拓扑,等化拓扑
indiscrete topology 密着拓扑
inductive topology 归纳拓扑

---

transformation [ˌtrænsfə'meiʃən, -fɔːr-] *n.*

Definition: a thorough or dramatic change in form or appearance 彻底的改变 chè dǐ de gǎi biàn,巨变 jù biàn,改观 gǎi guān

Origin: late Middle English: from Old French, or from late Latin transformatio(n-), from the verb transformare

Example:

the study of the flow and transformation of energy
力能学是研究能量的流动及转变的学科

Extended Terms:

continuous transformation 连续相变
reconstructive transformation 重建式相变
thermal transformation 热转变
abrupt transformation 突跃变换
adjoint linear transformation 伴随线性变换

T

---

transition [træn'ziʃən] *n.*

Definition: (Physics) a change of an atom, nucleus, electron, etc., from one quantum state to another, with emission or absorption of radiation(物理)跃迁,转变 yuè qiān, zhuǎn biàn

Origin: mid 16th century: from French, or from Latin transitio(n-), from transire "go across"

Example:

Novel Resonant Transition Soft-Switching Three-Phase PWM Inverters
新型谐振过渡软开关三相 PWM 逆变器

Extended Terms:

first order transition 一级相变
second order transition 二级相变
tricritcal order transition 三级相变
transition element[化]过渡元素
transition energy 转变能, 跃迁能
transition factor 过渡因素

---

translucent [træns'luːsənt, trænz-] *adj.*

Definition: (of a substance) allowing light, but not detailed shapes, to pass through; semi-transparent (物质)半透明的 bàn tòu míng de

Origin: late 16th century (in the Latin sense): from Latin translucent-"shining through", from the verb translucere, from trans-"through" + lucere "to shine"

Examples:

a sliver of rock thin enough to be translucent
一小片薄得半透明的岩石

When you first click a swappable element, a translucent rectangle appears underthe cursor.
当第一次单击可切换元素时,在光标下面会出现一个透明的矩形。

Extended Terms:

translucent beads 半透明球
translucent lighting 半透明照明
translucent surface 半透明表面
translucent scale 透明标度(标尺)
translucent glass 半透明玻璃

---

transparency [træns'peərənsi, -'pær-] *n.*

Definition: the condition of being transparent 透明 tòu míng,透明性 tòu míng xìng,透明度 tòu míng dù

Origin: late 16th century (as a general term denoting a transparent object): from medieval Latin transparentia, from transparent-"shining through"

Example:

We should take care of the transparency of ice.
我们须注意冰的透明度。

Extended Terms:

atmospheric transparency 大气透明度
seawater transparency 海水透明度
differential transparency of atmosphere 不同层次的大气透明度
self-induced transparency 自感应透明

---

transparent [træns'pεərənt] *adj.*

Definition: (of a material or article) allowing light to pass through so that objects behind can be distinctly seen (材料,物品)透明的 tòu míng de

Origin: late Middle English: from Old French, from medieval Latin transparent-"shining through", from Latin transparere, from trans-"through" + parere "appear"

Examples:

a transparent plastic used as a substitute for glass
一种可以代替玻璃的透明塑胶
Notice how the transparent color changes and blends with the background.
留意透明颜色如何变化并与背景融合。

Extended Terms:

transparent mineral 透明矿物
transparent material 透明材料
transparent plastic 透明塑料
transparent scale 透明度盘
transparent transmission 透明传送

---

triboluminescence [ˌtraibəuˌlumiˈnesns] *n.*

Definition: the emission of light from a substance caused by rubbing, scratching, or similar frictional contact 摩擦发光 mó cā fā guāng

Origin: 1885-1890; tribo-+ luminescence

Example:

synthesis, photoluminescence and triboluminescence of $[Eu(DBM)_4]MPy$
$[Eu(DBM)_4]MPy$ 的合成、发射光谱及摩擦发光性质

Extended Term:

phenomenon of triboluminescence 摩擦发光现象

---

tungstate [ˈtʌŋsteit] *n.*

Definition: (*Chemistry*) a salt in which the anion contains both tungsten and oxygen, especially one of the anion WO_4^{2-} [化]钨酸盐 wū suān yán

Origin: early 19th century: from tungsten + -ate

Example:

a mineral consisting of calcium tungstate; an ore of tungsten
一种由钨酸钙组成的矿物,是一种钨矿石。

Extended Terms:

tungstate molybdate and chromate minerals 钨钼铬酸盐矿物
ammonium tungstate 钨酸铵
barium tungstate 钨酸钡
bismuth tungstate 钨酸铋
cadmium tungstate 钨酸镉

---

turbidity [təːˈbiditi] *adj.*

Definition: (of a liquid) cloudy, opaque, or thick with suspended matter (液体)浑浊的

hún zhuó de,污浊的 wū zhuó de

Origin: late Middle English (in the figurative sense): from Latin turbidus, from turba "a crowd, a disturbance"

Example:

The slumping event deposits were mainly breccia dolostone (limestone) while the turbidity current event deposits were turbidity grain limestone (dolostone).
其中滑塌事件的沉积产物主要为滑塌角砾白云岩(石灰岩),浊流事件的沉积产物主要为浊积颗粒石灰岩(白云岩)。

---

twin [twin] *adj.*

Definition: (of a crystal) twinned (晶体)孪晶的 luán jīng de

Origin: late Old English twinn "double", from twi-"two"; related to Old Norse tvinnr. Current verb senses date from late Middle English

Example:

Research on the Mechanism of the Twin Boundary Motion of Ni-Mn-Ga Single Crystals and the Simulation of Stress-induced Martensitic Transformation
单晶 Ni-Mn-Ga 孪晶界迁动机制研究及应力诱发马氏体相变模拟

Extended Terms:

acline twins 底面双晶(水平双晶)
Baveno twin 巴韦诺双晶(曾用名:巴温诺双晶)
Brazil twin 巴西双晶(Carlsbad twin 卡尔斯巴双晶,即卡式双晶)
compound twinning 复合双晶
crosshatch twinning,grid twin,tartan twin 格子(状)双晶
cyclic twin 环状双晶(轮式双晶)
Dauphine twin 多菲内双晶(曾用名:道芬双晶)
elbow twin(knee shaped twin)肘状双晶(膝状双晶)
Emfola twin 恩福拉双晶
Esterel twin 埃斯特勒双晶
growth twin 生长双晶
Manebach twin 曼尼巴双晶
mechanical twin 滑移双晶(机械双晶)
Neustad twin 内乌斯塔德双晶
normal twin 垂直双晶(正交双晶)
parallel twin 平行双晶
penetrate twin 穿插双晶(贯穿双晶)
twin lamellae 双晶片

prism twin 柱面双晶
primary twin 原生双晶

twinning [ˈtwiniŋ] *n.*

Definition: the occurrence or formation of twinned crystals 晶体孪生 jīng tǐ luán shēng，孪晶形成 luán jīng xíng chéng

Origin: late Old English twinn "double", from twi-"two"; related to Old Norse tvinnr. Current verb senses date from late Middle English

Example:

effect of α_2 lamellars slip systems on yield stress of polysynthetically twinning crystals for γ-TiAl
α_2 相滑移系对 γ-TiAl 基多孪晶晶体屈服强度的影响

Extended Terms:

twinning striation 双晶条纹
chiral twinning（晶体）光学孪生
crossed twinning 十字双晶
polysynthetic twinning 多合孪生
secondary twinning 二次孪生

type [taip] *n.*

Definition: a category of people or things having common characteristics 类型 lèi xíng；种类 zhǒng lèi；品种 pǐn zhǒng

Origin: late 15th century (in the sense "symbol, emblem"): from French, or from Latin typus, from Greek tupos "impression, figure, type", from tuptein "to strike". The use in printing dates from the early 18th century: the general sense "category with common characteristics" arose in the mid 19th century

Example:

type classification and applied study of metallotectonic series
成矿构造系列的类型划分与应用研究

Extended Terms:

analog type 模拟型
antiresonant type 反谐振式；并联谐振式
assimilation type 同化类型
backhoe type 反铲式

balanced type 平衡型

typomorphic [ˌtaipəˈmɔːfik] *adj.*

Definition: 标型的 biāo xíng de

Origin: typo-+ morph + -ic

Examples:

The information on the zircon typomorphic characteristics indicate that the magma has originated from the boundary between the crust and the mantle, and the origin from the mantle is dominant.

锆石标型的信息表明,岩浆来源于壳幔边界处,且以幔源为主。

Typomorphic characteristics of pyrites in Au deposits, such as chemical elements, pyroelectricity and crystal form are closely related to the physio-chemical conditions under which they were formed.

金矿床中黄铁矿的化学元素、晶体形态、粒度等标型特征与其形成的物理化学环境及介质条件密切相关。

Extended Terms:

typomorphic characteristic 标型特征
typomorphic mineral 标型矿物
typomorphic peculiarities 标型特征

typomorphism [ˌtaipəˈmɔːfizm] *n.*

Definition: 标型性 biāo xíng xìng

Origin: typo-+ morph + -ism

Examples:

This paper deals with the typomorphism of alumino-chromian micas from the view-point of mineralogical phylogeny.

论文在晚期地壳演化尤其是成矿作用方面,提出了铬铝云母标型性特征,确立了铬铝云母在自然历史中固有的位置和作用。

The crystal form typomorphism of pyrite is one of instructing coefficient of mineralogcial mapping in the gold deposits.

黄铁矿晶体形态是金矿床矿物学填图的良好指示参数之一。

Extended Terms:

thermoluminescence typomorphism 热释光标型
exploration typomorphism 找矿标型

mineral typomorphism 矿物标型性
typomorphism of pyrite 黄铁矿标型特征
crystal morphological typomorphism 形态标型
quartz grain-size (typomorphism)石英粒度标型

ultramicroscopic [ˌʌltrəˌmaikrəˈskɔpik] *adj.*

Definition: too small to be seen by an ordinary optical microscope 超显微的 chāo xiǎn wēi de;超出普通显微镜可见度的 chāo chū pǔ tōng xiǎn wēi jìng kě jiàn dù de

Origin: mid 17th century: from modern Latin microscopium

Example:

During the study on uranium mineralization characteristics of the basin, pitchblende, uranium blacks and phosphuranylite are discovered in uranium deposit No. 382, and ultramicroscopic radioactive minerals are discovered in uranium deposit No.381.
通过对龙川江盆地铀矿化特征的研究,在 382 矿床内发现了沥青铀矿、铀黑和磷钙铀矿,在 381 矿床内发现了超显微状放射性矿物。

Extended Terms:

ultramicroscopic morphology 超显微形态学
ultramicroscopic granule 超显微粒
ultramicroscopic structure 超显微构造
ultramicroscopic dust 超微粉尘
ultramicroscopic observation 超微观察
ultramicroscopic method 超显微法
ultramicroscopic cracking 超微裂纹
ultramicroscopic organism 超显微微生物
macro-ultramicroscopic analysis 显微—超显微分析

unciform [ˈʌnsifɔːm] *adj.*

Definition: denoting the hamate bone of the wrist 钩状的 gōu zhuàng de

Origin: 1725-1735; from Neo-Latin unciformis, equivalent to Latin unc (us) a hook, barb (cognate with Gk ónkos)

Example:

Bent at the end like a hook, unciform.
钩形的端部弯成钩状的，钩形的。

Extended Terms:

unciform process 钩突
unciform fasciculus 钩束
unciform bone 钩骨

---

unconformity [ˈʌnkənˈfɔːmiti] *n.*

Definition: (*Geology*) a surface of contact between two groups of unconformable strata [地质]不整合接触 bù zhěng hé jiē chù

Origin: 1590-1600; un-+ conformity

Example:

Study on the play and the distribution of Ordovician unconformity in central east Ordos basin
鄂尔多斯盆地中东部奥陶系不整合面成藏组合及其分布规律

Extended Terms:

angular unconformity 角度不整合
blended unconformity 混合不整合
chemical unconformity 化学不整合
composite unconformity 复合(重叠)不整合
unconformity spring 不整合泉

---

uneven [ʌnˈiːvən] *adj.*

Definition: not level or smooth 不平坦的 bù píng tǎn de ;崎岖的 qí qū de;参差不齐的 cēn cī bù qí de

Origin: Old English unefen "not corresponding exactly"

Example:

The floors are cracked and uneven.
地板破裂不平。

Extended Terms:

uneven fracture 参差状断口,粗糙断口
uneven color 不匀染色

---

uranium [juə'reinjəm] *n.*

Definition: the chemical element of atomic number 92, a grey dense radioactive metal used as a fuel in nuclear reactors(Symbol:U) 铀(符号 U)yóu (fú hào U)

Origin: late 18th century:modern Latin, from Uranus:compare with tellurium

Example:

A certain scientist had discovered that a metal called uranium gave off a kind of radiation, which Marie Curie was later to call radio activity.
有一位科学家曾经发现了一种叫做铀的金属具有辐射性,后来玛丽·居里夫人把这种现象称作放射性。

Extended Terms:

uranium compound 铀化合物
uranium content 含铀量
uranium deposit 铀矿床
uranium helium method 铀氦法
uranium lead age 铀铅年龄
uranium lead method 铀铅法
uranium mine 铀矿
uranium mineral 铀矿物
uranium ore 铀矿石
uranium ore assay 铀矿石分析
uranium ore leaching 铀矿浸出
uranium separation 铀的分离
uranium uranium method 铀铀法
uranium xenon method 铀氙法

vacancy ['veikənsi] *n.*

Definition: the state of being vacant; emptiness 空缺(缺位)kòng quē (quē wèi)

Origin: 1570-1580; vacantia

Examples:

During the crystallization and growth of mineral, the liquid in clastic rock pores is trapped in the

crystal defects and holes, lattice vacancy and dislocation and microfissures of mineral to form fluid inclusions.
在这种矿物的结晶和生长过程中,碎屑岩孔隙中的液体会被捕获在矿物的晶体缺陷、空穴、晶格空位、位错及微裂隙之中,成为流体包裹体。
The results show that after partial substitution of La^{3+} with Sr^{2+} the oxygen vacancy concentration increases significantly, B^{3+}-B^{4+} system is formed and consequently the catalytic activity for simultaneous removal of NO_x and diesel soot particulates is significantly improved.
结果表明,Sr^{2+}部分取代La^{3+},催化剂表面氧空位浓度增加,同时形成B^{3+}-B^{4+}共存体系,催化剂对碳颗粒和NO_x催化活性显著提高。

Extended Terms:

lattice vacancy 晶格空位
subject vacancy 主体缺位
vacancy flow 空位流
vacancy center 空穴芯
vacancy loop 空位环
vacancy area 空地区域
vacancy concentration 空位浓度
vacancy concentration of oxygen 氧离子缺位浓度
vacancy condensation 空位凝聚
vacancy element 空位元素

---

valence [ˈveiləns] *n.*

Definition: relating to or denoting electrons involved in or available for chemical bond formation 价 jià,原子价 yuán zǐ jià,化合价 huà hé jià

Origin: 1865-1870; from Latin valentia strength, worth, equivalent to valent-(s. of valēns), prp. to be strong + -ia

Examples:

However, unlike the semiconductor silicon, graphene has no gap between its valence and conduction bands.
然而,与半导体硅不同,石墨烯的价带和导带之间没有带隙。
Hydrogen is a one-valence element.
氢是一价的元素。

Extended Terms:

active valence 有效化合价
chemical valence 化合价;化学原子价
chief valence 主要(化合)价

coordination valence 配(位)价
valence band 价带

vanadate ['vænədeit] *n.*

Definition: (*Chemistry*) a salt in which the anion contains both vanadium and oxygen, especially one of the anion VO_4^{3-} [化]钒酸盐 (huà) fán suān yán

Origin: mid 19th century: from vanadium + -ate

Example:

a mineral consisting of chloride and vanadate of lead; a source of vanadium.
一种含铅的氯化物和钒酸盐的矿物,是钒的来源。

Extended Terms:

ammonium vanadate 钒酸铵
ferric vanadate 钒酸铁
lead vanadate 钒酸铅
lithium vanadate 钒酸锂
potassium vanadate 钒酸钾

variety [və'raiəti] *n.*

Definition: the quality or state of being different or diverse; the absence of uniformity, sameness, or monotony 多样化 duō yàng huà; a thing which differs in some way from others of the same general class or sort; a type 品种 pǐn zhǒng

Origin: late 15th century: from French variété or Latin varietas, from varius

Example:

For a wide variety of lithologies from granite through gneiss, carbonate and amphibolite to basaltic rocks, it is shown that this relation is valid in the range 0~350API and 0.03~7$\mu W/m^3$ respectively with an error lower than 10%.
从片麻岩、碳酸盐岩、角闪岩、花岗岩到玄武岩,在这样宽的岩性变化范围内,显示出这种线性关系在 0~350API 和 0.03~7$\mu W/m^3$ 范围是有效的,其误差均小于 10%。

Extended Terms:

variety resources 品种资源
introducing variety 引种
differential variety 差别化品种
dynamic variety 动态变化
variety characteristic 品种特性

viscosity [viˈskɒsiti] *n.*

Definition: the state of being thick, sticky, and semi-fluid in consistency, due to internal friction 黏性 nián xìng;黏滞性 nián zhì xìng

Origin: late Middle English: from Old French viscosite or medieval Latin viscositas, from late Latin viscosus

Example:

The Elastic Viscosity Theory and Application of Polymer Solution
聚合物溶液的弹性黏度理论及应用

Extended Terms:

absolute viscosity 绝对黏度
anisotropic viscosity 各向异性黏性
anomalous viscosity 反常黏滞性
apparent viscosity 表观黏度
apparent shear viscosity 表观切变黏度
ash viscosity[地质]灰黏度
asphalt viscosity 沥青黏度
average viscosity 平均黏度
basic viscosity 基本黏度
bath viscosity 电介质黏度
differential viscosity 差分黏度
thixotropic viscosity 触变黏度

---

void [vɔid] *n.*

Definition: a completely empty space 孔隙 kǒng xì

Origin: Middle English (in the sense "unoccupied"): from a dialect variant of Old French vuide; related to Latin vacare "vacate"; the verb partly a shortening of avoid, reinforced by Old French voider

Example:

Soil-rock-mixture is a typical porous medium, and its seepage is closely related to the particle-size, void ratio and particle shape.
土石混合体属于典型的多孔介质,其渗透特性与颗粒的大小、孔隙比及颗粒形状关系密切。

Extended Terms:

void ratio 孔隙比
void test 空隙测定

void volume[物]空隙率
interstitial void 填隙空位
intragranular void 粒内空隙

volcanism [ˈvɔlkənizəm] *n.*

Definition: (*Geology*) volcanic activity or phenomena [地质]火山活动(dì zhì) huǒ shān huó dòng;火山现象 huǒ shān xiàn xiàng

Origin: 1865-1870; volcan(o) + -ism

Example:

During the late Archaeozoic, there was bimodal volcanism, alkalic magma eruption and intrusion, showing the tension-rifted characteristics of the valleys.
在晚太古代有“双峰态”火山活动、碱性岩浆喷发和侵入,反映了裂谷拉张特征。

Extended Terms:

sedimentary volcanism 沉积火山作用
lunar volcanism 月球火山作用
bimodal volcanism 双峰火山作用
external volcanism 外部火山作用
bimodal rift volcanism 双模式裂谷火山活动

water [ˈwɔːtə] *n.*

Definition: a cololess, transparent, odourless, tasteless liquid which forms the seas, lakes, rivers, and rain and is the basis of the fluids of living organisms 水 shuǐ

Origin: Old English wæter (noun), wæterian (verb), of Germanic origin; related to Dutch water, German Wasser, from an Indo-European root shared by Russian voda (compare with vodka), also by Latin unda “wave” and Greek hudōr “water”

Example:

a solid compound containing water molecules combined in a definite ratio as an integral part of the crystal
水合物结晶中有一定比例水分子成分的固体化合物,其中水分子与晶体的含量为一固定比例

Extended Terms:

water-repellent 防水剂
water-race 水道
crystal water 结晶水
crystallization water 结晶水
interlayer water 层间水

weathering [ˈweðəriŋ] *n.*

Definition: the various mechanical and chemical processes that cause exposed rock to decompose. 风化作用 fēng huà zuò yòng

Origin: 1655-1665; weather + -ing

Example:

Solid rock is broken down by weathering.
坚硬的岩石因风化而分解。

Extended Terms:

biological weathering 生物风化
concentric weathering 同心风化
deep-seated weathering 深层风化
double layer weathering 双风化层
frost weathering 冰冻作用

xenocryst [ˈzenəkrist] *n.*

Definition: (*Geology*) a crystal in an igneous rock which is not derived from the original magma (地质)异晶(dì zhì) yì jīng,捕获晶 bǔ huò jīng

Origin: late 19th century:from xeno-"foreign" + crystal

Examples:

Phengite Xenocryst in Mesozoic-Cenozoic Volcanic Rocks from Lhasa Block and Its Geological Implications
西藏拉萨地块中—新生代火山岩中多硅白云母捕虏晶特征及其地质意义

the discovery of deep Xenolith and Xenocryst in Mesozoic volcanic rock in Mingyuegou Basin of Yanbian area and their mineralogical chemistry
延边明月沟盆地中生代火山岩深源捕虏体和捕虏晶的发现及矿物化学

Extended Term:

accidental xenocryst 异源捕虏晶

---

xenolith [ˈzɛnə(ʊ)liθ] *n.*

Definition: (Geology) A piece of rock within an igneous mass which is not derived from the original magma but has been introduced from elsewhere, especially the surrounding country rock. (地质)捕虏体 bǔ lǔ tǐ,捕虏岩(指火成岩中与其无成因关系的包体)bǔ lǔ yán

Origin: 1900-1905; xeno- + -lith

Example:

The mantle events of 741 Ma and 78 Ma showed by the Sm-Nd reference isochron of these xenolith samples are in the ages consistent with the Jinning and Yanshanian movements on China continent, respectively.
地幔岩捕虏体的 Sm-Nd 计时获得 741Ma 和 78Ma 两条参考等时线,它们分别相当于晚元古代和晚中生代的地幔事件。

Extended Terms:

granulitic xenoliths 麻粒岩
mantle xenoliths 地幔捕虏体
lherzolite xenolith 尖晶石二辉橄榄岩
sandstone xenolith 砂岩捕虏体
xenolith granulites 捕虏体麻粒岩
peridotite xenolith 橄榄岩捕虏体
ultramafic xenolith 超镁铁质捕虏体
gabbro xenolith 辉长岩包体
granulite xenolith 麻粒岩捕虏体
pyroxenite xenolith 辉石岩捕虏体
composite xenolith 复合包体

---

xenomorphic [ˌzenəuˈmɔːfik] *adj.*

Definition: noting or pertaining to a mineral grain that does not have its characteristic crystalline form but has a form impressed on it by surrounding grains; anhedral 他形的 tā xíng de

Origin: 1885-1890; xeno- + -morphic

Example:

The automorphic, hypidiomorphic and xenomorphic granular textures are the main ore textures and vein and disseminated structures are the main ore structure.
矿石多具自形、半自形、他形粒状结构,以脉状、浸染状构造为主。

Extended Terms:

xenomorphic granular structure 他形粒状构造
xenomorphic-granular 变形粒状

X-ray ['eksrei] *n.*

Definition: an electromagnetic wave of high energy and very short wavelength (between ultraviolet light and gamma rays), which is able to pass through many materials opaque to light X 光 X guāng,X 射线 X shè xiàn

Origin: translation of German X-Strahlen (plural), from X-(because, when discovered in 1895, the nature of the rays was unknown) + Strahl "ray"

Example:

Also X-ray has got the extensive application in industry.
在工业中 X 射线也得到了广泛应用。

Extended Terms:

X-ray luminescence X 射线发光
hard X-ray 硬性 X 射线
heterogeneous X-ray 多色 X 射线
homogeneous X-ray 单色 X 射线
mesonic X-ray 介子 X 射线

zeolitic [ziːə'litik] *n.*

Definition: any of a large group of minerals consisting of hydrated aluminosilicates of sodium, potassium, calcium, and barium. They can be readily dehydrated and rehydrated, and are used as cation exchangers and molecular sieves (矿)沸石 (kuàng) fèi shí

Origin: late 18th century: from Swedish and German zeolit, from Greek zein "to boil" + -

lite (from their characteristic swelling when heated in the laboratory)

Example:

Advances in Synthesis Research of Meso-Microporous Zeolitic Molecular Sieves
中微孔结构沸石分子筛的合成研究进展

Extended Terms:

zeolitic water 沸石水
zeolitic cracking catalyst 沸石裂化催化剂
zeolitic ore deposit 含沸石矿床
zeolitic membrane 沸石膜
zeolitic soil 沸石土

zonal [ˈzəunəl] *adj.*

Definition: of or pertaining to a zone or zones 带状的 dài zhuàng de

Origin: late Middle English: from French, or from Latin zona "girdle", from Greek zōnē

Example:

Characteristic Zonal Wavenumber of Winter Circulation in Northern Hemisphere and Its Temporal and Spatial Evolutions
北半球冬季环流的特征纬向波数及其时空演变

Extended Terms:

zonal alteration 带状蚀变
zonal circulation 纬向环流
zonal distribution 分带分布
zonal index 纬向指数
zonal pegmatite 带状伟晶岩
zonal struture 环带构造

zone [zəun] *n.*

Definition: an area or stretch of land having a particular characteristic, purpose, or use, or subject to particular restrictions(有某种特点、特性、目的、用途等的)地带 dì dài,区域 qū yù

Origin: late Middle English: from French, or from Latin zona "girdle", from Greek zōnē

Example:

In the Cretaceous-Early Eocene (135-152 Ma), the zone was as a normal faulting with the component of dextral strike-slip (less than 100 km), the length of the fault zone increased considerably to 3,500 km, the depth of Penetration reached up to 30-40 km.
白垩纪—早始新世(135~152Ma),郯庐断裂呈现为略带右行走滑的正断层(走滑断距不超过 100km)，郯庐断裂带与其北部切割深度为 30~40 公里。

Extended Terms:

zone fossil 分带化石
zone melting 区域熔融
zone melting method 区域熔融法
zone of accumulation 淀积层
zone of contact 接触变质带
zone of faulting 断裂带
zone of fracture 破裂带
zone of investigation 甸区域
zone of leaching 淋滤带
zone of rainfall 降雨带
zone of rock flowage 岩柳
zone of rock fracture 岩石裂隙带
zone of saturation 饱水带
zone of subsidence 沉降带
zone of weathering 风化带